Lecture Notes of the Institute for Computer Sciences, Social Informatics and Telecommunications Engineering 378

More information about this series at http://www.springer.com/series/8197

Edgar Bisset Álvarez (Ed.)

Data and Information in Online Environments

Second EAI International Conference, DIONE 2021
Virtual Event, March 10–12, 2021
Proceedings

 Springer

Editor
Edgar Bisset Álvarez (iD)
Federal University of Santa Catarina
Florianópolis, Brazil

ISSN 1867-8211   ISSN 1867-822X (electronic)
Lecture Notes of the Institute for Computer Sciences, Social Informatics
and Telecommunications Engineering
ISBN 978-3-030-77416-5   ISBN 978-3-030-77417-2 (eBook)
https://doi.org/10.1007/978-3-030-77417-2

This Springer imprint is published by the registered company Springer Nature Switzerland AG
The registered company address is: Gewerbestrasse 11, 6330 Cham, Switzerland

Preface

It is a pleasure to present the proceedings of the 2nd EAI International Conference on Data and Information in Online Environments (DIONE 2021), which was planned to be held in Florianopolis, Brazil, during March 10–12, 2021, but, because of the COVID-19 pandemic, was celebrated in the cyberspace. In this second edition, we appreciated how the conference has become solid in terms of the number of presentations and the coverage of relevant and diverse topics. The conference kept its focus on the intersection between Computer Science, Information Science, and Communication Science. To this end, the presented papers covered topics related to text mining, big data environments, scientific data repositories, open data academic recommender algorithms, open science, research data sharing, scholarly publishing strategies, and fake news, among others.

DIONE 2021 was able to bring researchers, scholars, faculty members, and doctoral and post-doctoral students together to present and discuss innovative ideas related to the aforementioned topics. The conference consisted of 40 full papers selected from 86 submissions; 48% of the papers were from Brazilian authors, 41% were from Chinese authors, and the remaining 11% were from authors from other countries including Belgium, Canada, Peru, Germany, and Croatia. We highly appreciate the interest of all the authors who trusted DIONE to submit their research results. Special thanks go to the Organizing Committee, the members of the Technical Program Committee, and the external reviewers for their willingness to collaborate. In this edition, we were also privileged to have one keynote speaker: Prof. Pippa Smart from PSP Consulting (UK), who delighted us with a presentation on preprints and publishing.

We sincerely appreciate the guidance and support provided by EAI. Finally, we would like to thank the speakers and attendees for being a part of this conference.

We hope that these proceedings will be of interest for all the sectors involved with the aforementioned topics and that the research results will help to bring more insights in order to continue strengthening collaboration on these interdisciplinary issues.

March 2021   Edgar Bisset Álvarez

Organization

Steering Committee

Imrich Chlamtac	University of Trento, Italy
Carlos Luis González-Valiente	European Alliance for Innovation, Slovakia

Organizing Committee

General Chair

Edgar Bisset Alvarez	Federal University of Santa Catarina (UFSC), Brazil

General Co-chairs

Nancy Sánchez Tarragó	Federal University of Rio Grande do Norte, Brazil
Mario Antonio Ribeiro Dantas	Federal University of Juiz de Fora (UFJF), Brazil

Technical Program Committee Chair

Mirelys Puerta Díaz	Sao Paulo State University (UNESP), Brazil

Web Chair

Douglas Dyllon Jeronimo de Macedo	Federal University of Santa Catarina (UFSC), Brazil

Publicity and Social Media Chair

David Caldevilla-Dominguez	Complutense University of Madrid, Spain

Workshops Chair

Douglas Dyllon Jeronimo de Macedo	Federal University of Santa Catarina (UFSC), Brazil

Publications Chair

Genilson Geraldo	Federal University of Santa Catarina (UFSC), Brazil

Local Chair

Gustavo Medeiros de Araújo   Federal University of Santa Catarina (UFSC),
                                           Brazil

Technical Program Committee

Elías Sanz Casado   Carlos III University of Madrid, Spain
Gustavo de Araujo   Federal University of Santa Catarina, Brazil
Manik Sharma   DAV University, India
Ulises Orozco   CETYS University, Mexico
Edgar Bisset Alvarez   Federal University of Santa Catarina (UFSC),
                                           Brazil
Adan Hirales-Carbajal   CETYS University, Mexico
Enrique Orduna-Malea   Polytechnic University of Valencia, Spain
Genilson Geraldo   Federal University of Santa Catarina (UFSC),
                                           Brazil

Contents

Scholarly Publishing and Online Communication

Online Data Processing Technologies

Education in Online Environments

Evaluation of Science in Social Networking Environment

Depicting Recommendations in Academia: How ResearchGate Communicates with Its Users (via Design or upon Request) About Recommender Algorithms

Luciana Monteiro-Krebs[1,2]($\boxtimes$) (iD), Bieke Zaman[2] (iD), Nyi-Nyi Htun[3] (iD),
Sônia Elisa Caregnato[1] (iD), and David Geerts[2] (iD)

[1] Graduate Program in Communication and Information (PPGCOM), Federal University of Rio Grande Do Sul (UFRGS), Ramiro Barcelos, 2705/22201, Porto Alegre, RS, Brazil
luciana.monteiro@ufrgs.br
[2] Meaningful Interactions Lab (Mintlab), KU Leuven, Parkstraat 45 Bus 3605, Leuven, Belgium
[3] Human-Computer Interaction (HCI), KU Leuven, Celestijnenlaan 200A Box 2402, Leuven, Belgium

Abstract. Academic social networking sites (ASNSs) are increasingly using recommender systems to deliver relevant content to their users. Meanwhile, the profound opacity of the algorithms that filter information makes it difficult to distinguish among the elements that may suffer and/or exert influence over the interactions within ASNSs. In this article, we investigate how ResearchGate communicates with its users about the recommender algorithms used in the platform. We employ a walkthrough method in two steps (interface analysis and company inquiry using the General Data Protection Regulation (GDPR)) to investigate ResearchGate's communication strategies (via the design of the platform or upon request) regarding the use of recommender algorithms. The results show six main entities (Researcher, Institution, Research project, Publication, Job and Questions), along with the large amount of metadata involved in the recommendations. We show evidence of the mechanisms of selection, commodification and profiling, and in practice demonstrate the mutual shaping of the different actors (users, content, platform) in their interactions within the platform. We discuss the communication strategy of the company to shy away from providing details on automated profiling.

Keywords: Academic Social Networking Sites · Algorithmic transparency · Recommender systems · ResearchGate · Human-Computer Interaction · New materialism · GDPR

O presente trabalho foi realizado com apoio da Coordenação de Aperfeiçoamento de Pessoal de Nível Superior - Brasil (CAPES) - Código de Financiamento 001.
This paper is part of the research project Algorithmic Mediation in Academic Social Systems (AMASS) https://soc.kuleuven.be/mintlab/blog/project/amass/.

E. Bisset Álvarez (Ed.): DIONE 2021, LNICST 378, pp. 3–25, 2021.
https://doi.org/10.1007/978-3-030-77417-2_1

1 Introduction

The use of recommender systems in academia has been increasing, in part because of Academic Social Networking Sites (ASNSs) such as ResearchGate.net and Academia.edu. Because of the sheer amount of information available today, automated content filtering has become an essential feature of online platforms as a way to mitigate information overload. One way to filter content is using recommender systems, which have become popular across a variety of web-based services, including shopping, entertainment and social networking. By definition, recommender systems reduce the amount of information available to a user in a digital environment; this is done based on the predictions that dictate what the user may like. However, the selection criteria of the data used by the algorithm to recommend and how they are combined and weighted by the algorithm often appear to users as a black box [4, 6], varying across platforms and changing constantly. Some authors have started to raise awareness about recommender systems in attempts to increase algorithmic transparency [3, 6, 19, 20, 29, 32, 35, 42] or to improve the user's understanding of the recommendation mechanisms [7, 30].

Previous work on the topic of recommender systems has focused on improving their efficiency [28, 39, 40, 43], and on human factors, such as trust, privacy, robustness and serendipity [21, 31, 37, 39]. Although insightful, these works do not reflect on the effects of recommendations on the academic environment. At the same time, in the realm of ASNSs research, little has been explored regarding the functioning of recommendations. Previous research [12, 18, 22, 23, 33] has focused on the user's perceptions and behaviour. Studies that do consider the platform often scrutinise individual metrics on these platforms (such as RG Score) [34], here with a sharp topical focus. Current research on ASNS has yet to consider the agency of platforms in influencing users' decision making [13, 35].

ASNSs are socio-technical artefacts that shape and are shaped by human practices and economic, political and social arrangements. While the recommender algorithms mediate the interactions happening within those platforms, these systems remain neglected in literature about ASNSs and its mechanisms are referred to as black-boxes.

2 Background Literature

Many scholars recognize computational technologies as socio-technical artefacts [8–10, 25, 26]. Acknowledging platforms as (automated) mediators, they account for the *"[...] interplay and mutual shaping of technological tools, human action, and social/cultural formations"* [25, p. 44]. Within a connective social media environment, the practices of the users are mediated by the *"platform apparatus"* [9, p. 8]. This mediation determines how connections are taking shape, regardless of how much influence over the content the users can exert. The platform *selection* is *"[...] the ability of platforms to trigger and filter user activity through interfaces and algorithms, while users, through their interaction with these coded environments, influence the online visibility and availability of particular content, services, and people"* [10, p. 40].

The elements involved in the online platforms' dynamic are described in Lievrouw's mediation framework [25], which was created to shed light on the materiality of technology in communicative processes. This framework contains three elements of infrastructure (*artefacts, practices* and *arrangements*). The *artefacts* are the material devices and objects; the *practices* are the actions that people engage with; and the *arrangements* are the patterns of relations, organizing, and institutional structures. Aligned with the work of José van Dijck [8, 9], the mediation framework brings to light the *mutual shaping* of the different elements interplaying in the online performative environment. Therefore, the phenomenon of recommender algorithms on ResearchGate is to be understood at the level of the artefact, but also at the level of the practices and social, cultural and economic arrangements.

At the level of the *artefact*, when understanding recommender systems, we must understand their building blocks: they follow different techniques (similarity measures) to process data of users and items (products and/or content), according to the logic determined by the company that creates the system. Some of the most well-known recommendation techniques include *content-based filtering*, which recommends items that are similar to the ones consulted in previous interactions, and *collaborative filtering*, which selects content by similarity between users [11, 27, 39]. The exact variables used in this automatic inference and how they are weighted and combined is something that most companies keep as a commercial secret. This is why it is difficult to observe coding techniques (programmed by design) in social media platforms. However, this is possible through "*visible user interfaces and application programming interfaces (APIs), and sometimes though their (open) source codes*" [9, p. 6].

Recommendation techniques are used to process data about the items and users. In an information system, when data are used to represent entities (objects in the real world), these data are called metadata. Metadata "*[…] are descriptive elements or coded referential attributes that represent inherent or given characteristics to […] entities in order to uniquely identify it for later retrieval*" [2, p. 47]. For example, in the present study, a publication is an entity in ResearchGate, and it can be represented by many attributes, such as its authors, its keywords, the journal that published it, and its DOI. Not only do the inherent attributes of the entities (that is, attributes that describe the inherent properties of the entity) matter to the recommender system. Metadata that are somehow related to the entity, but rather describing the relations with other entities (such as when that publication was read by a certain user, who liked that publication, and how many keywords it has in common with yet another publication) are valuable to the recommender algorithm because they help to predict the relevance of that publication to the user. Therefore, in the current paper, we use recommendation attributes and metadata as interchangeable synonyms to refer to the classes of data that are used by the algorithm to form a recommendation.

The data used in platforms can also be categorized by where it comes from, that is, how the data are obtained, as pointed by privacy lawyer Simone van der Hof [15]. This typology distinguishes between the *data given, data traces* and *inferred data* [15]. *Data given* are provided by the user during the interaction; these data can be either about themselves or about other individuals, and normally, the user is aware of this provision (although not necessarily their intention to do so). Personal data about the users could be the email and password created by the users to login or their affiliation with an institution; hence, these data contain information about both the user and institution. *Data traces* are the data left behind, mostly unknowingly, by online interactions and are captured via data-tracking technologies such as cookies, web beacons and browser history. It meticulously documents the user's behaviour online, allowing the platform to know where the users spend more time on a page, where they click, and which paths they take to find certain information. Derived from the data given and data traces, it is possible to predict certain aspects about the users and/or their preferences. Van der Hof [15] calls this *inferred data.*

As for the logic of the recommender system at the level of the artefact, it is through the above mentioned machine understandable data that the recommendation algorithm can infer a certain preference of the user and predict what content might be desirable for a particular user at a particular time. The algorithm calculates the similarity between users and items and between users and items by processing metadata. It is possible to aggregate users by their attributes and past behaviour and then label the cluster with a certain profile (this process is referred to as *profiling*). For example, a user can be profiled as a heavy-user male, 36–55 years old, lecturer, and interested in astrophysics and mathematics. The profiling is often made by algorithms that employ data from within the platform, sometimes combined with other data sources, to find patterns and correlations [15].

At the level of *practices*, ASNSs provide an environment for researchers to expose their publications, projects and topics of interest and, to instantly connect users with other researchers and research groups with whom they wish to relate. Working like other social media platforms, it is possible to follow other people (i.e., academics), to recommend content (e.g., scientific papers, research projects), and post messages (e.g., intellectual output). Resembling the forums and groups on other social media platforms, users of ASNSs can engage in discussions around specific topics using the Q&A section (e.g., researchers may interact between different research groups, universities and countries through thematic affinity). A particular feature in these networks is the recommendation of new publications by those authors who have been previously cited by the user and who are included in their paper's references list. This way, the user can follow what the authors they cited in previous work are publishing at the moment.

Regarding *arrangements*, in general, it is considered that the recommendation algorithms on ASNSs are responsible for connecting researchers with common research interests, enabling collaboration and giving and receiving updates about the work in the field. However, the interests and motivations of the users are not the only ones to be considered. The development of a platform is guided by the economic and political interests of the company [8]. For example, ResearchGate positions itself under the paradigm of open science. Open science advocates for, among other things, providing unlimited and free-of-charge access to data sets and publications. This broader phenomenon is motivated by the belief that it would bring more fairness to science, countering the asymmetry between developed and developing countries. A feature in the platform that allows the researchers to upload and share their own intellectual output make this alignment tangible. Naturally, this policy has several implications on institutional and societal levels. Although researchers might use this feature by altruistic means and/or to boost their impact (by promoting their work to potentially increase the number of citations received), major publishers can engage in judicial disputes with the platform over copyright infringement.

The use of personal data in such platforms is also something to consider at the level of the arrangements. Because the recommender systems usually depend on the processing of personal data such as interests, location, gender and the browsing history of the user, the EU recently enforced a legal framework to ensure citizens' right to an explanation of how data are collected, processed, stored, shared and used on platforms. The General Data Protection Regulation (GDPR) [14] provides an access point for meaningful explanations about the functioning of ASNSs, and information requirements that are specifically geared towards automated decision making and profiling practices.

Still referring to arrangements, different pieces of communication form a whole that give a hint to the user regarding what the added value of such technology is, how it is meant to function or which risks it entails. Investigating the communication arrangement is an appropriate way to reveal how the platform is expected to be used [26]. From affordances and constraints on the interface to public announcements or private answers to requests, the manner in which that a platform communicates with its users plays an important role in how users perceive the technology.

Recent literature in Human-Computer Interaction (HCI) [4, 8, 9, 26] has shown how researchers can investigate the way platforms guide users through activities via their design and communication strategies. Some examples are information infrastructure studies [5], digital ethnography [36] and the walkthrough method [26]. In the intersection between science and technology studies (STS) and cultural studies, these approaches allow for *"[...] identifying the technological mechanisms that shape - and are shaped by - the app's cultural, social, political and economic context"* [26, p. 886].

Although great attention has been directed at social media platforms, to the best of our knowledge, there is no prior research investigating the role of the automated mediation of information (i.e., with the help of recommender algorithms), specifically in ASNSs. Against this backdrop, in the current paper we focus both on the artefact and on the communication arrangement of ResearchGate to answer the following research question: how ResearchGate communicates with its users (via design or upon request) about recommender algorithms? Our inquiry will be led by a few follow-up questions: What are the main entities involved in the recommendations on ResearchGate? Which mechanisms can be identified in the platform? How does ResearchGate communicate with its users about the recommender algorithms used in the platform?

In the present study, we chose to work with ResearchGate because of its popularity and outreach among researchers. The platform has been growing through the years in its number of users [41], quantity of documents that it holds [34] and the intensity of use [16], currently claiming to have 17 million users.

3 Methods

In the walkthrough method [26], we found a way to inspect the artefact (typical STS approach) while also expanding the analysis to arrangements (cultural studies approach, providing *"a frame from which to identify embedded cultural values"* [26, p. 888]).

Combining a technical walkthrough on the artefact with the analysis of the communication arrangement, we comprehensively delve into what is communicated via design (interface analysis - step i) and upon request (company inquiry - step ii) regarding the recommendations on ResearchGate. In the following subsections, we describe each step in detail.

3.1   (i) Interface Analysis

This analytical procedure consisted of two phases. First, on the interface, we identified all communicative elements (content labels) that are in one way or the other linked to recommendations. For the first phase, we looked for *visual evidence of a recommendation* in the interface, identifying content that was labelled as a suggestion or recommendation (e.g., when the word "suggested" appeared, or a button called "recommend" emerged). Through this search, we detected five content labels (header of a container) that we found were potentially showing recommended content. The labels are: "Suggested for you", "Who to follow", "Jobs you may be interested in", "Suggested projects" and "Questions we think you can answer". These content labels were above certain types of content, as can be seen in Fig. 1. Every time we found one of those labels, we clicked on the label's link, which then led us to a new page. If the communicative element led us to an independent page with further information and the attributes found there were also connected to other entities, we inferred that it was an entity.

   The second phase of the interface analysis, consisted of inspecting each of the entities in more depth and, also describing their corresponding (visible) attributes. For example, the entity *Researcher* has an independent page and is detailed by several attributes, such as name, RG Score, degree and current affiliation. We did this until we reached the saturation point (when there was no new entity found anymore, only repeated ones). This process resulted in the mapping of six main entities that are involved in recommendations on ResearchGate: *Researcher, Institution, Publication, Research Project, Job* and *Question.*

   We took screenshots and listed the entities, that can be seen in the results section. For the interface analysis, we accessed ResearchGate with the login of the first author using Google Chrome Version 75.0.3770.100 (Official Build) (64 bit) as a web browser. The data collection occurred in February 2019.

3.2   (ii) Company Inquiry

For the company inquiry, the first author sent a data access request to ResearchGate via their contact form on 2 April 2019, asking for the data they have on the user and an explanation of how they create the recommendations (based on what data and criteria). The company replied on the 9th of April via email (sender support@researchgate.net), as follows:

   We consider metadata we may have about you such as the names of published articles plus your past interaction with the site in order to present content that we think might be relevant and interesting to you. We partly use cookies to do this. To view our cookie policy or to opt out visit our cookie policy: https://www.resear chgate.net/cookie-consent-policy.

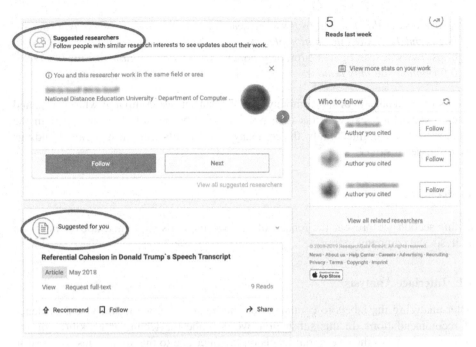

Fig. 1. Visual evidences of recommendation

The same day, on 9 April 2019, the first author wrote back with an extensive email citing the right to an explanation provided in GDPR [14] and specifying the exact information we wanted to receive. Art. 15 GDPR grants individuals the right to know whether or not their personal data have been processed (by the company and/or third parties); it also grants access to personal data and information when it is collected. Our response was structured around 12 questions that were developed with the assistance of a team of legal researchers[1]. ResearchGate did not respond to our inquiry within a reasonable time. Hence, we reinforced the request with another email on the 23 April 2019. On the 24th of April, ResearchGate (sender support@researchgate.net) responded and thanked us for contacting them whilst also informing us they were *"[...] in the process of responding to your request"*. On 13 May 2019, ResearchGate's Privacy department responded to our request. They sent us an introductory email with a seven-page plain explanatory text document as an attachment, and in the body text, they gave a reference to a URL to a set of 22 HTML files and 11 PDF files (see also the results section, below). The introductory email read as follows:

> *Thank you for your data subject access request dated 9 April 2019. Please find attached a document with more detailed answers to your questions and your data. If you wish to modify your privacy settings or update your personal data*

---

[1] Questions originally designed for the research project Algorithmic Transparency and Accountability in Practice (ATAP), in which Luciana Monteiro-Krebs and David Geerts participated. See more in https://soc.kuleuven.be/mintlab/blog/news/re-thinking_recommenders/.

please access this page https://www.researchgate.net/profile. *ProfilePrivacySettings.html. For more information please consult the Researchgate privacy policy at* https://www.researchgate.net/privacy-policy. *We remain available for any further information you may require.*

As for the analysis of the company inquiry data set, we compared what was stated in the explanatory text with the data set. We describe the information found in the categories of data pointed out by the company in the results section. Additional findings are discussed further.

4   Results

In this section we present the results of the two methods explained above: interface analysis and company inquiry.

4.1   Interface Analysis

After analysing the labels to containers on the home page, we found visual evidence of recommendations. In this subsection, we describe the technical walkthrough and how we went from the five initial labels on the interface to the six entities involved in recommendations and their respective attributes.

On ResearchGate's home page (see Fig. 2), the recommendations under the label **Suggested for you** refer to publications, such as articles, chapters, books, technical reports, theses, conference papers, data and preprints. We have identified the following attributes: title of the publication, the type of publication (paper, report, chapter, etc.), whether there is a full document available, the date of publication, and the number of reads. The item container under the label **Suggested for you** also shows the buttons "View", "Download" (or "Request full text" in case the file is privately archived), "Recommend", "Follow" and "Share". At the bottom of the item container, the number of researchers who follow or recommend this particular publication is made visible. Our findings further show that by clicking on the "View" button, we are led to the complete page of the publication within the platform. On that page, we could find specific information about that publication, such as title, author(s), abstract, editor/journal and date. At the bottom of the publication's page, a container recommending more research items under the label "Similar research" appears. Because publications appear under two labels that indicate recommendations ("Suggested for you" and "Similar research"), we identified *Publication* as a recommended entity. Additionally, the entity publication has a specific page for it: each paper, book chapter or preprint registered on the platform has its own page with all the metadata regarding that publication that can be retrieved and recommended from that metadata. On the home page, many recommended publications are from authors related to the user, either as coauthors, colleagues, or people the user follows or cites.

As for the label **Who to follow**, the interface analysis further shows that the main page gives concrete recommendations to follow other researchers. In the container under the label "Who to follow", there is a list of recommended researchers, showing a profile

Fig. 2. Publication recommended in the feed

picture, name, the connection between the recommended researcher and the user (e.g., if he/she is someone the user cited previously, or is a coauthor, etc.) and a "Follow" button. Three researchers are shown in this container on the home page.

At the bottom of that container, there is a link to "View all related researchers", which in turn leads to a new page with several options of researchers to follow. The recommended researchers to follow are introduced with short profiles that are ranked and separated by the following tabs: "Summary", "Your institution", "Your department", "Your coauthors", "Citations", "Similar interests", and "Your followers". The short profile features picture, name, institution/company, the connection with the recommended researcher and the suggested researcher's RG Score. The RG Score is a metric created by ResearchGate to, according to the platform, "*measure scientific reputation based on how your work is received by your peers.*" It is based on several aspects, including publications, citations and interactions within the platform. The RG Score is one of the few pieces of information that appears in the short summary of the researchers' profile on the interface. The interface also hints at the possibility of inspecting the connection in more detail, for example, by checking which publication the user and recommended researcher are co-authoring, the skills or expertise they share and the latest publication of the recommended researcher. Based on these attributes, we can infer two entities: *Researcher* and *Institution*. The *Researcher* has a specific page dedicated to it and is the entity that appears under the labels "Who to follow" and "View all related researchers". *Institution* appears to be relevant because it has its own specific page and because of two other reasons. First, in the list of recommended researchers, there are specific tabs for the people from "Your Institution" and "Your Department", to recommend colleagues for the user to follow. This suggests that it is because of the connection with the institution that other researchers are being recommended, and many suggestions on the home page are from people working in the same institution of the user. Second, in the container with Job offers, it is the logo of the institution that appears next to the job position. The institution can be a university or faculty, a research institute or a company.

Regarding the label **Jobs you may be interested in**, on a list of five job offers, ResearchGate first shows the title of the position and information about the institution or company: the logo, name and location (city and country). Depending on how recent the vacancy is, a label appears: "New job" or "Expiring soon". Two links appear at the bottom of this container on the home page in addition to the list of job positions: "Improve these suggestions" and "View more" (to visualize more suggestions). The link "Improve these suggestions", activates a pop-up box that allows for updating the list of skills and expertise. Some suggested skills also appear in this box, showing the importance of the keywords used in job recommendation. Clicking "View more" leads to a page with job positions with the exact same metadata that the item container on the home page brings, but instead of showing five options on the side bar, a page with dynamic scrolling is shown, with job offers appearing in a long list that occupies the entire page.

When clicking on one of the job positions in the recommended list, a complete register of the vacancy is provided on a new page, including more information on the title of the position, the date on which it was published, the institution, the location, the logo, a job description, areas of research (what knowledge fields that position encompasses), a list of other positions at the bottom of the page (link called "Discover more") and, on the right side of the screen, another list of recommendations: "Researchers also applied for". In the list "Discover more", which is at the bottom of the page, a list of 15 job positions is shown, but the only information on the link is the job title. In the list on the right side of the page, under the label "Researchers also applied for", the format is the same as that of the home page (i.e., job title, logo, name of the institution and location). The latest list ("Researchers also applied for") is shorter, with five positions only, and a link to "View more" appears at the bottom of this container. In this particular type of recommendation, we highlight the *Job* as an entity, because it appears under several recommendation labels ("Researchers also applied for", "Discover more" and "View more") and each job offer registered has a specific page. We confirmed the importance of the *Institution* because it appears prominently in job offers (logo and name of the institution). We also confirmed the importance of the entity *Researcher* through the label "Other researchers also applied for".

The interface analysis with respect to the label **Suggested projects** further shows that this label is above an item container that provides information on the title of the project, a brief project description and the name(s) of (a selection of) the researcher(s) who are involved in the project. The latest are ranked in a way that researchers with a shared connection are shown first independently of who is the project lead, along with the number of other researchers in the project. Figure 3 shows an example recommendation of a project. *Research project* is then mapped as an entity, because it have a specific page to describe its attributes and is closely connected to other entities, namely *Researcher* and *Publication*.

On the home page, there is a link to a section with questions and answers, which is labelled "Questions". Clicking on this link, a new page appears with, among other content labels, one called **Questions we think you can answer**. By clicking on this label, we are led to a page with open questions posed by other researchers. The attributes on this page are the name of the researcher, date, main topic, title of the question, the first sentences of the question, some keywords, the number of replies, the number of reads

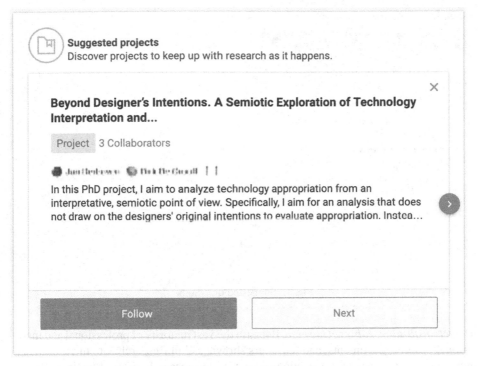

Fig. 3. Recommended projects

and the buttons "Reply", "Recommend", "Follow" and "Share". Figure 4 provides an example of a recommended question.

We identified *Question* as an entity, not only because it also has a specific page but because its importance is reinforced by yet another container in the home page with the label "Do you have a research question?". In this item container on the interface, the user is invited to ask questions to get help from experts in their field. The link for "Questions" also appears on the home page (feed) accompanied by the following sentence: "[Researcher] asked a question in [keyword]". From that sentence, we infer that the recommended question is influenced by the keywords list. This influence is reinforced by the item container with the user's skills and expertise in the right column of the "Questions" page. The container shows the sentence: *"We use your skills and expertise to show you relevant questions. You can edit your skills and expertise at any time."*, which is followed by a list of keywords that represent the user's skills and expertise. The association between questions and keyword (in the home page) is made even if the keyword is not present in the user profile or in their list of skills and expertise. In other words, users see recommended questions with the indication of a topic (e.g., communication) based on their profile, even if the users themselves did not list this specific keyword as a topic of interest. The inference made by the recommender algorithm might use co-occurrence as a similarity metric to suggest questions.

Summarizing the findings from the interface analysis (step i), the main entities involved in recommendations on ResearchGate are *Researcher, Publication, Research*

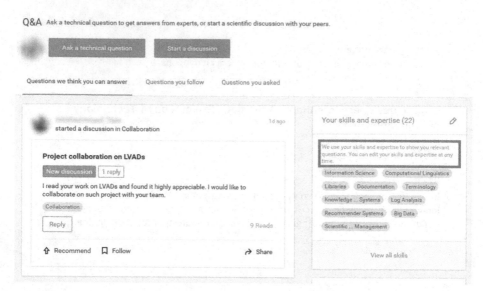

Fig. 4. Suggested questions

project, Institution, Job and *Question*. Because the vast majority of professional and institutional profiles, documents, and job offers shown within ResearchGate fit under these overarching categories of information mapped in our research, most of the interactions within the platform are somehow affected by the recommendation algorithm.

4.2 Company Inquiry

In this section, we describe the aggregated classes of data (with the quantity of attributes for each class) contained on the data set (seven-page text document attached to the conversation email, 22 HTML files and 11 PDFs). We further highlight some key findings before the Discussion section. The results of our company inquiry show that content on the ResearchGate platform is being recommended based on the processing of the following data:

Personal Data: According to the company, this information is used *"to understand more about the users, visitors and viewers, and how they interact with our platform"*. We identified 105 attributes (on the files "Account", "Account emails", "Your Privacy Settings" and "Your Notification Settings" describing user's personal data (including name, address and email) and preferences (such as if other researchers can see certain interactions of the user). When asked about the company's personal data sharing practices with a possible data processor, ResearchGate replied that they do share personal data with a partner (Lotame.com), but did not inform which data they consider personal.

Bibliographical Information: This includes information about academic content from different sources, including databases (e.g., PubMed) or the website of a publisher. The information includes, for example, title of the article, name of the journal, date of publication and names of the various authors of the content. In the data set, we found three attributes that fit this category (distributed on the HTML files "Coauthors", "Your Projects", "Your Publications"). The data set also contained the full publications in PDF format, although no information was provided on how the PDFs were indexed.

Information Pertaining to the User's Work: The HTML files "Project Collaborators" and "Profile information"[2] gather 29 attributes, including where and with whom the user works.

Historical Data: ResearchGate informed us that to recommend content, they process usage frequency, type of devices used, publications consulted, time spent on particular pages or parts of pages and number of clicks on a page of features. We identified 18 attributes on the HTML files "Login history", "Activity history"[3] and "Publication followings". These files contain the login data (when and from where the user logged in) and all sorts of interactions with different content (such as publications, researchers, advertisements, job offers and emails). The interactions include, but are not limited to, engage, view, react, update and open. For each interaction, the list shows the date, time, country, browser, operational system and (truncate) IP address[4] used in that activity.

ResearchGate states that, for security reasons, it keeps the number of pages viewed by the user to prevent data harvesting by third parties. Indeed, in the data set, we could see the number of read publications, the number of read projects and the number of citations. However, ResearchGate not only keeps the number of visited pages to avoid security attacks, but also registers every page and which type of interaction (view, engage, react) the user has with that specific content (see Fig. 5). This shows that the company not only keeps quantitative data about the accessed pages, but also keeps track, in great detail, of the users' interactions within the platform to observe their behaviour.

Authors the User May Have Chosen to Follow: The HTML files "Followers" and "Followings" (total six attributes) represent the network the user is in contact with.

The Subject Matter of Articles the User May Have Authored: As mentioned in the category bibliographical data, ResearchGate provided the PDF files of all publications registered in the platform by the user; however, it did not give information about the process of indexation of these content, that is, the extraction of the topics of the paper. We infer that there is a collection based on the PDF (and maybe that is why the platform is so insistent in asking the user to upload the full text). For example, there is no specific field on the interface to register the references of the papers. However, once the PDF is uploaded, metadata and links to the publications that are cited appear on the publication page. This could also be the case for automatic extraction of topics, which leads us to the keywords.

---

[2] The file "Profile information" could also be considered personal data because it includes email, phone and birthday.

[3] The files "Login history" and "Activity history" also fit the category of personal data because of the type of attribute they register. They registered two years of interactions within the platform.

[4] IP address is a numerical label assigned to each device connected to a computer.network, used to identify it individually.

- person:view:jobSuggestion @ 2019-04-16 09:12:46 from Belgium, Chrome 72 on macOS (ip: 1
- **person:view:recommendation** @ 2019-04-16 09:12:46 from Belgium, Chrome 72 on macOS (ip
- **person:view:jobSuggestion** @ 2019-04-16 09:12:46 from Belgium, Chrome 72 on macOS (ip: 1
- **person:view:jobSuggestion** @ 2019-04-16 09:12:46 from Belgium, Chrome 72 on macOS (ip: 1
- **person:view:jobSuggestion** @ 2019-04-16 09:12:46 from Belgium, Chrome 72 on macOS (ip: 1
- **person:view:jobSuggestion** @ 2019-04-16 09:12:46 from Belgium, Chrome 72 on macOS (ip: 1
- **person:view:activity** @ 2019-04-16 09:12:46 from Belgium, Chrome 72 on macOS (ip: 134.58.2
- **person:view:activity** @ 2019-04-16 09:12:46 from Belgium, Chrome 72 on macOS (ip: 134.58.2
- **person:view:activity** @ 2019-04-16 09:12:46 from Belgium, Chrome 72 on macOS (ip: 134.58.2
- **person:create:activity** @ 2019-04-16 09:12:46 from Belgium, Chrome 72 on macOS (ip: 134.58
- **person:update:device** @ 2019-04-16 09:12:45 from Belgium, Chrome 72 on macOS (ip: 134.58.
- **person:update:device** @ 2019-04-16 09:12:45 from Belgium, Chrome 72 on macOS (ip: 134.58.
- **person:update:device** @ 2019-04-15 14:52:08 from Belgium, Chrome 72 on macOS (ip: 134.58.
- **person:open:mail** @ 2019-04-15 14:52:08 from Belgium, Chrome 72 on macOS (ip: 134.58.253
- **person:receive:adslot** @ 2019-04-15 09:05:30 from Belgium, Chrome 72 on macOS (ip: 134.58.
- **person:view:adslot** @ 2019-04-15 09:05:23 from Belgium, Chrome 72 on macOS (ip: 134.58.25
- **person:receive:adslot** @ 2019-04-15 09:05:23 from Belgium, Chrome 72 on macOS (ip: 134.58.
- **person:view:jobSuggestion** @ 2019-04-15 09:04:59 from Belgium, Chrome 72 on macOS (ip: 1
- **person:view:jobSuggestion** @ 2019-04-15 09:04:59 from Belgium, Chrome 72 on macOS (ip: 1
- **person:view:jobSuggestion** @ 2019-04-15 09:04:59 from Belgium, Chrome 72 on macOS (ip: 1
- **person:view:jobSuggestion** @ 2019-04-15 09:04:59 from Belgium, Chrome 72 on macOS (ip: 1
- **person:view:jobSuggestion** @ 2019-04-15 09:04:59 from Belgium, Chrome 72 on macOS (ip: 1
- **person:view:recommendation** @ 2019-04-15 09:04:55 from Belgium, Chrome 72 on macOS (ip
- person:view:activity @ 2019-04-15 09:04:55 from Belgium, Chrome 72 on macOS (ip: 134.58.2

Fig. 5. Part of HTML file with historical data

Profiling Keywords: The data set also had a file called "Keywords and Skills", which contained a list of keywords, many of which were not added by the first author of the current paper. In total, the user profile in the platform had 22 keywords as skills and expertise at the time of the data gathering that were filled by the user and visible in the interface. The keywords HTML file of this same user had 67 keywords. Seventeen of them were classified as "Sciences" by the platform (e.g., Social Science, Semantics, Artificial Intelligence). As can be seen in step i (interface analysis), the section Questions normally uses skills and expertise to recommend questions to be answered by the user. Hence, if only 33% of the keywords (22 out of 67) in the user profile were actually provided by the user (data given), 67% of these recommendations are based solely on algorithmic inference (inferred data).

It is not clear, however, how the match between keywords and content is made to recommend publications to the users. The company makes recommendations based on *"The subject matter of articles the user may have authored"*, but no indication was found to identify how these topics are selected by the system. This vagueness is also reflected in the classification of information. Several attributes present in the data set could be classified as "Information pertaining to the user's work" or "Personal data", but the company did not make clear how this information was being used.

Content in the Platform: Content posted by the user within the platform fits this category. The HTML files contain "Messages" (three attributes, including the full content of the messages left on ResearchGate) and "Questions" (no attributes were mapped here because the first author did not publish any question within the platform at that time, hence the file came empty).

Scores/Stats: This refers to many aggregated metrics keeping track of the achievements of the user that are considered milestones. Examples include reads (the so-called *success stories* represent the number of reads the publications have across time), number of citations, likes on the user's publications and research projects (the button *Recommend* in the interface is equivalent to the *Like* button on other social media, and it is counted as a like in the HTML files). They also list the most relevant publications considering the h-index[5] on ResearchGate. The files "H-Index", "RG Score", "Account Stats", "Success stories" and "Top Publications by H-Index" total 18 attributes.

The HTML file "RG score" contains the composition of this metric (percentage distribution that comes from publications, questions, answers, and followers). For example, 99.48% of the first author's RG Score comes from publications, while 0.52% comes from followers. If the first author had posted or answered any question, this would also be part of the equation, but it was not the case. The platform dedicates a page to explain how the RG Score is built, but it does not show if and how it may influence the ranking of researchers recommended to the user (under the label "Who to follow").

As for the tailored advertising content (this content appears in the interface as "sponsored content"), ResearchGate listed the following information categories used: *"personal data provided by the user; personal data collected by the platform; and personal data inferred by the platform based on the use of the Service and the Internet"*. Some of this information is provided by the users themselves, some is collected by the platform, and some is inferred by the platform using a combination of data already in their possession. In the typology of privacy lawyer Simone van der Hof [15], those categories of data would fit as follows: *personal data provided by the user* is equivalent to *data given*; *personal data collected by the platform* is equivalent to *data traces*; and *data inferred by the platform based on the use of the Service and the Internet* is equivalent to *inferred data*. The three categories informed by the company are broad enough to include any kind of personal data used in the platform without necessarily specifying where the data come from and how the data are used in the advertisement. This classification is vague because it does not inform which data are considered personal by the platform.

The company also presented inconsistent information when answering the question on the logic involved in recommendations. ResearchGate denies automated decision making: *"We do not engage in automated individual decision making, including profiling, as proscribed in GDPR Article 22."* However, when asked about personal data usage, the company claims that: *"For the personal data where we provide a description of the data categories we cannot provide a copy of the data because the data is in an aggregated format."* The aggregated format can normally be seen as profiling [14, 15]. Hence, apparently, there is confusion about the meaning of profiling. In Art. 13(2)f and 14(1)f, the GDPR requires a platform to provide information regarding the existence of "solely automated decision making", including profiling. The law also requires that where such systems are deployed, meaningful information is given about the logic involved, as well

---

[5] The h-index is a bibliometric index given by the number of articles that have a number of citations equal to or greater than the number itself. If the researcher published 15 articles that obtained 15 or more citations each, then their h = 15. It was created in 2005 to assess the relevance of researchers, but rapidly spread across other entities and today it is applicable to researchers, institutions and journals.

as the significance and envisaged consequences of such processing for the individual. ResearchGate did not provide the explanation asked about the logic of the recommender system because of commercial secrecy. When explicitly requested to explain the logic behind the recommender system, the company responded with the following:

> *This request goes beyond your right to access deriving from Art. 15 GDPR, and an explanation of the set-up and specific functioning of our system would involve providing you with information we regard as business secrets.*

Summarising the results of the company inquiry (step ii), ResearchGate provided a long document explaining the recommendations together with a data set containing what the company claims to be all the data they have about the first author. However, this was received by the researchers at the third contact attempt, after an extensive request on behalf of the researchers with several specific and law-based questions. In the company's first answer (sent by the support team from Berlin), they provided two lines of explanation and a link to the cookies policy. We believe that the standard answer does not offer a complete overview about what is used to recommend content and how this process happens. At that moment, the company said they use data provided by the user to offer better personalised service and briefly referred to metadata (names of published articles and user's past interaction with the site) obtained partly by cookies, which are only two of the many attributes mapped in our rescarch. Their final answer, sent by the privacy team with no location given, was obtained six weeks after the first request. In the final answer, more information was provided, but the data set was still incomplete. For example, ResearchGate sent us PDF files with the publications authored and uploaded by the first author on the platform. However, the terms used to index the content of the publications were not revealed. It was not clear if this answer came from the data processor (Lotame) because there was no indication of location. From the absence of a nominal signature, we interpret that further contact on behalf of the user is not encouraged.

Even after receiving a document that is supposed to explain the logic of the rec- ommendations, the information we received was vague and sometimes inconsistent. Regarding examples of the former, there is a lack of accuracy in describing what is considered personal data (used to recommend regular and spon sored content); there is no information about how the inferences are made in the keywords used in recommen- dations; and there is no information regarding how much of the RG Score contributes to the ranking of researcher's recommendations, even though this is an important metric created to measure the reputation of researchers, as ResearchGate states on the interface. As for inconsistency, two occurrences were reported: first, the discrepancy between stat- ing how they keep quantitative data regarding the number of pages visited when they actually keep detailed data about the user's behaviour on the platform. And second, affirming that they do not use automated decisions while admitting to profiling users with a data processor.

5   Discussion

Analysing the artefact and the communication arrangement of ResearchGate in contrast to what usually interests ASNSs users, according to previous studies, we observe that most of what users seek is mediated by algorithms, even though this is not always clearly communicated to the user. For example, Jeng *et al.* [18] point out that ASNS platforms can facilitate scholarly information exchange, which in ResearchGate is possible through the section "Questions". Because *Question* is one of the entities shaped by recommender algorithms on this platform, the information-seeking that starts in this particular section receive algorithmic mediation. It is clear to the users that the information available in the "Questions" section receives automated curation (see Fig. 4), however it is unclear how much inference is happening behind the scenes. Moreover, the statement on the interface that the platform employs the users' skills and expertise to show relevant questions and that *the users can edit their skills and expertise at any time* can misinform the user. From these sentences, users can be misled to think that they are in control of the mediation, when in our study we found that nearly 67% of the keywords in the user profile were inferred and act behind the interface, not being available to manual edition by the user.

Previous studies identified practices of the ASNS users regarding contents usually available in these platforms. Our analysis can add to this knowledge showing the algorithmically mediated entities related to these specific contents. According to Nández and Borrego [33], researchers use ASNSs to follow and get in touch with other scholars. That involves the entity *Researcher*, which is used to recommend content and is also recommended in the platform. Users also disseminate their research results [33] via self-archiving (uploading one's own publications), motivated by accessibility [24]. The entities *Researcher*, *Research Projects* and *Publications* are directly involved in these actions. The recommendations used in *Job* positions are clearly based in collaborative filtering: "Other researchers apply for". However, differently from the other entities, the information provided here is generic, not disclosing which researchers applied for that specific vacancy, or how many. This might be an strategy from the platform to boost employment without jeopardize privacy and inner competition in academia. The competitive aspect of job seeking changes the configuration of the information provided, possibly to avoid endanger the negotiation between the *Institution* offering the position and the *Researcher*. This informs about how social arrangements, practices and artefact are mutually shaped [25], as a specific feature in the artefact reflects a social conduct and a certain "way of doing" that is professionally accepted by the academic community. By recommending a *Job* position this way, the platform protects the relations between the nodes of the network (imagine two colleagues knowingly running for the same vacancy) and the institution that offers the job (by not showing how many researchers have applied, the platform does not denounce how disputed that job vacancy - really - is).

5.1   Information Selection and Prioritization by the Platform

The company stated in their email that they use the recommendation engine to present content that they think might be relevant and interesting to the user. Analysing this, we see the platform selection [10] in practice. While the users are browsing, they are both actively providing data (on their profile or login history) and receiving recommendations that are based on these preferences. The users may provide information about their topics of interest (data given), publications (data given and data traces in case the metadata about the publications are collected by the platform in other databases) and what they like to see (by simply navigating and staying longer on a certain page these data are collected through the login history). However, it is the algorithm that decides how this information will be selected, processed and weighted, which ultimately defines what other content will be shown and in which order (prioritization) [6, 10]. In the context of recommendations in ASNSs, this means that although users have agency on the content that is published and consumed (e.g., uploading papers or inserting the institution name where the researcher works), how this content is used in the algorithm, namely, which attributes will be matched and which ones are more relevant in the ranking of recommendations, is a decision coming from the platform (automated decision making). The mutual shaping characteristic between the different elements (artefact, practices and arrangements) [25] is expressed in the way that users can freely engage with whatever content they want, while the platform nudges them to connect with certain items and forums through personalised recommendations (e.g., "you might like this" or "we think you can answer this question"). The user can decide what to click on, however, through automated filtering, the universe of choice is narrowed by the recommender algorithm. Aligned with Bozdag [6], Alvarado and Waern argue that, in promoting recommendations from the users' most active connections and downgrading the actions of the less active ones, *"[...] the system controls both the users' information and who they can reach"* [1, p. 2].

5.2   Commodification in ASNSs

ASNSs are designed in a way that the platform can benefit (economically and strategically) from the researchers practices, which is afforded by the artefact. For example, ResearchGate values the interactions within the platform, which is concretely expressed by the RG Score, that quantifies all these interactions. The interactions quantified in the RG Score (uploading full text publications, engaging in Q&A forums, and acquiring followers) are also beneficial to the platform: questions engage users, uploading full publications feeds into the database of the company with machine readable scientific content (that otherwise is protected by paywalls) and followers increase the trust on the digital environment. Probably not coincidentally, RG Score is one of the few elements shown in the short profile of the researcher. The platform also nudges researchers to invite their coauthors to become users as a way to "help their publication gain visibility" (sentence used in emails sent by the platform).

ResearchGate rewards the researcher that makes the upload of full text publications in two ways: adding up the users' RG Score and recommending new publications from "relevant" authors (that are inferred based on the list of references of the publications). By encouraging this practice (sharing the researchers' own work) the platform can then offer the publication free of charge to other users, which might increase the adherence of new users. This can be identified as the mechanism of commodification that "[...] involves platforms transforming online and offline objects, activities, emotions, and ideas into tradable commodities" [10, p.37].

Users receive reading recommendations based on a match between their own interests (keywords, readings) and the previous publications they wrote themselves. At the same time, they can also recommend content produced by others and, therefore, influence the way that publications are ranked in the platform (including for themselves). This recommendation is made through the button "Recommend", which is part of the design of the container for several types of content on the interface of ResearchGate, and it can be seen in previous figures (Fig. 1, Fig. 2 and Fig. 4). In a typical case of collaborative filtering, when user A recommends a certain content item, this item will appear to other users (users B, C, D) endorsed by user A. Prior research has shown that the collaborative filtering technique in recommender systems is inherently driven by social influence, as the "follow by example" pattern is automated by the algorithm [17, 38]. This may explain why ASNSs facilitate trust among users, as they consider the site to be an extension of their professional activities, therefore perceiving other members as trustworthy [22]. Hence, when a researcher connected to the user endorses a certain content, that content becomes more appealing and more likely to attract the user's attention and trust, which might be a strategy from the platform to increase their interest and to get users to trust more in the recommended content. This is aligned with what van Dijck [8] affirmed about how platforms are influencing human interaction on an individual, community and larger societal levels. This is happening *"while the worlds of online and offline are increasingly interpenetrating"* [8, p. 4].

5.3 Profiling in ASNSs

The three categories of personal data used in the digital environment are data given, data traces and inferred data [15]. Regarding the types of data used in recommendations, ResearchGate mentions to the use of personal data specifically when referring to tailored advertisement (sponsored content), as shown in step ii (company inquiry). However, data from all of these categories, particularly inferred data, are used for a number of different recommendations (not only sponsored content), connecting users and content on ResearchGate. For example, our research results have shown that only 33% of the keywords in the user's profile (in the HTML files) were stated by the first author in the field named *skills and expertise*. The company did not say how these inferences are made and what is the exact information used to generate them. Nevertheless, the results in step i (interface analysis) show that the entity *Question* is associated with keywords that are not listed in the user's profile or in their list of skills and expertise. Additionally, it is difficult to state which data are considered "personal" by ResearchGate because this was not detailed in the document. ResearchGate also did not specify which categories of information are shared with Lotame (data processor). Presenting customised content recommendations can be considered automated decision making, where recommendations depend on a profile that has been built out of the characteristics or interests of the user. Therefore, it would be desirable for the company to clearly explain to its users the process of automated profiling, either by design of the interface or upon request. The right to a meaningful explanation is ensured to the users by GDPR [14], and it is crucial to help them understand the mechanisms underpinning their interactions within the platform [30].

5.4 ResearchGate's Communication Strategy

Van Dijck and Poell [9] have stated that the technological mechanisms in social media are often invisible. This was possible to see in our data collection, with the delayed, vague and sometimes inconsistent answers to the company inquiry (step ii). This also goes towards Millecamp's claim: *"the rationale for providing individual recommendations remains unexplained to users"* [30]. Unfortunately, most people only have a vague idea of how recommender algorithms work, because these systems are often presented as a "black box" [6], which was also the result we got from company inquiry, when the company argued that the information asked regards business secrets.

The vagueness and inconsistency can have two motivations. On the one hand, it can be because of the recency of GDPR requirements and the lack of experience in providing detailed and meaningful explanations about the algorithmic mediation to users. On the other hand, it may be a conscious effort to keep the algorithmic logic safe from competitors (commercial secrecy). Nevertheless, transparency through design is a must regarding recommendations in ASNSs [32]. Providing tardy, unclear and discrepant explanation jeopardize the algorithmic transparency of ResearchGate and do not contribute to the user's understanding of the recommendation mechanisms.

6 Conclusion

In the current study, we conducted a socio-technical analysis of the recommendations on ResearchGate in light of the mediation framework [25]. Using the walkthrough method in two steps (interface analysis and company inquiry) we delved into what the platform communicates regarding the use of recommender algorithms via design or upon request. We identified the main entities involved in a recommendation: *Researcher, Institution, Publication, Research project, Job* and *Question*. Considering ASNSs are one type of social media, we analised how artefact, practices and arrangements mutually shape each other. We also verified how the mechanisms of platform selection, commodification and profiling [10] apply to the platform. We conclude that recommender algorithms mediate most of the content in the platform and that the mutual shaping characteristic of social media logic is also reflected in this particular ASNS. Even though the company denies automated decision making, our results point towards profiling (prediction based on inferred data). By reflecting on ResearchGate's communication strategies via visible interface elements and upon request, we suggest that the company implements tools to make sure the users are informed in a clear, agile and meaningful way about the algorithmic mediation.

7 Limitations and Future Work

Algorithms change constantly and are often protected by commercial secrecy. Hence, the present study is limited to the information we had access to in a particular period, including the data provided and what we identified following visual clues on the interface. At the time of publication, the platform, its logic and its effects in the interface might have changed. Future studies might investigate other ASNSs using the same approach to distinguish the potential patterns among platforms.

Acknowledgments. We thank Jef Ausloos and Pierre Dewitte (CiTiP/KU Leuven) for preparing the questionnaire which was adapted for this study, as well as the other ATAP project colleagues: Aleksandra Kuczerawy, Ingrid Lambrecht, Laurens Naudts, Oscar Alvarado Rodriguez and Elias Storms. We also thank Karin Hannes, Raquel Recuero, Kevin Sanders, Isa Rutten, Roos Voorend, Çisem Özkul and Niels Bibert for providing valuable feedback in the early stages of the research. This work was carried out with the support of the Coordination for the Improvement of Higher Education Personnel - Brazil (CAPES) - Financing Code 001.

References

1. Alvarado, O., Waern, A.: Towards algorithmic experience. In: Proceedings of the 2018 CHI Conference on Human Factors in Computing Systems, CHI 2018, pp. 1–9. ACM Press, Montreal (2018). https://doi.org/10.1145/3173574.3173860
2. Alves, R.C.V.: Metadados como elementos do processo de catalogação. Ph.D. thesis, Universidade Estadual Paulista, Faculdade de Filosofia e Ciências (2010). https://hdl.handle.net/11449/103361
3. Ausloos, J., Dewitte, P.: Shattering one-way mirrors - data subject access rights in practice. Int. Data Privacy Law **8**(1), 4–28 (2018). https://doi.org/10.1093/idpl/ipy001
4. Barassi, V.: Babyveillance? Expecting parents, online surveillance and the cultural specificity of pregnancy apps. Social Media + Society **3**(2) (2017). https://doi.org/10.1177/2056305117707188
5. Bowker, G.C., Baker, K., Millerand, F., Ribes, D.: Toward information infrastructure studies: ways of knowing in a networked environment. In: Hunsinger, J., Klastrup, L., Allen, M.M. (eds.) International Handbook of Internet Research, chap. 5, pp. 97–117. Springer, Dordrecht (2010). https://doi.org/10.1007/978-1-4020-9789-8
6. Bozdag, E.: Bias in algorithmic filtering and personalization. Ethics Inf. Technol. **15**(3), 209–227 (2013). https://doi.org/10.1007/s10676-013-9321-6
7. Brusilovsky, P., de Gemmis, M., Felfernig, A., Lops, P., O'Donovan, J., Semeraro, G., Willemsen, M.C.: Recsys'18 joint workshop on interfaces and human decision making for recommender systems. In: Proceedings of the 12th ACM Conference on Recommender Systems, RecSys 2018. ACM, New York, pp. 519–520 (2018). https://doi.org/10.1145/3240323.3240337
8. van Dijck, J.: The Culture of Connectivity. Oxford University Press, New York (2013)
9. van Dijck, J., Poell, T.: Understanding social media logic. Media Commun. **1**(1), 2–14 (2013). https://doi.org/10.17645/mac.v1i1.70
10. van Dijck, J., Poell, T., de Waal, M.: The Platform Society: Public Values in a Connective World. Oxford University Press, New York (2018)
11. Ekstrand, M.D., Riedl, J.T., Konstan, J.A.: Collaborative filtering recommender systems. Found. Trends Hum. Comput. Interac. **4**(2), 81–173 (2011). https://doi.org/10.1561/1100000009
12. Elsayed, A.M.: The use of academic social networks among Arab researchers: a survey. SSCR **34**(3), 378–391 (2016). https://doi.org/10.1177/0894439315589146
13. Frauenberger, C.: Entanglement HCI the next wave? . ACM Trans. Comput. Hum. Interact. **27**, 27 (2019). https://doi.org/10.1145/3364998
14. (GDPR), G.D.P.R.: Regulation (EU) 2016/679 of the european parliament and of the council of 27 April 2016 on the protection of natural persons with re gard to the processing of personal data and on the free movement of such data, and repealing directive 95/46/ec (general data protection regulation), April 2016. https://eurlex.europa.eu/legalcontent/FR/TXT/PDF/?uri=CELEX:32016R0679&from=FR

15. van der Hof, S.: I agree... or do I? A rights-based analysis of the law on children's consent in the digital world. Wis. Int'l L. J. **34**(2), 409–445 (2017). https://hosted.law.wisc.edu/wordpr ess/wilj/files/2017/12/van-der-Hof_Final.pdf

16. Jamali, H.R., Nicholas, D., Herman, E.: Scholarly reputation in the digital age and the role of emerging platforms and mechanisms. Res. Eval. **25**(1), 37–49 (2016). https://doi.org/10. 1093/reseval/rvv032

17. Jameson, A., et al.: Choice architecture for human-computer interaction. Found. Trends Hum. Comput. Interac. **7**(1–2), 1–257 (2014). https://doi.org/10.1561/1100000028

18. Jeng, W., DesAutels, S., He, D., Li, L.: Information exchange on an academic social networking site: a multidiscipline comparison on researchgate Q&A. J. Assoc. Inf. Sci. Technol. **68**(3), 638–652 (2017). https://doi.org/10.1002/asi.23692

19. Kleanthous, S., Kuflik, T., Otterbacher, J., Hartman, A., Dugan, C., Bogina, V.: Intelligent user interfaces for algorithmic transparency in emerging technologies. In: Proceedings of the 24th International Conference on Intelligent User Interfaces: Companion, IUI 2019, pp. 129–130. ACM, New York (2019). https://doi.org/https://doi.org/10.1145/3308557.3313125

20. Koene, A., et al.: Ethics of personalized information filtering. In: Tiropanis, T., Vakali, A., Sartori, L., Burnap, P. (eds.) Internet Science, pp. 123–132. Springer, Cham (2015)

21. Konstan, J.A., Riedl, J.: Recommender systems: from algorithms to user experience. User Model. UserAdap. Inter. **22**(1–2), 101–123 (2012). https://doi.org/10.1007/s11257-011-9112-x

22. Koranteng, F., Wiafe, I.: Factors that promote knowledge sharing on academic social networking sites: an empirical study. Educ. Inf. Technol. **24**(2), 1211–1236 (2018). https://doi. org/10.1007/s10639-018-9825-0

23. Laakso, M., Lindman, J., Shen, C., Nyman, L., Björk, B.C.: Research output availability on academic social networks: implications for stakeholders in academic publishing. Electron Markets **27**, 125–133 (2017). https://doi.org/10.1007/s12525-016-0242-1

24. Lee, J., Oh, S., Dong, H., Wang, F., Burnett, G.: Motivations for self-archiving on an academic social networking site: a study on ResearchGate. JASIST **70**, 563–574 (2019). https://doi.org/ 10.1002/asi.24138

25. Lievrouw, L.A.: Materiality and media in communication and technology studies: an unfinished project. In: Gillespie, T., Boczkowski, P.J., Foot, K.A. (eds.) Media Technologies: Essays on Communication, Materiality, and Society, p. 44–51. MIT Press, London (2014). https:// doi.org/10.7551/mitpress/9780262525374.003.0002

26. Light, B., Burgess, J., Duguay, S.: The walkthrough method: an approach to the study of apps. New Media Society **20**, 881–900 (2018). https://doi.org/10.1177/1461444816675438

27. Lops, P., De Gemmis, M., Semeraro, G.: Content-based recommender systems: State of the art and trends. In: Recommender Systems Handbook, pp. 73–105. Springer, Boston (2011). https://doi.org/10.1007/978-0-387-85820-3_3

28. Lorenzi, F., Abel, M., Loh, S., Peres, A.: Enhancing the quality of recommendations through expert and trusted agents. In: 2011 IEEE 23rd International Conference on Tools with Artificial Intelligence, pp. 329–335, November 2011. https://doi.org/10.1109/ICTAI.2011.56

29. Milano, S., Taddeo, M., Floridi, L.: Recommender systems and their ethical challenges. SSRN Electron. J. 1–20 (2019). https://doi.org/10.2139/ssrn.3378581

30. Millecamp, M., Htun, N.N., Conati, C., Verbert, K.: To explain or not to explain: the effects of personal characteristics when explaining music recommendations. In: Proceedings of the 24th International Conference on Intelligent User Interfaces, IUI 2019, pp. 397–407. ACM, New York (2019). https://doi.org/https://doi.org/10.1145/3301275.3302313

31. Montaner, M., López, B., de la Rosa, J.L.: Developing trust in recommender agents. In: Proceedings of the First International Joint Conference on Autonomous Agents and Multiagent Systems: Part 1, pp. 304–305. ACM (2002)

32. Monteiro-Krebs, L., Alvarado Rodriguez, O.L., Dewitte, P., Ausloos, J., Geerts, D., Naudts, L., Verbert, K.: Tell me what you know: GDPR implications on designing transparency and accountability for news recommender systems. In: Proceeding CHI EA 2019 Extended Abstracts of the 2019 CHI Conference on Human Factors in Computing Systems, CHI EA 2019, pp. 1–6. ACM, New York (2019). https://doi.org/10.1145/3290607.3312808

33. Nández, G., Borrego, A.: Use of social networks for academic purposes: a case study. Electron. Libr. **31**(6), 781–791 (2013). https://doi.org/10.1108/EL-03-2012-0031

34. Orduña-Malea, E., Martín-Martín, A., Delgado-López-Cózar, E.: Researchgate como fuente de evaluación científica: desvelando sus aplicaciones bibliométricas. Profesional de la Información **25**(2), 303 (2016). https://doi.org/10.3145/epi.2016.mar.18

35. Pariser, E.: The Filter Bubble. Penguin Books Limited, New York (2012)

36. Pink, S., Horst, H., Postill, J., Hjorth, L., Lewis, T., Tacchi, J.: Digital Ethnography: Principles and Practice. SAGE Publications Ltd., London (2016)

37. Pu, P., Chen, L., Hu, R.: Evaluating recommender systems from the user's perspective: survey of the state of the art. User Model. User Adap. Inter. **22**(4–5), 317–355 (2012)

38. Ramos, G., Boratto, L., Caleiro, C.: On the negative impact of social influence in recommender systems: a study of bribery in collaborative hybrid algorithms. Inf. Process. Manage. **57**, 1–18 (2020). https://doi.org/10.1016/j.ipm.2019.102058

39. Ricci, F., Rokach, L., Shapira, B.: Recommender Systems Handbook. Springer, Boston (2015). https://doi.org/10.1007/978-0-387-85820-3

40. Tsai, C.H., Brusilovsky, P.: Leveraging interfaces to improve recommendation diversity. In: Adjunct Publication of the 25th Conference on User Modeling, Adaptation and Personalization, UMAP 2017, pp. 65–70. ACM, New York (2017). https://doi.org/10.1145/3099023.3099073

41. Van Noorden, R.: Online collaboration: scientists and the social network. Nature News Feature **512**(7513), 126–129 (2014). https://www.nature.com/news/online-collaboration-scientists-and-the-social-network-1.15711

42. Vedder, A., Naudts, L.: Accountability for the use of algorithms in a big data environment. Int. Rev. Law Comput. Technol. **31**(2), 206–224 (2017). https://doi.org/10.1080/13600869.2017.1298547

43. Amine, Abdelmalek, Bellatreche, Ladjel, Elberrichi, Zakaria, Neuhold, Erich J.., Wrembel, Robert (eds.): CIIA 2015. IAICT, vol. 456. Springer, Cham (2015). https://doi.org/10.1007/978-3-319-19578-0

A Study on the Process of Migration to Training in Brazil: Analysis Based on Academic Education Data

Higor Alexandre Duarte Mascarenhas[1]([⊠])(iD),
Thiago Magela Rodrigues Dias[1](iD), and Patrícia Mascarenhas Dias[1,2](iD)

[1] CEFET-MG, Divinópolis, Minas Gerais, Brazil
thiagomagela@cefetmg.br
[2] UEMG, Divinópolis, Minas Gerais, Brazil

Abstract. In recent years, a fact that stands out at the national level is the movement of individuals to other locations at some point in their lives. There are several causes that motivate these types of displacement, among them, one of the main reasons for training, especially at the level of academic training. Given this scenario, this work aims to carry out an analysis of how Brazilian academic mobility occurred, through data extracted from currencies and institutions registered in the Lattes Platform. Thus, extracted from the curricula of Brazilians, resulting in 308,317 records. From the extraction, the data was filtered, obtaining the following relevant research items, performed, treatment of the analysis items, removing irrelevant and incomplete terms and, subsequently, improving the data with information on the geographical location of each institution. At each level of education the entire group analyzed, from the place of birth to the individual's current professional performance. Soon after, with the set of detected data, it was possible to carry out the monitoring in the measurements of the social networks, in which the networks were characterized considering the displacements between the places in the academic formation process used. As a result, an image of how the scientific formation process took place after a long process of capacitation was used, making it possible to measure the migratory flow of individuals and trends in the formation processes.

Keywords: Platform Lattes · Brazilian scientific exodus · Migratory flow · Data analysis

1 Introduction

The emigration of Brazilians to other countries and to other states has increased significantly, so that in Brazil, studies show that some cities have rates of 10 to 30% of migrants who do not live in their home state [2]. In many cases, Brazilians go out for a job or study, always looking for quality of life.

© ICST Institute for Computer Sciences, Social Informatics and Telecommunications Engineering 2021
Published by Springer Nature Switzerland AG 2021. All Rights Reserved
E. Bisset Álvarez (Ed.): DIONE 2021, LNICST 378, pp. 26–40, 2021.
https://doi.org/10.1007/978-3-030-77417-2_2

Among the main reasons for applying is the need for training at a high level of training. One of the main causes for the mobility option in the Brazilian territory, refers to the quality of higher education in other states, the search for new opportunities and more experiences in their areas [9]. Another refuge for these students is directed to other countries, thus seeking cultural exchange and better investment in research grants. The student's departure to other countries is not only interesting for the student, but also for the institutions of origin, as it the same return in most cases more productive, with the most extensive contact network, greater experience and available in the future with the sharing their experiences with other students from the home institution.

According to [6], every day it has become more difficult to produce scientific research in Brazil, due to investment cuts using education. One of the main reasons for the migration of Brazilian researchers to other countries can be pointed out by the lack of government support. Therefore, with this scenario of Brazilian researchers go out in the country, thus making it difficult to return due to the lack of opportunities. Most of the Brazilian scientists who return to Brazil do not work in their area of training, so they do not progress in their careers.

A program that facilitated and helped a lot or the entry of students in institutions with institutions in other countries is Ciência sem Fronteiras, for being a program that supports students, that offers scholarships. In 2015, the government planed to reach 101,000 scholarships for researchers, graduates, PhD students, post-doctoral students, encouraging students to capacitate in institutions of recognized relevance [4]. Today, with only 5,000 scholarships available, the Program has lost a lot of influence in the entry of students to other countries, due to investment cuts.

As a motivation for the study to have an understanding of the Brazilian scientific performance aiming to obtain an opportunity to understand the current scenario, and to adopt measures to promote possible openings of new undergraduate or postgraduate courses in areas where a deficit in specific areas of the knowledge. Another issue related to the economic issue, is a better exploration of the specific area of education in a region of Brazil.

Given this scenario, this work presents a study on the exod of individuals that left your state/city of birth to other states/cities and/or those that went to other countries in search of training. To use the data of Brazilian students analyzed in this study, the framework LattesDataXplorer [7] was used, a tool responsible for extracting and processing CVs of users registered on the Lattes Platform. Currently, the Lattes Platform curricula repository that records academic/scientific and professional information, has approximately 6,750,000 registered curricula. Therefore, a set of components created for the purposes of this study and incorporated into the *framework*, thus enabling a broad and unprecedented view of the Brazilian scientific exod.

2   Related Works

According to [8] the shifts in the researcher's training demonstrate a correlation as characteristics of the individual, one of the main characteristics is the degree of international cooperation or scientific production.

[1] in his study carried out an analysis of the migratory flow of people born in 196 countries on all continents, his research carried out an analysis in the mid-1990s until the year 2010, with a study to understand of patterns and trends in flow of immigration from countries and continents selected by the authors. The authors were able to identify the migration flows of individuals registered in the study according to the level of development of the countries.

[5] conducted a study with data from the Lattes Platform of postgraduate researchers, collecting data on the researcher's trajectory from birth to his last formation degree. Having analyzed the group of PhD, reaching the conclusion of 95% are from the South, Southeast and Northeast states. It was mentioned that 40% of the first PhD training courses were carried out in the cities of origin and 87% of those individuals displaced to other cities not exceeded or within the limit of 1,000 km. It was also highlighted that the city with the highest number of PhD in São Paulo.

Already [10] analyzes the mobility of Brazilian researchers and students throughout their academic training. It is worth mentioning that 20% of the researchers work, however, more than 500 km from the institution where they entered the academic trajectory, in contrast, the majority work at about 100 km. This mobility made the interviewed researchers involved in several lines of research, making their work more known, in the places of their trajectory. The study indicates that the states in the southeastern region, mainly São Paulo, are those that most researchers are natural from there; the other states have a temporary migratory pattern.

[3] carried out a study to analyze the circulation of people throughout the academic journey of individuals, as well as their workday. In your study the author, mentioned which researchers from different nationalities choose to engage in migration or obtain more experience in your area and expand a network of contacts at other universities. Subsequently, it is mentioned that in Brazil there is no form of incentive facilitated for Brazilian students emigrated to other countries, and this fact can harm even the same Brazil, as this way hinders the network of contacts between Brazilian researchers and researchers from other nationalities. However, Brazil invests in foreign researchers to study to Brazil, to make contacts networks, however, it is often difficult to find an attraction for foreigners, since there was a reduction in scholarships in the country. Other countries, such as China, invest in obtaining a network of contacts between other countries, and concluded that the United States is one of the countries in which more Chinese choose to perform the exodus.

[11] proposed a study with analysis of intrastate use in Rio Grande do Norte, with occurrences between the Metropolitan Region of Natal and the interior of the state, and between the interior of the state and the Metropolitan Region of Natal and the data obtained for this research, based on two periods, from 1995/2000 and 2005/2010, with data provided by the Instituto Brasileiro de Geografia e Estatística (IBGE). From the data obtained, the authors carried out statistical analyzes to compare the results and point out the main figures on the flows that occurred, and found that individuals choose to exercise the flow from

interior to metropolis, but in both cases in the Metropolitan Region of Natal shows a decline in population gains. About the analysis of migrations from cities that they observed in Natal, there is a greater volume of migrations "from" and "to" the interior of Rio Grande do Sul, because the capital concentrate as the with activities related to the sectors of service, commerce, tourism and education.

Therefore, it is notorious that a large scale of individuals choose to obtain capacitation at a high level of formation left your home city to another, and with a smaller scale, part of your home country seeking capacitação abroad. It is worth mentioning that many works related to this project prefer to extract data from the curricula registered in the Lattes Platform, as it is a repository of great importance for the study of Brazilian scientific production.

3   Methodology

In the present work, the main source of data used was the curriculum repository available on the Lattes Platform. Initially, it was necessary to use the *Lattes-DataXplorer* [7] to extract the data, given the difficulty of obtaining them, since the interface to query the Lattes Platform curricula allows access to only one curriculum per time, so the analysis of large groups of individuals becomes a limiting factor. Data extraction was carried out in May 2019, totaling 308,317 resumes from individuals with completed PhD degree, considering all PhD regardless of the date of completion of formation.

Soon after the data extraction was carried out, treatments were carried out with the objective of obtaining formatted data extracts in order to facilitate future analyzes. Thus, steps such as "Data Selection" and "Data processing" were performed according to the scheme shown in Fig. 1.

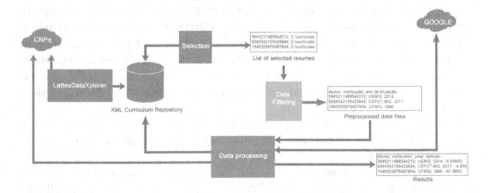

Fig. 1. General aspect of the set of components used. Source: Authors.

In the "Selection" Step, the XPath query language (XML Path Language) is used for research and subsequent generation of the subgroups to be analyzed. The XPath language allows the construction of expressions that will go through an XML document in a similar way to the use of regular expressions. Therefore, it allows the grouping of a set of curricula with desired parameters, such as academic training or areas of expertise.

The list stores the identifiers for each curriculum and the path in which it is stored locally, so it will be possible to analyze only the selected curricula. In view of the above, only curricula were collected from individuals with completed PhD degree, as this is the group with the highest level of academic education; since these are curricula that are frequently updated and most of the parameters required for the present work are registered in their curricula.

After selecting the set to be analyzed, the "Data filtering" module, which is responsible for analyzing the resumes in XML files in order to obtain relevant information for research, features an extract of formatted data (Preprocessed data files). The curriculum information registered in the file has: curriculum identifier; individual's state and city of birth; institution code, name and zip code of the individual's current employment relationship, in addition to the identification code and name of the institution for each level of education completed, considering since undergraduate to PhD.

Afterwards, the module "Data processing" (Fig. 2) is executed, in which four steps are performed: Obtaining the institution's CEP; Search by geographic location; Data cleaning and grouping and Data normalization. The first step carried out is the "Obtaining CEP of the institution" in which, from the institution's code retrieved from the curriculum, it is consulted in the institutions directory of the Lattes Platform, in order to obtain the institution's data and, thus, retrieve it from the address section, the institution's characteristics data, from then on, the website will return the institutions' information and thus obtain the institution's zip code.

The "search for geographic location" stage is a task performed with the purpose of geolocating an institution. Accessing the Google geolocation API (Application Programming Interface), the institution's address will be sent, to later have the institution's geographic location (latitude and longitude) returned.

In the "Data cleaning and grouping" stage, exclusion of possible terms that are irrelevant to the search occurs, in order to reduce the volume of data to be processed and analyzed. As an example: removing *stopWorlds* in city names; normalization to extract accented words, and replace them with their equivalent without accent.

The "Data normalization" stage, on the other hand, aims to reduce the redundancy of information, discarding attributes with the absence of data, such as CEP with no digits.

Subsequently, the "Results File" is generated, representing a summary of all the data obtained in the curricula of Brazilian PhD degree, without needing to consult the XML files of the extracted curricula, having all the specific data for carrying out the analyzes of this research.

Soon, after all the steps described above have been carried out, several metrics are applied to understand how the mobility of Brazilian PhD degree has occurs throughout their academic training process.

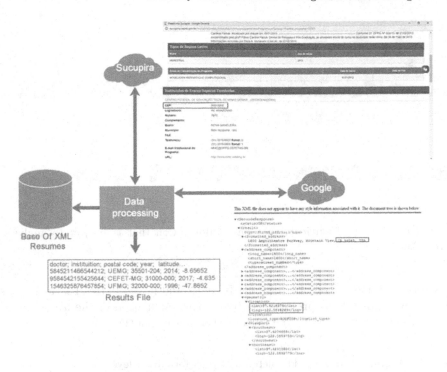

Fig. 2. Data processing. Source: Authors.

4 General Characterization

Initially, it was possible to characterize the analyzed set. As it is a group of individuals who have the highest level of education, to be considered, only those formations whose status in the curriculum are "completed" were included in the analysis, resulting in a total of 308,317 individuals. In order to evaluate the analysis potential of the extracted data, Fig. 3 shows the number of curricula that have attributes registered, such as: city of birth, institution of activity and formation institutions.

Most individuals have their birth city registered with a total of 293,340 (95%) records, as this is a mandatory field when registering on the Lattes Platform. Those individuals who do not have a birth city registration are considered to be older curricula, in which city registration was not mandatory. Of the other institutions shown in the graphic, the one that identifies itself as superior in quantitative data is the *institution of completion of the PhD degree*, totaling 297,815 (96%) registrations, as it is the group selected to carry out the study. The *postdoctoral institution* is the one with the lowest number of registrations, with an amount of 70,405 (22%), this fact is justified, because the selection of the groups was directed to individuals with a completed PhD degree; for this reason, an individual who holds a PhD degree does not always have a postdoctoral degree.

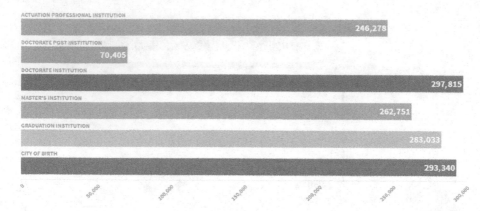

Fig. 3. Quantitative of curricula that have information that is subject to analysis. Source: Authors.

After obtaining the number of records registered in the Lattes Platform curricula, it was possible to present in Table 1, the calculation of the average distances in kilometers between one level of training to another.

Table 1. Average distance in kilometers between formation levels. Source: authors.

Distance (km)	Graduation	Master's degree	PhD degree
Birth	291.58	548.59	1,000.75
Graduation	–	432.38	901.26
Master's degree	–	–	619.00

It is possible to observe the result of the average distance of all stages of training of Brazilian PhD, during their academic training. It can be observed that the average distance between the stages has a variation considered. Initially, analyzing the average distance from the place of birth for graduation it is noticed that this is the shortest calculated distance. One of the factors that influence this phenomenon is that a large part of Brazilian cities have institutions that provide undergraduate courses to students, and those that do not, in most cases, are close to other cities that hold courses at this level of training of interest to students. The greatest distances, on the other hand, are between the place of birth and formation at the PhD level, followed by the undergraduate/PhD degree in which the displacement is greater than the other levels of training.

Figure 4 aims to demonstrate a characterization of the number of bonds obtained by PhD's at birth, at each level of education, as well as their professional performance. In order to explore better data visualization, the representation of the links has been separated by states, and the colors treated in the heat map vary according to the number of links the state has, with values ranging from 0 links to 40,000 links, aiming to address the discrepancy of the data in relation to the way it was distributed. In the image, there are five different graphics, dealing with the ties of PhD's at each level of education, they are: birth (a); graduation (b); master's degree (c); PhD (d) and professional performance (e).

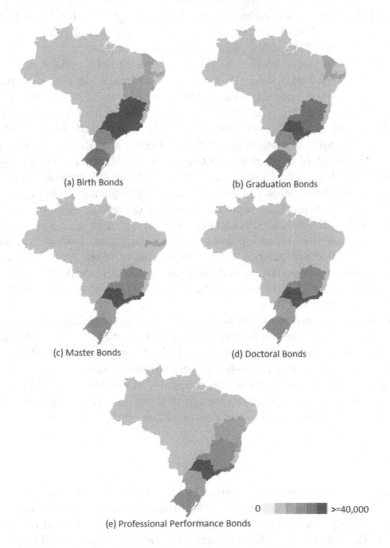

(a) Birth Bonds

(b) Graduation Bonds

(c) Master Bonds

(d) Doctoral Bonds

0   >=40,000

(e) Professional Performance Bonds

Fig. 4. Bonds of PhD's in Brazilian states. Source: authors.

In all the graphs obtained, the state of São Paulo has values on average well above the limit rated by the authors, because if the number of links obtained by the state of São Paulo was used as the highest value, data visualization would be unfeasible from heat maps.

It is observed that the region that stands out most at all levels present, is the Southeast region, which have greater influence in the states of São Paulo, Rio de Janeiro and Minas Gerais, a possible justification is the high concentration of institutions and universities federal governments in this region, in the state of Espírito Santo, do not have a high concentration of bonds in any of the graphs. With less influence, but highlighted in all charts, the state of Rio Grande do Sul also stands out at all levels of education, representing one of the states with a large concentration of Brazilian universities and institutions.

When it comes to the level of birth, in addition to the greater prominence in the southeastern region and the state of Rio Grande do Sul, it is possible to observe the highest concentration in the state of Paraná, and with the least amount of links, there are some states in the northeast region, such as: Bahia, Pernambuco and Ceará respectively. Unlike the graph with birth level, it is possible to observe that the states of Minas Gerais and Rio de Janeiro stand out less in the graph of undergraduate ties and that the state of Bahia does not stand out in this level of education, a possible justification for the lower assiduity of links from the three states, may be the search for capacitation in other states.

Analyzing the graphs that represent master's degrees, it is clear that in the Northeast region, the state of Pernambuco stands out with a greater number of degrees present at this formation level. When taking into account the number of PhD degrees, it confirms once again the hegemony of the state of São Paulo and Rio de Janeiro respectively, and right after Rio Grande do Sul and Minas Gerais, due to the amount of post-graduate courses offered is higher in the four states compared to the others.

When observing professional performance, the state of Bahia stands out compared to the other states, excluding the quartet of states (São Paulo, Rio de Janeiro, Minas Gerais and Rio Grande do Sul) characterizing one of the states with the highest number of professional performance bonds.

It is observed how less frequent are the states of the North and Center-West regions, due to the fact that these regions have fewer universities and institutions, and consequently a lower number of postgraduate courses on offer.

5   Results

This chapter presents a characterization of the formation based on the networks of PhD's throughout their academic capacitation and also their professional performance, through their links between Brazilian states and several countries in the world. The networks are directed, and the nodes are represented by the location where the individual was trained and the edge represents the interaction carried out between places where the individuals displaced. The diameter of each knot characterizes the number of degrees it has.

Figure 5 represents the interaction between nodes (Brazilian states/other countries), constituting five different networks: birth - graduation (a); graduation - master's degree (b); master's degree - PhD degree (c); PhD degree - professional performance (d) and birth - professional performance (e).

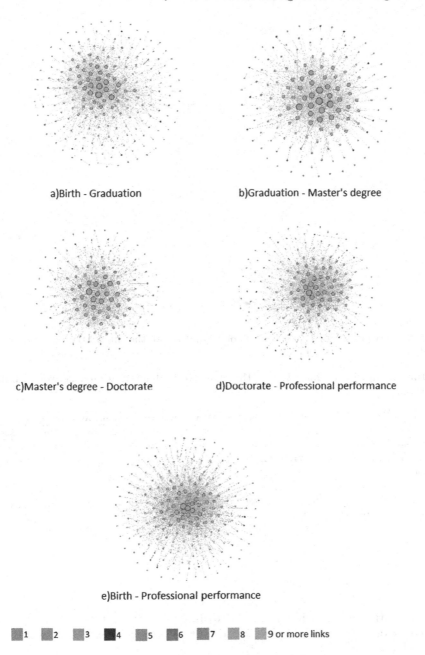

a)Birth - Graduation

b)Graduation - Master's degree

c)Master's degree - Doctorate

d)Doctorate - Professional performance

e)Birth - Professional performance

1 2 3 4 5 6 7 8 9 or more links

Fig. 5. International link networks. Source: authors.

It is possible to observe that some networks have a lower number of nodes in comparison to the others, for example the networks characterized in Figs. 5b and 5c, since generally individuals choose to remain in the same location in

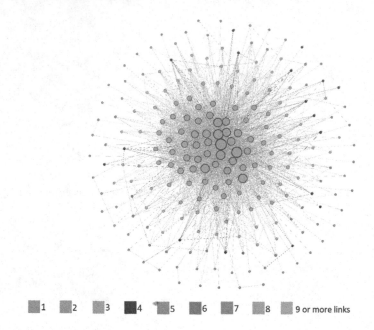

1 2 3 4 5 6 7 8 9 or more links

Fig. 6. Network of links at all levels of formation at the international level. Source: authors.

Table 2. Metrics extracted from the characterized international networks. Source: Authors.

Metrics	Birth-Grad	Grad-Mest	Mest-Doct	Doct-Actuation	All	Birth-Actuation
Number of nodes	201	144	124	174	217	210
Number of Edges	2,246	1,814	1,840	2,297	4,039	3,389
Average degree of nodes	11.174	12.597	14.839	13.201	18.613	16.138
Nodes of G.C.	189	135	119	171	217	206
% of nodes in the G.C.	94.03	93.75	95.97	98.28	100	98.1
Edges of G.C.	2,237	1,805	1,835	2,295	4,039	3,385
% of edges in the G.C.	99.99	100	99.72	99.91	100	99.88
Network Density	0.056	0.088	0.121	0.076	0.086	0.077
Network Diameter	5	5	5	5	6	5
Average path length	2.321	2.106	2.009	2.182	2.261	2.179

Note: G.C.: Giant Component

the formation levels, when it comes to undergraduate - master's degrees, and master's degrees - PhD degree, as a consequence of this, the networks have fewer links.

It is also observed that in the networks characterized by Figs. 5a and 5e, they present greater amounts of links, due to the presence of many countries that do not appear in other networks, mainly due to the fact that some PhD's registered in the Lattes Platform were born in other countries, not being Brazil, and/or in some cases, choose to work professionally in other countries.

Some dyads stand out in the link networks, in which when analyzing them, it is noticed that these dyads are represented by links between us with spelling errors, such as: the individual inserted the origin of the "Butsuana" link and the destination of the link "Botswana"; in another case, representing incomplete data, the individual inserted Bosnia and Hezergovina-Bosnia, also causing a dyad. However, really representing a dyad we have a fact that a PhD inserted the bond in Timor-Leste and the bond in Kosovo.

The networks do not have isolated components, since all links have a node characterizing the entry (being represented as the emigration location) and another node as the destination (representing the immigration location).

A network was created with all the links obtained at birth, using all levels of education, and professional performance (Fig. 6), just not taking into account the birth level for professional performance.

In this network, all nodes belong to a giant component, having no isolated component, and the greatest number of nodes with degrees whose degree value is greater than 9 stands out, it was observed that all Brazilian states are present in this network of links, with a degree higher than 9. It was noted that nodes with degree values greater than 60 have links with most nodes with amount of nodes greater than itself.

The Table 2 is intended to represent a summary of the characterized networks and calculations of metrics adopted.

When comparing the number of nodes and edge rate in the Giant Component, with the global network, it is noticed that the network that stands out the most is the network that has all the links, since it has the largest connected Giant Component, corresponding in 100% of the edges and 100% of the nodes, different from the other networks, in which the values of the rates of the nodes are also high, but with a percentage lower than the network of all links. As far as density is concerned, the network with all links has the third lowest value, different from the master's and PhD network that has the highest density, since the nodes are quite connected, and a possible justification for this is the networks of collaborations belonging to individuals who are in the transition between these levels of education.

It is observed that the largest diameter is represented by the network of all links, with a value equal to 6, where there is a distance of 5 nodes between the two most distant nodes. The other networks have the same diameter value totaling the value of 5, in which the two most distant nodes need to travel 4 nodes to meet.

The network from birth to graduation has the lowest average degree of nodes, being justified because some nodes have a low degree value because they do not have large amounts of bonds, thus decreasing the average degree value. Unlike the network with all links that has the highest average degree value, where the nodes with the highest grades increase the average. With regard to the average minimum path, it is possible to observe that the networks have very close values (approximately 2), on average it only takes two edges to reach a given node from any other node that makes up the network.

To expand the analysis and obtain a better view of the states and countries that have the highest degrees of entry, Fig. 7 was characterized.

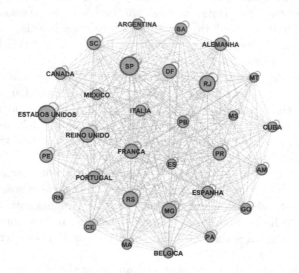

Fig. 7. Link networks with higher degrees of entry. Source: authors.

It is possible to highlight that the network is dense with a value of 1, since all nodes are connected to each other. It was also observed that all the states of the South, Midwest and Southeast regions are present, with a hypothesis for this the number of universities belonging to these regions. The North and Northeast regions, on the other hand, have lower numbers of states present, due to the inferior job offers and inferior qualities of graduate courses compared to the other regions.

When observing the countries present in the network, it is clear that Argentina is the only South American country. Still in analyzes aimed only at American countries, Cuba, Mexico, Canada and the United States stand out as the largest nodes, which are those with the greatest number of ties. The United States country stands out with the highest degree value, being inferior only to the state of São Paulo. Some of the possible reasons why the United States is the country that has the highest degree among nodes are: it is the country that has the largest world economy; have a universal language; have a large number of scholarship offers, among others.

When it comes to countries in Europe, seven countries are present, the United Kingdom, France, Portugal, Germany, Spain, Belgium and Italy with the highest degrees of entry of individuals respectively.

6   Conclusion

From the results obtained, it was possible to verify the feasibility of adopting the curricula registered in the Lattes Platform as a source of data for analysis on how the Brazilian Scientific Exodus occurs.

The choice of the group of PhD's is characterized as a significant portion of the entire set of data registered in the Lattes Platform, considering that they are the individuals with the highest level of academic education completed. It was also noticed that in general their resumes are recently updated and most of them have a registered professional address.

In addition, it also became clear how the southeast region and the state of Rio Grande do Sul concentrate the vast majority of Brazilian PhD's throughout the academic education process, a fact directly influenced by the concentration of the main public universities in the country.

After carrying out the general characterization of the set of individuals to be analyzed, the next stage of the work was to obtain the geolocation data from all the institutions in which the PhD's were trained at some level of academic training.

After it was possible to find the geolocation data of the institutions, it was possible to identify the average distance traveled by individuals throughout their academic training.

Also noted that the country United States stands out from the rest of the countries, since it corresponds to the country with the largest number of Brazilian fellows enrolled in the country. There is also a greater interaction between European countries compared to South American countries, because European countries have a better quality of education.

References

1. Abel, G.J., Sander, N.: Quantifying global international migration flows. Science **343**(6178), 1520–1522 (2014)
2. Almeida, G.Z.R.: Fluxos migratórios: a distribuição da população de cada estado pelo país (2017). https://www.nexojornal.com.br/grafico/2017/12/01/Fluxos-migratorios-a-distribuicao-da-populacao-de-cada-estado-pelo-pais. Accessed 1 Jul 2020
3. Andrade, R.O.: O impacto da circulação de cérebros. Revista Fapesp, pp. 18–25 (2019)
4. Aveiro, T.M.M.: O programa ciência sem fronteiras como ferramenta de acesso à mobilidade internacional. Tear: Revista de Educação Ciência e Tecnologia **3**(2) (2014)
5. Chaves, L.C.R., et al.: Analisando a mobilidade de pesquisadores através de registros curriculares na Plataforma Lattes. Dissertation – Universidade Federal da Paraíba (2016)
6. Demartini, M.: Falta de oportunidades mantém cientistas brasileiros no exterior (2017). https://exame.abril.com.br/ciencia/falta-de-oportunidades-mantem-cientistas-brasileiros-no-exterior/. Accessed 1 Jul 2020

7. Dias, T.M.R.: Um estudo da produção científica brasileira a partir de dados da plataforma Lattes. Thesis (Ph.D. in Mathematical and Computational Modeling)-CEFET-MG (2016)
8. Jonkers, K., Tijssen, R.: Chinese researchers returning home: Impacts of internationalmobility on research collaboration and scientific productivity. Scientometrics **77**(2), 309–333 (2008)
9. Lombas, M.L.D.S.: A mobilidade internacional acadêmica: características dos percursos de pesquisadores brasileiros. Scielo **19**(44), 308–333 (2017)
10. de Pierro, B.: Circulação limitada. Pesquisa Fapesp, pp. 36–39 (2016)
11. de Silva, P.S., de Queiroz, S.N.: Migração intraestadual no rio grande do norte. Idéias, **11**, e020008 (2020)

Indexes for Evaluating Research Groups: Challenges and Opportunities

Areli Andreia dos Santos[(✉)] and Moisés Lima Dutra

PGCIN, Federal University of Santa Catarina, Florianópolis, Brazil
{areli.santos,moises.dutra}@ufsc.br

Abstract. Several indexes have been proposed to evaluate the scientific research productivity. These indexes are useful for several management decisions, such as career development and distribution of financial capital to scientific researches. In the last two decades, several works have proposed indexes to measure how productive and relevant the work of a researcher is. Some of these indexes aim to be applicable to evaluate an individual researcher and/or groups of researchers, applying the same formula for both situations. However, in some cases, this application is not straightforward. In this work, we studied the main indexes used to evaluate productivity of scientific researchers, discuss their aspects and organize them according to their characteristics, by pointing up the variables that are used to compose each index. We also analyzed works that applied the h-index for a group of researchers, highlighting the different ways adopted for this application. Furthermore, we discussed the opportunities to expand these metrics in a fairer way to evaluate quantitative and qualitative aspects for groups of researchers. After a literature review of these indexes, we conclude that the h-index can be used in many ways to evaluate groups of scientific researchers. Hence, there are other challenges and opportunities in the proposal of indexes to evaluate these groups, such as performing experiments for smaller groups, evaluating social aspects and defining better ways for selecting articles to evaluate the research productivity of groups.

Keywords: Scientific research evaluation · Indexes · Metrics · Group of researchers

1 Introduction

In the last decades, the access to scientific publications such as articles and journals was facilitated by the use of the Internet. In addition, several indexes [1–3] were proposed to compare and evaluate how relevant a publication is and what are its impacts on society. These indexes are useful specifically to evaluate quantitative aspects, measuring how much a researcher, a group of researchers and even a university or country produces in science. One of the main uses of these statistical data is for a fairer distribution of funds and other resources among scientific entities [4].

E. Bisset Álvarez (Ed.): DIONE 2021, LNICST 378, pp. 41–53, 2021.
https://doi.org/10.1007/978-3-030-77417-2_3

Expressing such an intangible value as scientific productivity is not an easy task, and there will always be distortions of the reality. The most primary idea is to use the number of papers, but it only measures quantity, not quality. Another common metric is the number of the citations, that can express how much relevant a scientific paper is to other scientists. The number of citations is the most relevant and traditional metric to evaluate the impact and relevance of a scientific article.

In order to combine both quantity and quality, several indexes (i.e. h-index, e-index, i10index, etc.) had been proposed by using both the number of papers and citations, aiming to measure not only the quantity, but how useful a paper is. After that, self-citation became an issue, and other indexes [2, 3, 5, 6] were proposed to avoid this distortion. More recently, aspects considering the online environment have also been applied to these indexes, bringing new metrics such as clicks and downloads, i.e. the altmetrics. But the number of papers and citations still remain as one of the most relevant metrics for measuring production and relevance in science.

While some of the proposed indexes are used to calculate data for an individual researcher [1, 7, 3], others are used in the calculation for a group of researchers [8–10]. In this work, we consider a group of researchers as any group of two or more researchers, which could be students, professors, and other academics who work together. This group setting can vary from a small group composed of a few research colleagues who are applying for a small and non-financed project to a large research group working on an international project with many researchers and resources from different countries. Besides, a group of researchers can also be considered on a larger scale, such as a group of all the researchers of a country or even all the researchers in a given field considering a global scale.

Some of the proposed works claim that the application of existing metrics originally created for evaluating individual researchers to a group of researchers is a straightforward task, e.g. there are several approaches that use the h-index to evaluate research groups [11, 12]. However, these approaches use specific h-index extensions to analyze groups. The main extensions of h-index for groups are: i) the h_1-index [11], which considers an entire institution or department, taking into account all the papers of this institution/department in its calculation; ii) [13, 14] calculate an average of the h-indexes for a group of researchers; iii) a so-called successive h-index [15] calculates the h-index of an institution by identifying as top researchers all those who have a number of publications greater than or equal to a predetermined limit value for the h-index. According to this proposal, to calculate the h-index for an entire country, it is necessary to identify the top institutions that have an h-index greater than or equal to this predetermined limit value.

Nonetheless, even the established ways to extend the h-index for evaluating groups raise questions: All the papers of an author who worked in several institutions should be considered, or only the ones published by him/her while working in his/her present institution? Shall we consider all the citations of all authors, regardless of their co-authors? Or should we focus only on the papers in which all members of the group participate? Could we also focus on the papers in which only a few members of the group participate? The number of authors in each paper must be considered? All these questions are still not answered and can be grouped in one big question: which articles should be selected to measure how relevant the scientific research productivity of a group is?

There are in the literature some surveys that address the indexes used to measure scientific researchers [4, 16, 17]. While some of them focus on analyzing a specific knowledge area [16, 17], some others are more generalists [4]. These works are focused on metrics to analyze papers and researchers, but there is no specific emphasis for a group of researchers. Still, some other works proposed the application of the h-index for a group of researchers [11, 13–15]. These works mostly used a transformation, such as averaging, to apply the h-index to analyze a specific field, by comparing different institutions. However, to the best of our knowledge, none of these works summarized the two approaches: indexes to evaluate both individuals and groups of researchers. Considering the problem of applying metrics that consider both quantitative and qualitative aspects for a group of researchers, the objective of this work is to study the main indexes used to measure science, by means of evidencing which values are used to compute each index. Subsequently, based on a literature review, we present the main challenges and opportunities for research in this area.

The remainder of this article is organized as follows. In Sect. 2, we summarize the research methodology that was used to search and select the main works related to scientific measurement. In Sect. 3, we summarize the main aspects of each index in a table, and present a more detailed explanation for each one. In Sect. 4, we discuss the challenges and opportunities for measuring the scientific production of research groups. Finally, in Sect. 5 we present the conclusions of this work.

2 Research Methodology

In this section we present a state-of-the-art on indexes used to evaluate productivity in science. We conducted an exploratory search, and found the main papers through Google Scholar. An initial search in SCOPUS database was performed, but since no relevant results were found, we focused on the first source.

The terms used in this exploratory search were: "Measurement scientific research", "Index scientific research", "Bibliometric scientific research", "Index Group of researchers", and "h-index for groups". Several false positive results were shown in Google Scholar, however an initial group of ten papers was selected. From them, new works were included. After reading each paper, we could identify some surveys, and most of these papers were focused on highlight pros and cons for each metric. However, none of them presented a summary combining the two indexes for individual and groups of researchers. Subsequently, we evaluated each description for the indexes in order to identify and simplify the way in which these indexes are presented. Thus, we summarized the description of each index in Table 1, which presents the main outcome of this work, and depicts the application of each index for groups of researchers. Finally, we analyzed the existing works in order to identify the challenges to assess the productivity of a group of researchers. The focus of our analysis is to propose research opportunities for this area.

In the next section, we present a literature review summarizing the main indexes and providing a more detailed explanation of each index.

Table 1. Metrics to evaluate research productivity

Metric	Used for	Source	Short description
Number of papers	Groups and Individuals	Number of papers	Counter of the number of papers
Number of citations	Groups and Individuals	Number of citations	Counter of the number of citations
h-index (Hirsch 2005)	Individuals	N. of papers and citations	Total of h papers with at least h citations each
h(2)-index (Kosmulski 2006)	Individuals	N. of papers with n^2 citations	Total of h papers with at least h^2 citations each
g-Index (Egghe 2006)	Individuals	N. of top papers	Total of g papers with at least g^2 citations altogether
A-index (Jin et al. 2007)	Individuals	Average citations	Average of citations of an author's publications
R-index (Jin et al. 2007)	Individuals	N. of citations	$\sqrt{}$A-index (square root of A-index)
AR-index (Jin et al. 2007)	Individuals	N. of papers and citations, age of papers	$\sqrt{}$A-index normalized by the age of the article
RP-Index (Altmann et al. 2009)	Individuals	N. of papers and citations, weight of author contribution and age of the article	NC_{ij} considers all variables described. Similar to h-index, but considers the NC_{ij} not h
e-index (Zhang CT. 2009)	Individuals	N. of citations	Prunes the total number of citations by a number of h^2
Index h_m (Schreiber 2009)	Individuals	N. of citations, papers and authors	Counting the papers fractionally according to the inverse of the number of authors
i10-index (Google 2011)	Individuals	N. of papers and citations	Number of papers with at least 10 citations each
m-Quotient (Hirsch 2011)	Individuals	h-index and age of 1st paper	Divide the h-index by the number of years since the first publications
Crown Indicator	Groups and Individuals	N. Citations	Compares a paper with the average citations of other similar articles

(continued)

Table 1. (*continued*)

Metric	Used for	Source	Short description
AIF - Author Impact Factor (Pan and Fortunato 2014)	Individuals	Average of citations, time window	Average number of citations from an author's papers in the last years
PageRank-Index (Senanayake 2015)	Individuals	PageRank of Google Search applied to an article	The higher the PageRank of a papers is, the higher is its relevance
h_α-index (Hirsch 2019)	Individuals	h-index	The number of papers where an author is the main author
h_1-index (Mitra 2006)	Groups	N. of papers and citations	Total of h papers with at least h citations each
h_2-index (Mitra 2006)	Groups	N. of researcher and h-index of researchers	Total of n researchers with an h-index of at least n each
Successive h-indices (Shubert 2007)	Groups	h-index	Successive application of h-index formula using other values
CP-Index (Altmann et al. 2009)	Groups	RP-Index	Number of researchers n with an Total of n researchers with an RP-index of n least each
h_2/h_1−ratio (Rousseau et al. 2010)	Groups	h_1-index, h_2-index	h_2-index /h_1-index

3 Literature Review

Table 1 presents the source for the calculation of each one of the indexes that are used to measure quantity and/or quality for individuals and groups of individuals. After, we present a more detailed review of each index that is currently being used to assess scientific outputs.

3.1 Summary Table of Indexes for Evaluating Researchers

In order to provide a simpler way to identify each index, we organized them chronologically, accordingly to the time of each proposal, as shown in Table 1. Metrics to evaluate research productivity. The first column of Table 1 provides the index names. The second one depicts whether the index is used for groups or individuals. The third column presents the data source used to calculate each index. Finally, the fourth column provides a short description of how to calculate the index.

A more detailed explanation of Table 1 is presented in the next section.

3.2 Main Indexes to Evaluate Scientific Research

The first presented metrics, the counters, brings the simplest idea to evaluate productivity. More specifically, the number of papers to evaluate quantity and the number of citations to evaluate quality, i.e. how relevant a research is to others. It is the most seminal idea of an index and most of the subsequent works are based on these counters. Then, we explain the h-index itself, as well as other indexes to evaluate research productivity for individuals and groups of individuals.

Number of Papers. It is the simpler way to evaluate scientific productivity, by straight-forwardly counting all papers by a researcher or a group of researchers. The number of papers can also be specified, for instance, to consider only articles from a given area or only papers published in high impact factor (IF) journals. However, this metric cannot be used to evaluate quality, and so, it is usually classified as a quantitative measure [4].

Number of Citations. It is a most sophisticated counter, that considers the number of times where a certain paper is cited, and thus used by another scientific work [16]. This metric can be used to evaluate researchers, a group of researchers and even journals, by the sum of citations for a certain group of papers. However, this metric cannot be used to evaluate a solid contribution in a certain field, because it could be biased if a single paper is highly cited. Another bias of the number of citations is that authors that are focused on surveys can have a higher number of citations than a scientist who is effectively producing new contributions.

h-index. One of the most famous metrics that combines number of papers and number of citations [1]. If one paper has at least one citation, the h-index is 1. If two papers have at least two citations each, the h-index is two. Consequently, if ten papers have at least ten citations each, the h-index is ten, and so on. This became one of the most used indexes in the last decade, being used in Web of Science Indexing, Scopus Indexing, Google Scholar Indexing, among others.

h(2) index. "A scientist's h(2) index is defined as the highest natural number such that his h(2) most-cited papers received each at least $[h(2)]^2$ citations." (Kosmulski 2006) [5]. For instance, a h(2) of 10 express the value of 10 papers that were cited at least 100 times each. Accordingly to Kosmulski, this index is useful to the chemistry area, and can be adapted to a h(n)-index, accordingly to another specific field, even using the h-index as an h(1)-index. This index was one of the first ones to try overcome the self-citation issue.

g-index. This metric uses an input of papers ranked by citations. The higher the number of citations, the higher in the list will be the article. As a result, the g-index delivers the largest number of g-articles that had g^2 citations altogether. This index gives a higher value than the h-index and can be biased by a few articles with a high number of citations [2].

A-index. Proposed by Jin et al. in 2007 [18]. It is named A-Index because it is calcu-lated with the average of citations of an author's publication and takes into consideration the same papers used to evaluate the h-index. This index might not be ideal to compare researchers with different number of papers, since the average can be biased by a high number of papers with only a few citations, even when the individual has solid contributions with other works.

R-index. Proposed by Jin et al. in 2007 [18], it was presented as a solution to the A-index problem. In order to normalize the number of papers, the R-index is simply calculated by the square-root of the A-index.

AR-index. Also proposed by Jin et al. 2007 [18], this index is claimed as a solution for one of the h-index problems: when a top-researcher retires or even reduce the number of his/her publications, his/her h-index can still have a high value. Thus, this index is a variation of R-index, but instead of use the square root for calculating the average of citations, the number of citations is normalized by the age of a paper. The older the paper, the less this paper will impact the index.

RP-Index. Proposed by Altmann et al. in 2009 [19]. RP stands for Research Productivity. This index considers the NC_{ij}, that is calculated by the number of citations, divided by the age of the paper and multiplied by a factor of collaboration of an author for the paper, which is a number varying from 0 to 1. The NC_{ij} factor is then used similarly to the h-index as "the largest natural number x such the top x publications have at least in average a value of x for their NC_{ij}" [19].

Index h_m. Schreiber in 2009 [20] proposed an index that considers co-authorships. Through this index, the papers are counted fractionally according to the inverse of the number of an author's rank.

e-index. Proposed in 2009 by Zhang CT [3]. This index was created as a criterion for when more researchers have the same h-index value, but a different number of citations. Must be used only for highly cited authors, it prunes the number of citations by a number of h2 citations to calculate a modified h-index.

i10-index. Proposed by Google in 2011 and used in the Google Scholar platform. This metric corresponds to the number of papers from a researcher that has, at least, 10 citations each [21].

m-Quotient. Hirsch, the creator of h-index, proposed in 2011 his own improvement to the famous index. The m-quotient is based on the h-index, which is divided by the number of years since the first publication of an author [17].

Crown-Indicator. Proposed by Centre for Science and Technology Studies at Leiden University, this index is also known as field citation score (Joshi 2014). It considers only the citations from publications by a researcher or a group of researchers that can be equivalent to the world average of citations of works at the same time span, document type, subject area and age, configuring as a subset of comparable works. For instance, if the Crown-indicator of a paper is 0.8, that means this article has 20% less citations than the average. And if a paper has a Crown-indicator of 1.1, that means that this paper has 10% more citations than the average of other papers that fit the same criteria [4].

Author Impact Factor. In 2014, Pan and Fortunato [22] proposed the Author Impact Factor given by "AIF of an author A in year t is the average number of citations given by papers published in year t to papers published by A in a period of Δt years before year t". This metric is another proposal to overcome the h-index issue of researchers that are no longer actively working, but have kept a high h-index score over several years.

PageRank Index. This index was proposed in 2015 by Senanayake [23]. In order to calculate its value, it is required to analyze a citation network by using the Google's PageRank algorithm. The higher the PageRank, the more relevant the article is.

h_α-index. Hirsh [24], in 2019, proposed the $h\alpha$-index to quantify an author's leadership. Its value is obtained considering the same papers used to calculate the h-index, i.e. it is a counter. In other words, the $h\alpha$-index is the number of papers where an author is the main contributor of the paper.

The aforementioned indexes focus on evaluating an individual researcher, more specifically his/her productivity, and also how relevant a certain set of publications is. One of the most relevant indexes, the h-index has variations to evaluate groups of researchers. Some of the proposals to extend the h-index to evaluate a group of researchers are shown in the following.

h_1 index. Mitra in 2006 [11] proposed two extensions of h-index to evaluate research institutions. The first one, the h_1-index corresponds to $h_1 = h$ when the institution has published h papers, with at least h citations each. As long as the h-index has as an input of papers of a given author, the same idea is used by Mitra to evaluate an institution. This is a very simple metric that can be easily applied. However, as for the h-index, when the publication is an effort of two or more institutions, the same paper is integrally counted for both of them.

h_2 index. The second index proposed by Mitra [11] is the h_2 index. This metric takes into account the already calculated h-index of each researcher and gives an idea of the top individuals from some institution. In this scenario, an institution possesses an h_2 index $= 1$ if one of its researchers has an h-index of 1, another institution has an h_2 index $= 2$ if two of its researchers have an h-index of 2, ..., and, consequently, an institution will have an h_2 index $= 20$ when 20 of its researchers have an h-index of at least 20, and so on. However, this index can be biased by one group of researchers of the same institution that publish altogether, since in this case the same papers will be considered two or more times in the calculation.

Successive h-indices. It is a model for successive h-indexes proposed by Shubert in 2007 [15]. An author's h-index is calculated by using his/her number of papers and citations. In an institutional level, the idea is to use the h_2-index proposed by Mitra. For higher levels, e.g. a whole region or country, the same idea based on the h_2-index of the lower levels is applied. The modeling of successive h-indexes proposed that a country can be represented by the h_2-indexes of its institutions, e.g. with that approach being applicable in many different levels.

CP-Index. Proposed by Altmann et al. in 2009 [19]. It is based on the RP-Index. The CP-Index is defined as "the largest natural number y such that the top y researchers of this research community have at least in average a value of y for their RP-Index, given that the researchers are sorted according to their RP-Index in decreasing order". It is similar to the Successive h-index model, but instead of being based on the h-index of the researchers, it uses its own RP-Index metric.

(h_2/h_1)-Ratio. Proposed by Rousseau et al. in 2010 [25]. In this paper, the authors highlight that the indexes proposed by Mitra [11] can be calculated in different ways: either considering all the papers of a given author, or only the ones that possess the institutions address in the body of the paper. They also point that two institutions with the same h_2-index might have very different profile of publications. Besides, authors with medium h-index might feel not valued in their institutions. They proposed then dividing h_2/h_1 in order to give a structural indicator of the institution.

The h-index and other standard bibliometric indicators for chemistry research groups was compared by Raan in 2006 [26]. In order to apply h-index for groups, he used all papers of the research group. The study was performed for 147 chemistry research groups in Dutch universities. The study observed a correlation between h-index and citation, however for smaller groups with "less heavy citation traffic" the crown indicators were considered more appropriate.

In 2009, Jacsó [12] applied the h-index for South American countries in Web Of Science and Scopus. He compared the top 10 rank h-indexes of universities in both databases. The study considered that the h-index for groups in Scopus and Web of Science was robust, since both ranks showed almost the same results, except for one position.

An application of h-index average was proposed by Rad et al. in 2010 [13] to evaluate research in radiology. In this study, they selected radiology programs and obtained the h-indexes and the numbers of citations and publications for selected radiologists. They performed a regression analysis to determine which variables were best associated with the academic ranking. A correlation was identified between unrated growth and the h-index of the institution's researchers.

Another application of h-index average to evaluate neurosurgery departments was performed by Khan et al. in 2013 [14]. They evaluate the h-index average by considering sex, academic rank, years of practice, subspecialty, and institution. Their conclusion is that the h-index average can distinguish productivity for academic rank, subspecialty and years of practicing. They also point out that the application of h-index average is not reliable to compare neurosurgery departments.

In this section, we presented several indexes to evaluate a researcher, as well as h-index variations to evaluate research groups, by showing different applications and results accordingly to each field. In the next section we present the challenges and opportunities for measuring groups of researchers.

4 Discussion

This chapter is divided in two sections. In the first one, we present the challenges identified by means of analyzing the current proposed indexes. In the second section, we point out opportunities for evaluating research groups.

4.1 Challenges for Evaluating Groups of Researchers

The h-index is the most popular metric to combine quantitative and qualitative aspects for scientific evaluation of research productivity. The main idea behind it is very simple and practical. It provides a number that combine a quantitative measure assessed of productivity with the relevance of the work according to the number of citations received by it. However, even being a very useful index, a few issues have been brought to light in the last years.

One of the first issues is related to the "quality versus quantity" problem. Authors that are focused on survey papers can have a higher number of citations than others that are producing novel knowledge. The most relevant index to overcome this issue is the Crown-indicator, which only considers papers identified as equivalent by certain characteristics, such as: area of application, publication type, publication period, among other aspects, in order to provide a fairer comparison.

Another issue related to h-index is that an author could have a very high h-index even after years of his/her retirement, while new researchers with new contributions might be hampered, even when presenting quality and quantity in a still short career. Consequently, a few indexes were proposed to overcome this problem, by means of using a time gap to take papers into account, such as the AIF metric [22].

Authors with very different number of papers and citations can have the very same h-index, so the use of self-citation to boost their h-indexes also has become a relevant issue. Indeed, the h-index in certain fields does not represent quite well the reality. A few indexes were proposed by applying a mathematical transformation on the h-index, such as considering citations on squared values [6, 27].

When addressing the evaluation of groups, many other problems become clear. The first one is related to which articles should be used to calculate the indexes. Many proposals focus on the institutional level and so, consider all the papers of a given institution. However, it is not clear if all the papers of each research of the institution are considered, or if there is any kind of special processing for the papers of some researcher who possesses multiple ties or who has changed his/her workplace over the years. This question is important to be considered in the variations of the h-index applied to groups: h_1-index, successive or average, since all of them depend on a set of papers as an input.

Other relevant issue not addressed in the previous works is the strength of a group as a whole. The works focus on institutions or university departments, but not on two or three researchers that are producing science together. There is somehow a union of separated parts and not a whole group. Some recent works use co-authorship [28, 29] and co-citation [30] metrics to evaluate groups of researchers. However, these works focus on analyzing specific areas, not in providing an index to evaluate the scientific weight or value of a group.

In the next section, we present some opportunities to overcome these issues.

4.2 Opportunities for Evaluating Groups of Researchers

Based on the issues highlighted in the last section, it is possible to observe that some of the gaps pointed out in the existing indexes have already been overcome. However, we believe that one of the most remarkable issues is related to the evaluation of the prestige, quality and quantity of the knowledge production of a group of researchers based on their publications.

Scientific papers and even projects usually are an effort of many people. It is commonplace to evaluate new proposals of research projects based on the fundamentals and criteria related to the project itself. Thus, when several project proposals are well formulated a new set of more objective criteria must be used, in order to provide a fairer evaluation. New measures must take into account the number of researchers, the quality of their work as a group, and the aspects of each researcher and the lead researcher. Another relevant aspect is how connected the researchers in this group are, or how linked their research fields are. These questions are important for evaluating research projects and defining, based on the objectives, which group are more suitable for each line of work.

As for future work, we intend to propose new metrics to evaluate: 1) How scientifically relevant a research group is, by proposing a single metric to evaluate both productivity and quality of a group of researchers, when considering their work together; 2) How strong is the connection from one researcher to another, so that is possible to estimate how close their research fields are. Consequently, the idea is to assess whether a group is more likely to become a joint effort in the same field, or an innovative group by merging two or more initiatives from different fields, and; 3) A new different metric for a group of researchers that combine two aspects: how relevant a scientific group is and how strong their connection as researchers is.

In this chapter, we presented the most relevant challenges and opportunities. Moreover, we talked about how the indexes have evolved to overcome the h-index limitations. Finally, we envisaged the proposition of new possible future metrics to overcome the issues presented in the evaluation of a group of researchers. The paper's conclusions are presented below.

5 Conclusions

In this work we presented and discussed the main indexes to evaluate the productivity of a researcher, in qualitative and quantitative ways. First, we summarized the literature review in a table, evidencing how each index is comprised and organizing them into two big groups: (i) the ones that evaluate a single researcher and (ii) others that apply the h-index in different ways to evaluate groups of researchers.

Based on the review of literature, we analyzed three different ways to expand the h-index for the group evaluation scenario: using all of the institution's work, averaging the h-index, and applying the successive h-index approach. Then, we pointed out some challenges to evaluate groups of researchers, such as self-citations and the difficulty in select which papers should be used to evaluate groups in an institutional level, considering that a researcher can change his/her place of work and also have more than one affiliation.

In the end, we also identified some opportunities for assessing research groups, such as evaluating small groups of researchers who work together and considering how linked they and their research fields are among themselves. This is an ongoing work and the opportunities addressed earlier are being worked on and will be more detailed in experiments to be presented in future works.

References

1. Hirsch, J.E.: An index to quantify an individual's scientific research output. Proc. Natl. Acad. Sci. **102**, 16569–16572 (2005). https://doi.org/10.1073/pnas.0507655102
2. Egghe, L.: Theory and practise of the g-index. Scientometrics **69**, 131–152 (2006). https://doi.org/10.1007/s11192-006-0144-7
3. Zhang, C.-T.: The e-Index, complementing the h-index for excess citations. PLoS One **4**, (2009). https://doi.org/10.1371/journal.pone.0005429
4. Joshi, M.A.: Bibliometric indicators for evaluating the quality of scientific publications. J. Contemp. Dent. Pract. **15**, 258–262 (2014). https://doi.org/10.5005/jp-journals-10024-1525
5. Kosmulski, M.: A new Hirsch-type index saves time and works equally well as the original h-index. ISSI Newslett. **2**(3), 4–6 (2006)
6. Flatt, J., Blasimme, A., Vayena, E.: Improving the measurement of scientific success by reporting a self-citation index. Publications **5**, 20 (2017). https://doi.org/10.3390/publications5030020
7. Hirsch, J.E.: An index to quantify an individual's scientific research output that takes into account the effect of multiple coauthorship. Scientometrics **85**, 741–754 (2010). https://doi.org/10.1007/s11192-010-0193-9
8. Valles, M., Injante, R., Hernández, E., Riascos, J., Galvez, M., Velasco, J.: An altmetric alternative for measuring the impact of university institutional repositories' grey literature. In: Mugnaini, R. (ed.) DIONE 2020. LNICST, vol. 319, pp. 222–234. Springer, Cham (2020). https://doi.org/10.1007/978-3-030-50072-6_17
9. Torres-Salinas, D., Robinson-Garcia, N., Jiménez-Contreras, E.: Can we use altmetrics at the institutional level? A case study analysing the coverage by research areas of four Spanish universities, p. 8 (2016)
10. Rousseau, R., Yang, L., Yue, T.: A discussion of Prathap's h2-index for institutional evaluation with an application in the field of HIV infection and therapy. J. Informetr. **4**, 175–184 (2010). https://doi.org/10.1016/j.joi.2009.11.007
11. Mitra, P.: Hirsch-type indices for ranking institutions scientific research output. Curr. Sci. **91**, 1439 (2006)
12. Jacsó, P.: The h-index for countries in web of science and scopus. Online Inf. Rev. **33**, 831–837 (2009). https://doi.org/10.1108/14684520910985756
13. Rad, A.E., Brinjikji, W., Cloft, H.J., Kallmes, D.F.: The h-index in academic radiology. Acad. Radiol. **17**, 817–821 (2010). https://doi.org/10.1016/j.acra.2010.03.011
14. Khan, N., Thompson, C.J., Choudhri, A.F., et al.: Part I: the application of the h-index to groups of individuals and departments in academic neurosurgery. World Neurosurg. **80**, 759–765.e3 (2013). https://doi.org/10.1016/j.wneu.2013.07.010
15. Schubert, A.: Successive h-indices. Scientometrics **70**, 201–205 (2007). https://doi.org/10.1007/s11192-007-0112-x
16. Agarwal, A., Durairajanayagam, D., Tatagari, S., et al.: Bibliometrics: tracking research impact by selecting the appropriate metrics. Asian J. Androl. **18**, 296 (2016). https://doi.org/10.4103/1008-682X.171582

17. Gasparyan, A.Y., Yessirkepov, M., Duisenova, A., et al.: Researcher and author impact metrics: variety, value, and context. J. Korean Med. Sci. **33**, (2018). https://doi.org/10.3346/jkms.2018.33.e139
18. Jin, B., Liang, L., Rousseau, R., Egghe, L.: The R- and AR-indices: complementing the h-index. Chin. Sci. Bull. **52**, 855–863 (2007). https://doi.org/10.1007/s11434-007-0145-9
19. Altmann, J., Abbasi, A., Hwang, J.: Evaluating the productivity of researchers and their communities: the RP-index and the CP-index, p. 15 (2009)
20. Schreiber, M.: A case study of the modified Hirsch index h_m accounting for multiple coauthors. J. Am. Soc. Inf. Sci. Technol. **60**, 1274–1282 (2009). https://doi.org/10.1002/asi.21057
21. Kozak, M., Bornmann, L.: A new family of cumulative indexes for measuring scientific performance. PLoS ONE **7**, (2012). https://doi.org/10.1371/journal.pone.0047679
22. Pan, R.K., Fortunato, S.: Author Impact Factor: tracking the dynamics of individual scientific impact. Sci. Rep. **4**, 4880 (2015). https://doi.org/10.1038/srep04880
23. Senanayake, U., Piraveenan, M., Zomaya, A.: The pagerank-index: going beyond citation counts in quantifying scientific impact of researchers. PLoS ONE **10**, (2015). https://doi.org/10.1371/journal.pone.0134794
24. Hirsch, J.E.: hα: an index to quantify an individual's scientific leadership. Scientometrics **118**, 673–686 (2019). https://doi.org/10.1007/s11192-018-2994-1
25. Ye, F.Y., Rousseau, R.: Probing the h-core: an investigation of the tail–core ratio for rank distributions. Scientometrics **84**, 431–439 (2010). https://doi.org/10.1007/s11192-009-0099-6
26. van Raan, A.F.J.: Comparison of the Hirsch-index with standard bibliometric indicators and with peer judgment for 147 chemistry research groups. Scientometrics **67**, 491–502 (2006). https://doi.org/10.1556/Scient.67.2006.3.10
27. Egghe, L.: Theory and practise of the g-index. Scientometrics **69**, 131–152 (2006). https://doi.org/10.1007/s11192-006-0144-7
28. Affonso, F., Dias, T.M.R., de Oliveira Santiago, M.: A strategy for co-authorship recommendation: analysis using scientific data repositories. In: Mugnaini, R. (ed.) DIONE 2020. LNICST, vol. 319, pp. 167–178. Springer, Cham (2020). https://doi.org/10.1007/978-3-030-50072-6_13
29. Fernandes, D., David, N., Cortinhal, M.J.: A distributed tool for online identification of communities in co-authorship networks at a university. In: Mugnaini, R. (ed.) DIONE 2020. LNICST, vol. 319, pp. 179–189. Springer, Cham (2020). https://doi.org/10.1007/978-3-030-50072-6_14
30. Marcelino, L.V., Pinto, A.L., Marques, C.A.: Intellectual authorities and hubs of green chemistry. In: Mugnaini, R. (ed.) DIONE 2020. LNICST, vol. 319, pp. 190–209. Springer, Cham (2020). https://doi.org/10.1007/978-3-030-50072-6_15

Expert Bibliometrics: An Application Service for Metric Studies of Information

Adilson Luiz Pinto[1]([envelope]) [iD], Rogério de Aquino Silva[1] [iD], André Fabiano Dyck[1,2] [iD], Gustavo Medeiros de Araújo[1] [iD], and Moisés Lima Dutra[1] [iD]

[1] PGCIN, Federal University of Santa Catarina, Florianópolis, Brazil
adilson.pinto@ufsc.br
[2] SETIC, Federal University of Santa Catarina, Florianópolis, Brazil

Abstract. Following the current trend of developing software as a service, we propose a software application to handle and support metric studies of information by applying the main laws of Bibliometrics, such as Lotka, Bradford, and Zipf, by means of integrating content and format in a single analysis process. Our application manages metrics according to the relevance of data, dispersion, rule of three and square root, in order to generate new indexes by relying on theories and laws already established among the scientific community. In addition, the developed tool has a pleasant aesthetic, along with a low cognitive effort for the user. To achieve such a scenario, a standardization of the interface combined with the fluidity of navigation within the application were used. The proposed application is suitable for those who work with academic and/or scientific issues, offering quick results compared to manual work, in which a lot of time is spent, either creating a system or analyzing spreadsheets. The result is a beta tool, available online at http://expertsbibliometrics.ufsc.br/, with login and password, respectively: 'ebbc2020@ufsc.br' and 'trocar123'.

Keywords: Bibliometric software · Metric studies application · Lotka application · Bradford application · Zipf application

1 Introduction

The development of tools to support the various types of metric studies presents a successful case among the scientific environment, since every knowledge area makes use of them to organize and understand its specific body of knowledge. These tools help facilitate the understanding of technical and scientific scenarios and trends. Typically, they are shaped in the form of software applications, which handle metric criteria according to specific needs. In this sense, there are some initiatives, such as Vantage Point [1]; CopalRed [2]; IN-Spire [3]; InCites [4]; SciMAT [5]; CiteSpace [6]; BibExcel [7]; SciVal [8]; Sci2 Tool [9]; Publish or Perish [10]; Network WorkBench [11]; VOS Viewer [12]; and Sitkis [13] that stand out in the global scope of metric studies of information. Those tools work with content-clustering systems, frequency of occurrences and average values for processing their calculations.

© ICST Institute for Computer Sciences, Social Informatics and Telecommunications Engineering 2021
Published by Springer Nature Switzerland AG 2021. All Rights Reserved
E. Bisset Álvarez (Ed.): DIONE 2021, LNICST 378, pp. 54–63, 2021.
https://doi.org/10.1007/978-3-030-77417-2_4

This group of applications usually calculate frequency for co-words, authority citation, journal citation, bibliographic coupling, clustering, h-index, co-authorship, performance of institutions, geolocation, among others. However, to the best of our knowledge, there is a lack of tools that apply the laws of Bibliometrics and some bibliometric standards that gave rise to these studies, when considering the complexity of representing data and images, for example. The Bibliometric laws were defined based on studies carried out by **Lotka** [14], **Bradford** [15], and **Zipf** [16].

Lotka [14] proposes the 80/20 rule (inverse distribution law), which establishes there is a core group of highly productive authors in a given theme or area of knowledge and their function is to assess the regularity of scientific productivity in that area. The 80/20 rule works with a constant for each theme (C) divided by the square of the total number of publications by author (n^2). In general, this rule first ranks the 20% of most productive authors identified by the amount of their publications. In fact, nowadays it is quite complex to keep this 80/20 ratio up to date. This is due to the fact that it is extremely difficult to keep an updated count of the authors' publications. There are more and more journal editions being published, new journals being created (especially the open access ones), and new sub-areas of knowledge being defined. In addition, there is the fact that one begins to quantify the efficiency of the authors from other types of documents, such as books, works presented at events, and book chapters. This is because not all areas of knowledge give exclusivity to the publication of articles in scientific journals.

$$Y_{(n)} = \frac{C}{n^2}$$

Bradford [15] determines the core group of the most productive journals in a given theme and whose representation is organized based on the number of titles (both within the same theme and within the same database), by zones or subsets (e.g.: zone of 33%) multiplied by the proportionality factor of the number of journals in each of the defined zones. Often the sum of the number of journals is calculated by a decreasing ranking, by means of using three equal dimensions of 1/3 each or 33% in each zone. The calculation is a division of the total by 3, forming a zone with a large percentage (1st scale), but with few titles; another zone with a median representative percentage (2nd scale), but with a considerable number of journals; and a third zone with a small representative percentage (3rd scale), but with so many journals that, as a rule, only count a single appearance on the topic;

$$p : p1 : p2 : 1 : n : n^2$$

Zipf [16] proposes metrics for supporting thematic issues of publications and revealing what the interdisciplinary contents of documents are. The focus of its applicability relies on the analysis of journal keywords, in order to identify which journals are the most representative ones in the complete universe of existing journals. However, this law's initial proposal aimed to identify the ranking position of frequent words in a given text in relation to the most frequent ones. At that time, all words were counted, including articles and prepositions. In current applications, only the keywords of the articles are worked on, thus facilitating an analysis more directed to the understanding of the syntheses created by the authors to shortly and objectively represent their texts via keywords. Zipf's Law works with a constant extracted from the principle of least effort (c), which is obtained by multiplying the ranking of the most frequent words in the text (r) by the frequency of occurrences of the words in the text (f). Recently, this same law was applied, with variations, in "stability zones of appearance of terms", with the first zone being reflected as "trivial", the second zone as "interesting data" and the third zone as "noise". It is possible to apply this same approach in the way we understand the information resulting from searches on search engines.

$$(r)(f) = c$$

In this paper, we propose an application service to work with metric studies of information that uses not only the three basic laws of Bibliometry, but also that includes the application of other techniques. This proposal includes the application of two other techniques for analyzing scientific impact, which use other metrics in addition to calculating the frequency of terms. These techniques are the ones proposed by **Price** [17] and **Platz** [18].

Price [17] is an analysis model that identifies the elite of most cited authors, by using the sum total of citations received by the authors. To identify the amount of these elite authors, the square root of the total number of authors is calculated.

$$E = \sqrt{n}$$

Platz [18] works with a visibility index related to the presence of journals in their scientific contexts, which is calculated by dividing the total number of citations received (from other journals) by the total number of articles published by this same journal.

$$V = \frac{In\_cb}{A}$$

The proposed analytical tool has the potential to become one of the most relevant in its niche, whether by calculating scientific productivity or by calculating the impact of publications. Moreover, it comprises the fusion of content for analysis extracted from different databases, such as Web of Science, Scopus, LISA, and Google Scholar, for example. The proposed tool, Expert Bibliometrics, can be accessed at http://expertsbibliometrics. ufsc.br/,with login and password, respectively:**'ebbc2020@ufsc.br'** and **'trocar123'**.

2   Goals

The need for an software application that could integrate content and different formats of corpora in a single analytical dashboard, along with the need for an analytical tool that checks the laws and the main standards of metric studies of information, were the main motivating factors behind this proposal. Furthermore, our expectation is to go beyond the mere representation of indicators by analyzing frequency of terms. We intend to do this by means of applying dispersion techniques, rules of three and square roots, in order to generate new indexes based on theories and laws already established in the scientific environment, but little represented in software applications.

Thus, the main objective of this proposal is to develop a software tool to analyze data through the application of the main laws and bibliometric standards. Specifically, the proposal aims to:

a) Analyze content extracted from several extentions of databases, like CSV format separated by comma and in BIBTEXformat, and aggregate them into common fields;
b) Apply bibliometric laws (Lotka, Bradford and Zipf), depending on the relevance of the data;
c) Apply other bibliometric techniques, such as Price and Platz's theories, for authority and journal citations; and
d) Generate analysis results by more than one representation, such as data clustering based on the graph theory.

3 Development

The main features of the Expert Bibliometrics tool are:

a) It is a software application made available as a service, accessible from any browser, on any operating system;
b) The input data to be processed are the result of the search in several journal databases (Web of Science, Scopus, LISA, IEEE, etc.), in which the user undertook a previous search. To achieve this scenario, the same search terms are used by Expert Bibliometrics. When this occurs, that is, when analyzing data from several journals, the user needs to ensure that his previous searches in each of the databases have used the same search terms. After that, the raw data is downloaded to the user's local computer. The user needs to be aware of the standards defined for the column names in the downloaded data, so that it is possible to select which columns will be analyzed together, when applicable, e.g.: Author, Authors, and/or AU. Through this combination, the system will be able to uniformly reformat the data;
c) It provides an option called "Organize Table", which standardizes the representation of the uploaded files. For example, one file could use "semicolons" to separate fields, another one could use "comma space-space" or "comma space", and still other ones could use "semicolon" to separate different fields and "comma" within the same field. Care must be taken not to separate the surname from the Author's first name, for example. This software feature makes a quantification of commas within each field, to know whether to separate the columns with a comma or semicolon. This arrangement is done before the file is effectively separated into columns;
d) It allows uploading files in CSV format separated by commas and in BIBTEX format. The user needs to know the content of these files, in order to be able to select the correct columns to be analyzed;
e) It can analyze files with only a few dozen records, as well as files with thousands of records. The processing time is proportional to the number of records to be analyzed;
f) It is able to analyze a wide variety of files to apply the laws of Lotka, Bradford and Zipf. The results are presented in the form of: (i) a synthetic summary; (ii) graphical visualizations of the percentages of publications; and (iii) in the form of graphs, in which the clusters of authors are shown.

g) In the next version of the tool, still under development, it will also be possible to analyze visibility applications.

The features of Expert Bibliometrics can be classified into three categories (Figs. 1 and 2):

(a) **Import**. Feature that receives raw data from different periodical databases.

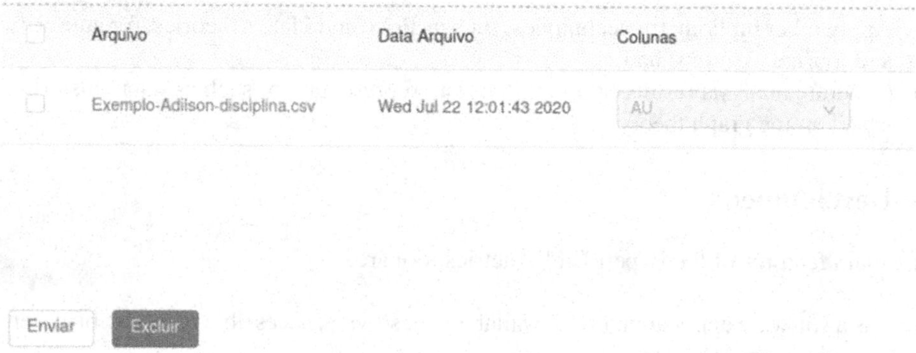

Fig. 1. Files uploaded to be analyzed

(b) **Analysis**. A application of the laws of Lotka, Bradford and Zipf on imported data, and.

Fig. 2. File and column selection (AU) for analysis of authors

(c) **Report**. Presentation of the results.

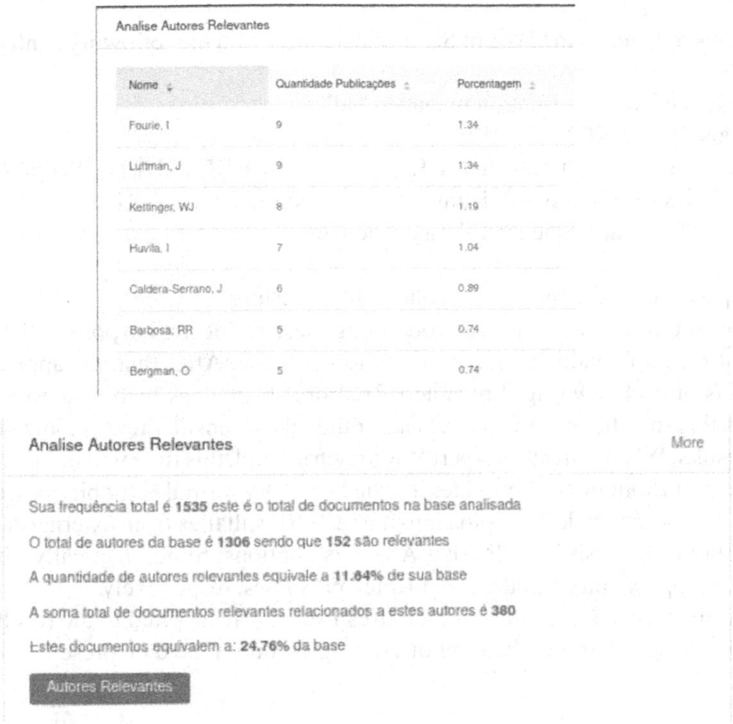

Nome	Quantidade Publicações	Porcentagem
Fourie, I	9	1.34
Luftman, J	9	1.34
Kettinger, WJ	8	1.19
Huvila, I	7	1.04
Caldera-Serrano, J	6	0.89
Barbosa, RR	5	0.74
Bergman, O	5	0.74

Analise Autores Relevantes

Analise Autores Relevantes   More

Sua frequência total é **1535** este é o total de documentos na base analisada

O total de autores da base é **1306** sendo que **152** são relevantes

A quantidade de autores relevantes equivale a **11.64%** de sua base

A soma total de documentos relevantes relacionados a estes autores é **380**

Estes documentos equivalem a: **24.76%** da base

Autores Relevantes

Fig. 3. Summary report of the analysis of relevant authors.

4 Results

The tool's dashboard was developed with the focus on offering the user a standardized interface with pleasant aesthetics, low cognitive effort, with the least possible number of clicks per functionality, and with smooth navigation in the software.

The proposed application service, Expert Bibliometrics 1.0, has the following modules:

a) User authentication module;
b) Module for loading multiple files in CSV format separated by comma and in BIBTEX format, extracted from different databases;
c) Module for analyzing author relevance;
d) Module for analyzing the more representative journals;
e) Module for analyzing keyword relevance;
f) Analysis report module, in table format (Fig. 3);
g) Analysis report module, in graph format (Fig. 6); and
h) Authors' analysis report module, in graph format (Fig. 6).

5 Discussion

We undertook a search in the Web of Science database, with the following configuration:

- search expression: "information management";
- field: Topic; range: 2008 to 2017;
- collections: Science Citation Index Expanded (SCI-EXPANDED), Social Sciences Citation Index (SSCI), Arts & Humanities Citation Index (A & HCI);
- category: Information Science Library Science

When performing the search, we obtained 672 results.

In a non-automated scenario, after obtaining these results and exporting them, there is a manual work of handling on author and journal data. After that, the application of the formulas of each law (e.g. Lotka and Bradford) begins, as well as a co-authoring analysis. In this specific scenario, both tasks could take a considerable amount of time to generate results. When using the Expert Bibliometrics tool, this time will be considerably reduced. Expert Bibliometrics provides: (i) the necessary formulas for bibliometric laws integrated in its source code; (ii) uploading the search result files from external databases; (iii) the "Author Analysis" or "Journal Analysis" options, to automatically obtain the results of the application of Lotka's and Bradford's laws, respectively.

The "Author Analysis" feature, which uses Lotka's law applied to the results of the aforementioned search, for a given set of criteria, produces the data presented in Fig. 4 as a result.

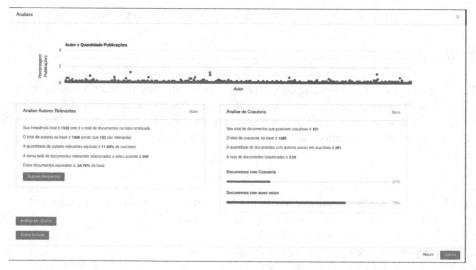

Fig. 4. Analysis of relevant authors and analysis of co-authorship

The analysis of relevant authors shows that the total number of documents analyzed in the database was 1535. The total number of authors in the database is 1306, of which

152 are relevant. The number of relevant authors is equivalent to 11.64% of the total number of documents within the database. The total sum of relevant documents related to this total number of authors is 380, which is equivalent to 24.76% of the base documents. In addition, it is also possible to click on the "Relevant Authors" button to obtain a list of the names of the relevant authors.

The co-author analysis shows that the total number of documents with co-authors is 421. The total number of co-authors in the database is 1285. The number of documents with single authors is 251. The rate of co-authors is 3.05%. The "Journal Analysis" feature, which uses Bradford's law applied to the search results, with the above criteria, produces the data presented in Fig. 5 as a result.

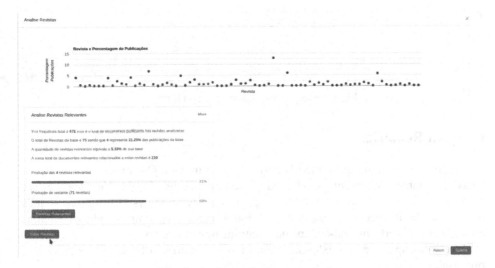

Fig. 5. Analysis of relevant journals

The analysis of relevant journals shows that the total number of documents published in the analyzed journals is 672. The total number of journals in the database is 75, with only 4 of them representing 31.25% of the publications in the database. The number of relevant journals is equivalent to 5.33% of the total base. Finally, the total sum of relevant documents related to these journals is 210.

In addition to the analysis of bibliometric laws, the system is also programmed to present the results in a graph form. This presentation, in addition to being more visually pleasing, also provides a range of indicators of co-authorship, co-relationship, proximity, intermediation.

The use of the proposed tool on the research undertaken in the Web of Science database allowed us to verify the speed and ease in obtaining the desired results. Consequently, we believe that its public availability on an open access portal will bring enormous benefits to researchers in the area of metric studies of information.

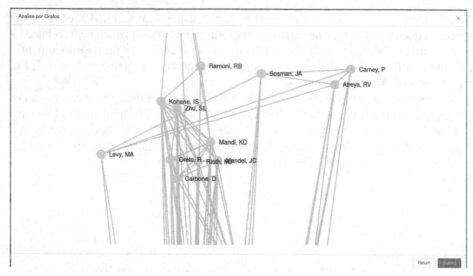

Fig. 6. Analysis of authors in graph format

6 Final Remarks

The proposed software application is currently still in beta. Our expectation is that a tested and stable version will be released in the first months of 2021. New features are being added, such as the analysis of Price's elite authority (in analysis of citation), Platz's theory (for the impact of journals), in addition to processing simpler indexes, such as co-authorship, works by authorship, and citations by authorship.

The following is a non-exhaustive list of new features to be added to the initial proposal:

a) The "Graph Analysis" report will show the power of influence of each author (analysis of force), based on changes in the diameters of the nodes.
b) The "Analysis by Graphs" report will allow the selection of authors to be analyzed, which will allow a faster result to be obtained.
c) The option to select authors will also be offered in the main menu of the tool.
d) The "Graph Analysis" report will allow a click on the author's name to display his/her affiliation.
e) From the identification of affiliation, the analysis of authors will show which universities cooperate and collaborate with each other. In addition, it will be possible to know which authors interact with which universities.
f) A feature will be implemented that will allow the results of the analyzes to be sent by e-mail. Some possible export formats that will be used: PDF, CSV, or LaTeX.
g) The user will be offered the option to define filters by year or other time intervals, so that it is possible to analyze the number of publications per year, for each author.
h) Analysis of citations, citations of the most cited authors, and the most cited journals will be made available.
i) The user will be able to save his analyzes, in order to be able to access them later.

References

1. Porter, A.L.; Cunningham, S.W.: Tech Mining: Exploiting New Technologies for Competitive Advantage. Wiley-Interscience (2004)
2. Bailón-Moreno, R.: Ingeniería del conocimiento y vigilancia tecnológica aplicada a la investigación en el campo de los tensio activos: desarrollo de un modelo cienciométrico unificado. Doctoral Tesis. Universidad de Granada, Granada (2003)
3. PNNL. IN-SPIRETM: Visual document analysis. U.S. Department Energy (2019). https://inspire.pnnl.gov/index.stm. Accessed 07 Aug 2020
4. Clarivate Analytics. InCities Help (2020). http://help.incites.clarivate.com/inCites2Live/indicatorsGroup/aboutHandbook.html. Accessed 07 Aug 2020
5. Cobo, M.J., López-Herrera, A.G., Herrera-Viedma, E., Herrera, F.: SciMAT: a new science mapping analysis software tool. J. Am. Soc. Inf. Sci. Technol. **63**(8), 1609–1630 (2012). https://doi.org/10.1002/asi.22688. Accessed 07 Aug 2020
6. Chen, C.: CiteSpace II: detecting and visualizing emerging trends and transient patterns in scientific literature. J. Am. Soc. Inf. Sci. Technol. **57**(3), 359–377 (2006)
7. Perssonl. O., Danell, R., Schneider, J.W.: How to use Bibexcel for various types of bibliometric analysis. ISSI Newslett. **5**(1), 5–24 (2009). https://homepage.univie.ac.at/juan.gorraiz/bibexcel/ollepersson60.pdf. Accessed 07 Aug 2020
8. ELSEVIER. SciVal (2020). https://www.scival.com/landing. Accessed 07 Aug 2020
9. Linnemeier, M.: Science of Science (Sci2) tool manual (2014). http://sci2.wiki.cns.iu.edu. Accessed 07 Aug 2020
10. Harzing, A.W.: The Publish or Perish Book. Tarma Software Research Pty Ltd, Melbourne (2010)
11. Börner, K., Chen, C., Boyack, K.: Visualizing knowledge domains. Ann. Rev. Inf. Sci. Technol. **37**(1), 179–255 (2003)
12. van Eck, N.J., Waltman, L.: VOS: a new method for visualizing similarities between objects. In: Decker, R., Lenz, Hans -J. (eds.) Advances in Data Analysis. SCDAKO, pp. 299–306. Springer, Heidelberg (2007). https://doi.org/10.1007/978-3-540-70981-7_34
13. Singh, N.: Complementing Bibliometrics with Network Visualization to Support Scientific Spheres. IFLA WLIC, Atenas (2019)
14. Lotka, A.J.: The frequency distribution of scientific productivity. J. Washington Acad. Sci. **16**(12), 317–323 (1926)
15. Bradford, S.C.: Sources of information on specific subjects. Eng. Illustrated Weekly J. **137**, 85–86 (1934)
16. Zipf, G.K.: Human Behaviour and The Principle of Least Effort. Addison-Wesley Press, Boston (1949)
17. Price, D.J.S.: Networks of scientific papers. Science **149**(july), 510–515 (1965)
18. Platz, A.: Psychology of the scientist: XI Lotka's law and research visibility. Psychol. Rep. **16**(2), 566–568 (1965). https://doi.org/10.2466/pr0.1965.16.2.566. Accessed 07 Aug 2020

The Role of Artificial Intelligence in Smart Cities: Systematic Literature Review

Ivana Dominiković, Maja Ćukušić(✉) ⓘ, and Mario Jadrić ⓘ

Faculty of Economics, Business and Tourism, University of Split,
Cvite Fiskovica 5, 21000 Split, Croatia
idomin01@live.efst.hr, {mcukusic,jadric}@efst.hr

Abstract. The increase in urban population has brought climate, technological and economic changes that may negatively affect the quality of life in cities. In response, the concept of a smart city has emerged referring to use of novel ICTs to reduce the adverse effects on cities and its inhabitants. Among other technologies, Artificial Intelligence (AI) is used in that context, evolving rapidly and playing an essential role in supporting intelligent city-wide systems in different domains. It is thus beneficial to identify current research advances and get a better understanding of the role the AI plays in this particular context. Consequently, there is a need to systematically study the connection between AI and smart cities, by focusing on the findings that uncover its role, possible applications, but also challenges to using the concepts and technologies branded as AI in smart cities. Therefore, the paper presents a systematic literature review and provides insights into the achievements and advances of AI in smart cities pertaining to the mentioned aspects.

Keywords: Smart city · Artificial intelligence · Systematic literature review

1 Introduction

Growing urban populations, environmental pollution, climate change, infrastructure breakdown, and lack of resources are some of the challenges cities face. Such challenges require novel solutions, giving rise to an encouraging concept of a "Smart City" (SC). By employing smart solutions, existing day-to-day scenarios in the cities can be significantly improved, and new ones could be created whereby the aim is to increase the quality of life of its residents.

With the rapid development of Artificial Intelligence (AI), opportunities emerge to improve public services, some of which are presented in the paper. As AI develops faster, new and smart solutions for cities and its citizens are developing at the same rate, but also leading to the development of associated tools used for malicious purposes. Cyber-attacks can compromise the privacy of data collected through the Internet of Things (IoT) devices containing not only public but also personal data of individuals. In addition to privacy concerns, if the safety of citizens and systems that use smart solutions is compromised, it can lead to potentially disastrous situations such as attacks on traffic, health, surveillance and security systems etc.

E. Bisset Álvarez (Ed.): DIONE 2021, LNICST 378, pp. 64–80, 2021.
https://doi.org/10.1007/978-3-030-77417-2_5

Ethical issues remain another significant challenge. A well-known example is a dilemma involving autonomous vehicles in SCs, where, in an accident scenario, the vehicle decides whether to sacrifice the passenger, a pedestrian or hit another vehicle (Dennis and Slavkovik 2018). Such issues require human participation in decision-making and novel standards and ethical norms. Further to this, there are ethical issues concerning facial recognition, behavioural traits identification, voice recognition, and other trends. The accelerated development of AI in the future is likely to increase the number of ethical challenges.

The rise in interest in AI and SCs, its potentials and issues, has led many researchers to contribute to the field, resulting in a sharp increase in the number of published studies. Although this increase provides more insights and greater body-of-knowledge in a specific area, we as users need more time to identify relevant studies in a particular case. Thus, this paper presents an attempt to systematically review the relevant literature on the topic of AI in SCs, and can consequently serve as a starting point for further research in the field.

The methodology for the review is presented in Sect. 2, categorisation of papers in Sect. 3, followed by the review in Sect. 4, and concluding remarks in Sect. 5.

2 Methodology for Literature Review

A systematic literature review was done to formally synthesise primary studies relevant to the already described area of interest (Kofod-Peterson 2015). It differs from non-systematic studies as it is methodologically framed, with a series of well-defined steps. A predetermined protocol includes identification, selection, evaluation and analysis of relevant studies that are eligible for review, going beyond limitations of traditional reviews (Palka et al. 2018). During the process, relevant recommendations (Wolfswinkel et al. 2013) were followed, and standard databases (Web of Science – WoS and Scopus) were selected as sources. Although other databases may be suitable for this purpose, they contain a large number of papers that are already indexed in the two databases so this would significantly increase the number of duplicates. The keywords used to search for the papers are "artificial intelligence" AND ("smart city" OR "smart cities"), but limited to scientific papers written in English. Another limitation is that alternative and related keywords such as "computational intelligence" were not considered at this stage.

Visualisation of the terms from the papers from the WoS database search is presented in Fig. 1 to get a general overview of the results. It was created using the VOSviewer tool based on a set of general data of the papers and abstracts. Four clusters are revealed, each marked with a different colour. The size of a circle indicates the frequency of occurrence of a particular term in a dataset, and the length of the line between two terms their relationship. Understandably so, the term AI appears in the dataset most frequently. The shorter length of the line between the two terms, for example, between the terms AI and big data, indicates a stronger connection confirming the relevance of big data technologies in AI.

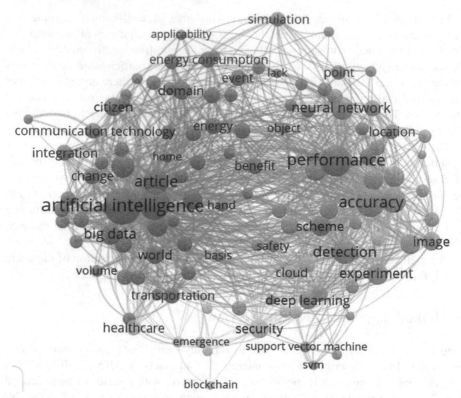

Fig. 1. Network visualisation of the terms from WoS database search results (Source: Authors)

The literature sampling procedure is illustrated in Fig. 2, with the numbers of papers handled in each phase. Out of the results from WoS (360) and Scopus (230), there was less than 5% of the papers that could not be accessed (some indexed in both databases) that were then excluded from further analysis. Further to that, the papers indexed in both databases (duplicates) were excluded from one. In the first round of selection, papers that met the following criteria were kept: abstracts precisely corresponding to the research topic, abstracts containing both keywords, or even some of the known subcategories, methods or algorithms. For example, all papers that did not contain "artificial intelligence" and "smart city(s)" in the abstract, but instead contained keywords such as "smart healthcare" or "smart transportation system" in addition to one of the keywords related to AI such as "supervised learning", "reinforcement learning", and similar – were left in the sample. The second round of selection eliminated all false-positive papers that passed the first round of selection, but after reviewing the rest of the text and graphics did not meet the stated criteria. The process of forward- and backwards tracking citations, i.e. searching for papers that cited papers in the sample, or that were cited by the authors of sampled papers, resulted in including 9 additional papers in the sample for review. Since duplicates were removed from the Scopus list, to make the graphical representation representative, papers are subsequently grouped and striped of database classification.

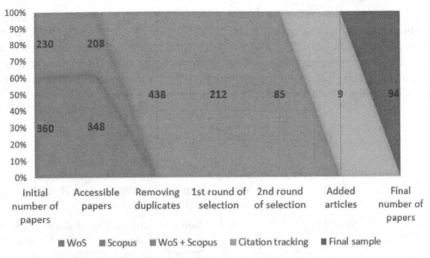

Fig. 2. Literature sampling procedure (Source: Authors)

The final number of papers after the sampling procedure is 94. The span of these papers is only five years, starting from 2015 to 2020. Although both terms (AI and SC) are represented in the database earlier than 2015, and the initial list of papers contained papers published earlier (though not before 2012), many papers did not meet the presented criteria for the 1st and 2nd selection rounds. Looking at the final list of selected papers, in 2019 a total of 39 were published, i.e. 3.25 per month, while in 2020 a total of 25 papers were published during the first six months, i.e. 4.17 per month. The growth in the number of papers is not surprising, given that AI is a rapidly evolving discipline. Out of the selection, the largest number of papers (17) was published in the journal IEEE Access. This is followed by the IEEE Internet of Things Journal (with 9 papers), Applied Sciences (6 papers), Sensors (5 papers) and Sustainability (3 papers). The remaining 54 papers were published in other journals (45 different outlets), one or two papers per journal. In the five journals listed here, 40 out of 94 papers were published, accounting for 43%.

3   Categorising the Papers for the Literature Review

Due to the diversity in the content of the final list of papers focusing on different aspects of the connection between AI and SCs, several categories were created based on relevant literature reviews (Rjab and Mellouli 2018, 2019). The three main categories are General role, Application areas and Challenges. Each of these categories refers to AI in SCs, i.e. the role of AI in SCs, the application of AI in SCs and the challenges that may arise from the use of AI in SCs. These categories are further divided into subcategories, as presented in Table 1. In particular, a large number of papers explores the use of AI in mobility (e.g. traffic safety, smart parking) so these were grouped under the subcategory Mobility of category Application areas. Other subcategories were defined similarly. Consequently, and in line with good review practices (Webster and Watson 2002), a cluster of references was created in the form of a concept matrix. This way, it is easy to see the most represented concepts/subcategories (in terms of content), as well as the ones containing fewer studies so it can serve as a good starting point for future reviews. Several papers fall into two different categories, for example (Liang et al. 2019) discussing both the role and challenges of AI for the security in SCs (thus bringing the total number of papers across categories to $N = 109$).

Table 1. The categorisation of selected papers

Cat.	Subcategories	No. of papers	Authors
General role	Ethical aspects	3	Shen et al. (2019), Sholla et al. (2020), Zhou et al. (2020)
	Security	7	Chen et al. (2019a), Diro et al. (2017), Falco et al. (2018), Kim et al. (2020), Liang et al. (2019), Roldan et al. (2020), Xu et al. (2020)
	Data management	9	Al Zamil et al. (2019), Aydin et al. (2015), Chen et al. (2019b), Dilawar et al. (2018), Ferrara et al. (2019), Gong et al. (2019), Iqbal et al. (2020), Shu et al. (2019), Yao et al. (2019)
	Other	3	Anthony (2020), Austin et al. (2020), Gomez et al. (2018)
Application areas	Mobility	20	Chen et al. (2019b), Iqbal et al. (2020), Asad et al. (2020), Aymen et al. (2019), Cai et al. (2019), Choudhury et al. (2018), El-Wakeel et al. (2018), Hossen et al. (2019), Huang et al. (2016), Huang et al. (2019), Hwang et al. (2019), Ke et al. (2020), Ke et al. (2020), Li et al. (2018), Mannion et al. (2015), Martinez Garcia et al. (2018), Niu et al. (2015), Qiu et al. (2020), Wan et al. (2018), Wang et al. (2020)
	Environment	11	Chen et al. (2019b), Ahmed et al. (2019), Cao et al. (2019), Jung et al. (2020), Khan et al. (2019), Mo et al. (2019), Park et al. (2019), Ping et al. (2020), Rojek et al. (2019), Wu et al. (2020), Zhang et al. (2019)
	Surveillance	12	Chen et al. (2019b), Jung et al. (2020), Iqbal et al. (2020), Almeida et al. (2018), Castelli et al. (2017), Eldrandaly et al. (2019), Medapati et al. (2019), Liu et al. (2019a), Miraftabzadeh et al. (2018), Qin et al. (2018), Xiong et al. (2017), Zhao et al. (2019b)

(continued)

Table 1. (*continued*)

Cat.	Subcategories	No. of papers	Authors
	Energy	12	Aymen et al. (2019), Austin et al. (2020), Abbas et al. (2020), Almeshaiei et al. (2019), Hurst et al. (2020), Idowu et al. (2016), Liu et al. (2019c), Park et al. (2019), Serban et al. (2020), Le et al. (2019a), Le et al. (2019b), Vazquez-Canteli et al. (2018)
	Smart home	6	Hurst et al. (2020), Xu et al. (2020), Lin et al. (2017), Ponce et al. (2018), Sanam et al. (2020), Yassine et al. (2017)
	Health	11	Chen et al. (2019b), Iqbal et al. (2020), Zhou et al. (2020), Ajerla et al. (2019), Alhussein et al. (2019), Alhussein et al. (2019), Amin et al. (2019), Mohanta et al. (2019), Obinikpo et al. (2017), Venkatesh et al. (2018), Zhang et al. (2017)
	Education	3	Gomede et al. (2020), Gomede et al. (2018), Wang S. (2019)
	Other	8	Iqbal et al. (2020), Dilawar et al. (2018), Liu et al. (2020), Liu et al. (2019b), Manzanilla-Salazar et al. (2020), Shousong et al. (2019), Talamo et al. (2020), Zhao et al. (2019a)
Challenges	Security	1	Liang et al. (2019)
	Ethical issues	3	Calvo (2020), Dennis et al. (2018), Etzioni et al. (2016)

Providing a better view of representation per subcategories, Fig. 3 emphasises the popularity of AI use in mobility systems with as many as 20 papers. The gap between the prevalent subcategory in terms of the Application area – Mobility, and the two succeeding – Surveillance and Energy (equally represented by 12 papers per subcategory) demonstrates the interest in AI in the context of smart mobility well. This is followed by Health and Environment application areas with the same number of papers (11), Smart home (6), Education (3), and 8 papers on various topics in the subcategory Other.

4 Overview of Research Studies Focusing on Artificial Intelligence in Smart Cities

4.1 Studies on the General Role of AI in Smart Cities

Many (ethical) considerations have to be taken into account in the implementation of SC solutions. Collecting large amounts of data, public and personal, using a variety of

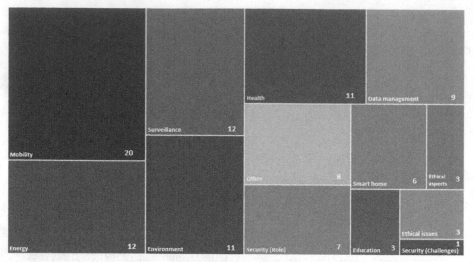

Fig. 3. Visual representation of subcategories with number of papers (source: authors)

IoT devices brings attention to issues related to the privacy of collected data. Papers presenting IoT-collected data protection solutions are based on blockchain and machine learning (ML) (Shen et al. 2019), and the human-in-the-loop model (Zhou et al. 2019). Furthermore, it is given that smart devices are deprived of moral, religious and legal responsibilities which can lead to situations where human rights may be compromised; here Sholla et al. (2020) propose a system for checking the ethical compliance of smart things.

Security solutions based on AI algorithms are devised to ensure safe and reliable use of smart solutions, so, e.g., Chen et al. (2019a) propose an algorithm that provides smart control of wireless communications and intelligent applications. Also, the use of deep learning (DL) algorithms has been proposed to detect cyber-attacks more effectively (Diro and Chilamkurti 2017), similar as recommending other AI techniques for tackling data that cannot be processed in time and can thus pose a potential threat, to detect anomalies (Xu et al. 2020). Falco et al. (2018) present the development of a tool for assessing the risk of critical infrastructure that could potentially encounter a cyber-attack. Similarly, there are examples of good practices for using AI in the context of cybersecurity in SCs, e.g., a solution that uses ML and proposes an intelligent architecture that can detect different types of IoT security attacks in real-time (Roldan et al. 2020), and an integrated security system presented by Kim and Ben-Othman (2020). Numerous benefits of using ML to increase the security of IoT devices are identified by Liang et al. (2019).

On a related note, issues associated with *data management* stem from the fact that IoT devices collect audio, video, images or text data that is difficult to manage with traditional algorithms. The role of ML is inspected (Aydin et al. 2015) to facilitate data management with applications in image classification (Shu and Cai 2019) and by employing DL algorithms (Chen 2019b) to classify, among other, audio data (Al Zamil et al. 2019). Ferrara et al. (2019) present models that use AI to analyse data collected through sensors in real-time. It is also deemed necessary to improve management of data traffic in IoT-networks, by, e.g., classifying data traffic using DL techniques to improve network security and service quality (Yao et al. 2019). Other examples of using ML algorithms are focusing on social networks data to get a better understanding of citizens' preferences (Dilawar et al. 2018). On the note of participation, it is essential to involve nonprofits in SC projects, and the factors that drive them to engage are analysed through the use of ML successfully (Gong et al. 2019). The concept of big data plays a vital role in building functional SCs since traditional data analysis cannot cope with the sheer volume of data, which is why AI methods are used (Iqbal et al. 2020).

There are *other* problems where AI can bring relevant solutions to SCs. For example, to improve the capacity of network structures in SCs, Gomez et al. (2018) suggest optimal network load balancing techniques. For strategic decision-making in the implementation of SC initiatives, AI is used to develop a recommender system for SC planning (Anthony 2020). Another option is, with the help of AI and ML, to build city's digital twin that will play a role in data collection and processing, event identification and automated decision making (Austin et al. 2020).

4.2 Studies on the Application Areas of AI in Smart Cities

As presented already, (smart) mobility is where AI is used frequently as a part of SCs initiatives. Traffic data is analysed using DL algorithms to enable safer and smarter use of transport networks with applications to transport flow, personal mobility and parking in particular (Chen 2019b). With the help of sensors and ML techniques (Hossem 2019), such as neural networks (Cai et al. 2019), and a combination of AI algorithms and edge computing paradigms (Ke et al. 2020), drivers can get helpful assistance in finding a parking space. Radar images are also implemented and analysed as an excellent option for SCs as these minimise privacy issues (Martinez Garcia et al. 2018). An increasing number of electric vehicles in the streets necessitates research on new approaches to energy management, e.g. based on two-way communication between vehicles and buildings using neural networks (Aymen and Mahmoudi 2019). One of the most significant challenges in urban mobility management is stakeholder safety. In that context, an improved framework for detecting pedestrians using neural networks was presented (Choudhury et al. 2018). Poor roads cause vehicle damage and congestion and can affect traffic safety. In response to this problem, a road anomaly monitoring framework was introduced that uses ML techniques and collects data from motion sensors available in most vehicles, allowing the detection of road irregularities and their impact on vehicle movement (El-Wakeel et al. 2018). An intelligent system focused on forecasting taxi demand was also presented (Iqbal et al. 2020), and tested in the city of Shanghai (Huang et al. 2019). Another example from China is the use of DL in transportation systems (Li et al. 2018). Vehicle identification also plays a vital role in forensic research; DL techniques are used for improved model detection capability (Ke and Zhang 2020). Differences in the datasets used to identify vehicles can lead to inadequate solutions, and a specific learning method has been proposed to address it (Wang et al. 2020). With solutions for traffic management using DL queues and waiting times could be reduced effectively (Mannion et al. 2015; Wan and Hwang 2018). An innovative route search mechanism based on traffic prediction and DL techniques can also improve driving conditions and reduce time-to-destination (Niu et al. 2015). Precise determination of paths using DL to optimise travel times is proposed (Qiu et al. 2020). Energy-wise optimisation of train routes, as a part of a train management framework using ML, was presented (Huang 2019). In this context, passenger data from sensors installed at railway stations have also been used to optimise passenger traffic (Asad et al. 2020; Hwang et al. 2019).

Effective environmental monitoring is one of the critical tasks in SCs. Using neural networks, scientists can identify the causes of problems, issue prior warnings, and organise resources for rescue (Chen et al. 2019b), e.g. for wildfires, extreme waves, cold, and heat, a faster decision-making framework based on AI has been presented (Jung et al. 2020). Air pollution is becoming a severe problem; and a system for early detection of changes in air quality is of great importance (Mo et al. 2019), as is the analysis of water quality trends and detection of anomalies (Ahmed et al. 2019). A system for detecting water leaks from pipes has been developed based on neural networks to minimise water losses (Rojek and Studzinski 2019). Noise detection is another essential aspect of SC management; a method based on neural networks improves the current performance of urban sound recognition (Cao et al. 2019). Early detection of smoke can prevent large disasters where neural network-based smoke detection systems can be effective (Khan et al. 2019). Also, a system with a mechanism to minimise data transmission delays in the described contexts was presented (Liu et al. 2019c), while other AI methods are proposed to detect and predict fire sources in inconspicuous places such as tunnels (Wu et al. 2020). Furthermore, street waste detection systems based on DL are also valuable (Ping et al. 2020; Zhang et al. 2017).

Surveillance cameras can be used for various purposes in sustainable cities; new AI algorithms are introduced to adjust camera orientation and improve view coverage (Eldrandaly et al. 2019). It is becoming common to surveil and monitor whether a large number of people in one place represents traffic congestion or a traffic jam (Liu et al. 2019a). In that regard, for network congestion management, an algorithm for network load balance is designed (Zhao et al. 2019b). Video surveillance for fire detection using neural networks is also a welcomed solution (Jung et al. 2020) for increased safety, as well as systems that analyse data on crime from large datasets (Castelli et al. 2017). A model that describes user behaviour based on DL predicts subsequent actions and recognises behavioural anomalies (Almeida and Azkune 2018). Similarly, it is possible to identify abnormal behaviour in mass video surveillance data (Qin et al. 2018). Other examples include face recognition systems that use AI (Chen et al. 2019b) and compare face images with those from databases (Medapati et al. 2019), identifying people using their biometric and behavioural traits (Iqbal et al. 2020), and re-identifying individuals (Xiong et al. 2019), especially in crowds (Miraftabzadeh et al. 2018).

Energy management is one of the most demanding problems in SCs, this being a reason why the topic is getting a lot of attention. Here, an energy management design based on IoT devices and ML is proposed (Liu et al. 2019c) and a platform for the development of an energy-wise sustainable city based on AI (Park et al. 2019). Renewable energy is vital for the development of cities where significant progress is made, as confirmed by analyses on the use of AI in the renewable energy sector (Serban and Lytras 2020; Almeshaiei et al. 2019; Abbas et al. 2020). A digital twin architecture and a simplified analysis of energy consumption in buildings in the Chicago area using ML (Austin et al. 2020) and a similar purpose model (Idowu et al. 2016) and techniques (Le et al. 2019a) have been proposed, e.g. predicting heat load in buildings (Le et al. 2019b). Also, a simulation environment is presented, combining an energy simulator and a platform for the implementation of ML algorithms to plan and manage energy consumption in buildings (Vazquez-Canteli 2018). A study exploring the use of AI to optimise energy use in electric vehicles (Aymen and Mahmoudi 2019) has already been mentioned above.

In smart homes, ML can be used to analyse data from smart gas meters and predict fuel consumption (Hurst et al. 2020). Data on indoor air trends and quality can be used to detect correlations with peoples' behaviour (Lin et al. 2017), and predict indoor climate conditions (Ponce and Gutierrez 2018). Precise localisation of different subjects indoors using wireless signals based on ML techniques (Sanam and Godrich 2020) is also of relevance in this context. Collecting data from smart devices in home environments can reveal behavioural patterns of individuals who live there and use ML techniques to obtain information used in healthcare (Yassine et al. 2017). Smart devices, in combination with DL algorithms, can serve to detect other anomalies in households (Xu et al. 2020) as well.

With the development of AI technologies, many applications in healthcare have been presented, including a model that uses neural networks and enables biometric identification of a person based on electrocardiograms (Zhang et al. 2017). By collecting data from sensors, it is possible to (remotely) monitor the condition of patients (Chen et al. 2019b; Iqbal et al. 2020; Venkatesh et al. 2018; Ajerla et al. 2019) and with the implementation of new 5G networks, the benefits could be even more significant (Mohanta et al. 2019). Other notable examples include a DL platform that records voice using smartphones and sends information about voice changes to reduce visits to the doctor in particular cases (Alhussein and Muhammad 2019). A system has also been proposed that automatically detects possible diseases using DL to examine brain activities using electroencephalographic (EEG) data (Alhussein et al. 2019). EEG sensors and others can record patient data such as facial expressions, speech, movements, gestures to be processed by AI. In this way, the patient's condition is monitored in real-time, and emergency care is provided as needed (Amin et al. 2019). Patient data collected through sensors are considered sensitive data; therefore, DL techniques can be used for its protection (Obinikpo and Kantarci 2017) in addition to prediction and decision-making purposes. A study combining AI with the human-in-the-loop model to preserve privacy in smart healthcare (Zhou et al. 2020) has been mentioned already.

Optimising educational content that would match the abilities of learners is one of the main tasks of smart education. To that end, an AI model was presented (Wang et al. 2019). AI methods were used in another study to improve e-learning systems with regard to different learning styles (Gomede et al. 2018), later extended to include more effective ways to leverage AI for developing individual learner profiles in individualised approach and goal-setting (Gomede et al. 2020).

Other uses of AI that cannot be grouped straightforwardly in the categories above are listed here. For example, inadequate and small data samples and errors in quantifying the features needed to assess the value of residential land make it difficult to estimate its value adequately. Here, a value estimation method using AI and transfer learning was studied (Shousong et al. 2019). Another example is selecting the optimal location of new stores based on DL (Liu et al. 2019b). Conservation and restoration of cultural assets is a complex process that requires the participation of individuals from different fields, including data scientists as the processes include the use of IoT and AI (Talamo et al. 2020). A framework based on ML can be used to successfully detect failures on wireless network base stations (Manzanilla-Salazar et al. 2020). Another area where AI is used is robot intelligence; to ensure the timely response of robots to stimuli from the external environment, DL and the Internet of Robotic Things (IoRT) have been proposed to monitor and control their behaviour (Liu et al. 2020).

4.3 Studies on the Challenges of AI in Smart Cities

Smart solutions require storing large datasets collected from IoT devices and often remain unprotected from cyber-attacks. Although ML is frequently used in supporting the security of cybernetic systems (as presented already with the studies that tackle the role of AI in security management), full protection cannot be guaranteed. Equally, AI is also used when performing cyber-attacks (Liang et al. 2019), and for that reason, it is worth analysing it from the perspective of lasting challenges.

Due to the ability of AI to analyse large datasets, in specific cases, some control and decision-making processes have been taken over by algorithms, some of which can contribute to unequal opportunities and treatment of individuals, contributing to social inequalities (Calvo 2020). Also, the number of smart systems such as autonomous vehicles that make independent decisions is growing, which is why there is a need to eliminate possible unethical behaviour through controlled models of AI (Etzioni and Etzioni 2016). In response to ethical questions related to the use of AI in SC solutions, a new discipline of artificial intelligence has emerged - machine ethics that develops intelligent systems with ethical concepts. Since this discipline is still in its infancy, it faces many challenges (Dennis and Slavkovik 2018).

5   Concluding Remarks

By selecting scientific papers from the interest area following a standard systematic process, a total of 94 papers was identified as relevant. All papers from the selected sample were published from 2015 to mid-2020, with the largest number in the last two years. Therefore, the review contains a novel overview of AI in SCs, complementing recent studies (Rjab and Mellouli 2018; Rjab and Mellouli 2019).

Categories and subcategories have been defined, and the papers categorised considering its primary focus. There is an observable disproportion between the number of papers in different categories, with the majority representing specific solutions for the application of AI in SCs, and a minimum number of those that cover the area of challenges and barriers to using AI in SCs. A popular topic is smart mobility, addressing improvement of parking, traffic management and safety. On the other hand, the lack of papers dealing with ethical and safety issues does not mean that such problems do not exist, but that either there is no interest from researchers, or that these problems have not yet been sufficiently researched. Although AI provides many benefits for the development of smart solutions, it has been used for cyber-attacks and manipulation of personal data collected from various sensors, making this problem non-negligible.

The main shortcomings of the paper are the limited number of databases searched (WoS and Scopus) and possible bias of the authors in selecting the final list of papers for the review. Even though increasing the number of sources (beyond the two standard ones) would significantly increase the number of duplicates, it would improve the quality of the literature review by offering a more representative view of publications on the topic. The paper can thus serve as a starting point for further research.

Acknowledgment. This work has been supported by the Croatian Science Foundation (project No. IRP-2017-05-7625).

References

Abbas, S., et al.: Modeling, simulation and optimisation of power plant energy sustainability for IoT enabled smart cities empowered with deep extreme learning machine. IEEE Access **8**, 39982–39997 (2020)

Ahmed, U., Mumtaz, R., Anwar, H., Mumtaz, S., Qamar, A.: Water quality monitoring: from conventional to emerging technologies. Water Supply **20**(1), 28–45 (2019)

Ajerla, D., Mahfuz, S., Zulkernine, F.: A Real-time patient monitoring framework for fall detection. Wirel. Commun. Mobile Comput. (2019). https://doi.org/10.1155/2019/9507938

Al Zamil, M.G.H., Samarah, S., Rawashdeh, M., Karime, A., Hossain, M.S.: Multimedia-oriented action recognition in Smart City - based IoT using multilayer perceptron. Multimedia Tools Appl. **78**, 30315–30329 (2019)

Alhussein, M., Muhammad, G.: Automatic voice pathology monitoring using parallel deep models for smart healthcare. IEEE Access **7**, 46474–46479 (2019)

Alhussein, M., Muhammad, G., Shamim Hossain, M.: EEG pathology detection based on deep learning. IEEE Access **7**, 27781–27788 (2019)

Almeida, A., Azkune, G.: Predicting human behaviour with recurrent neural networks. Appl. Sci. **8**, 305 (2018). https://doi.org/10.3390/app8020305

Almeshaiei, E., Al Habaibeh, A., Shakmak, B.: Rapid evaluation of micro-scale photovoltaic solar energy systems using empirical methods combined with deep learning neural networks to support systems' manufacturers. J. Clean. Prod. **244**, 118788 (2019). https://doi.org/10.1016/j.jclepro.2019.118788

Amin, S.U., Shamim Hossain, M., Muhammad, G., Alhussein, M., Abdur Rahman, M.D.: Cognitive smart healthcare for pathology detection and monitoring. IEEE Access **7**, 10745–10753 (2019)

Anthony Jnr, B.: A case-based reasoning recommender system for sustainable smart city development. AI Soc. **22**, 1–25 (2020). https://doi.org/10.1007/s00146-020-00984-2

Asad, M.S., Ahmad, J., Hussain, S., Zoha, A., Abbasi, Q.H., Imran, M.A.: Mobility prediction-based optimisation and encryption of passenger Traffic-Flows using machine learning. Sensors **20**, 2629 (2020)

Austin, M., Delgoshaei, P., Coelho, M., Heidarinejad, M.: Architecting smart city digital twins: combined semantic model and machine learning approach. J. Manag. Eng. **36**(4) (2020). https://doi.org/10.1061/(asce)me.1943-5479.0000774

Aydin, G., Hallac, I.R., Karakus, B.: Architecture and implementation of a scalable sensor data storage and analysis system using cloud computing and big data technologies. J. Sens. (2015). https://doi.org/10.1155/2015/834217. paper 834217

Aymen, F., Mahmoudi, C.: A novel energy optimization approach for electrical vehicles in a smart city. Energies 12(929) (2019). https://doi.org/10.20944/preprints201901.0214.v1

Cai, B.Y., Alvarez, R., Sit, M., Duarte, F., Ratti, C.: Deep learning-based video system for accurate and real-time parking measurement. IEEE Internet Things J. 6(5), 7693–7701 (2019)

Calvo, P.: The ethics of Smart City (EoSC): moral implications of hyperconnectivity, algorithmisation and the datafication of urban digital society. Ethics Inf. Technol. 22, 141–149 (2020)

Cao, J., Cao, M., Wang, J., Yin, C., Wang, D., Vidal, P.P.: Urban noise recognition with convolutional neural network. Multimedia Tools Appl. 78, 29021–29041 (2019)

Castelli, M., Sormani, R., Trujillo, L., Popovič, A.: Predicting per capita violent crimes in urban areas: an artificial intelligence approach. J. Ambient Intell. Humaniz. Comput. 8, 29–36 (2017)

Chen, M., Miao, Y., Jian, X., Wang, X., Humar, I.: Cognitive-LPWAN: towards intelligent wireless services in hybrid low power wide area networks. IEEE Trans. Green Commun. Networking 2(3), 402–417 (2019a)

Chen, Q., Wang, W., Wu, F., De, S., Zhang, B., Huang, X.: A survey on an emerging area: deep learning for smart city data. IEEE Trans. Emerg. Top. Comput. Intell. 3(5), 392–410 (2019b)

Choudhury, S. K., Padhy, R. P., Sangaiah, A. K., Sa, P. K., Muhammad, K., Bakshi, S.: Scale aware deep pedestrian detection. Trans. Emerg. Telecommunications Technol. 30(3552) (2018)

De Paz, J.F., Bajo, J., Rodriguez, S., Villarrubia, G., Corchado, J.M.: Intelligent system for lighting control in smart cities. Inf. Sci. 372, 241–255 (2016)

Dennis, L.A., Slavkovik, M.: Machines that know right and cannot do wrong: the theory and practice of machine ethics. IEEE Intell. Inform. Bull. 19(1), 8–11 (2018)

Dilawar, N., et al.: Understanding citizen issues through reviews: a step towards data informed planning in smart cities. Appl. Sci. 8(1589) (2018). https://doi.org/10.3390/app8091589

Diro, A.A., Chilamkurti, N.: Distributed Attack detection scheme using deep learning approach for internet of things. Future Generation Computer Systems 82, 761–768 (2017). https://doi.org/10.1016/j.future.2017.08.043

Eldrandaly, K.A., Abdel-Basset, M., Abdel-Fatah, L.: PTZ-Surveillance coverage based on artificial intelligence for smart cities. Int. J. Inf. Manage. 49, 520–532 (2019)

El-Wakeel, A.S., Li, J., Noureldin, A., Hassanein, H.S., Zorba, N.: Towards a practical crowdsensing system for road surface conditions monitoring. IEEE Internet Things J. 5(6), 4672–4685 (2018)

Etzioni, A., Etzioni, E.: AI assisted ethics. Ethics Inf. Technol. 18, 149–156 (2016)

Falco, G., VisWanathan, A., Caldera, C., Shrobe, H.: A master attack methodology for an AI-Based automated attack planner for smart cities. IEEE Access 6, 48360–48378 (2018)

Ferrara, E., et al.: An AI approach to collecting and analysing human interactions with urban environments. IEEE Access 7, 141476–141486 (2019)

Gomede, E., de Barros, R.M., de Suoza Mendes, L.: Use of deep multi-target prediction to identify learning styles. Appl. Sci. 10(5), 1756 (2020). https://doi.org/10.3390/app10051756

Gomede, E., Gaffo, F.H., Brigano, G.U., de Barros, R.M., de Suoza Mendes, L.: Application of computational intelligence to improve education in smart cities. Sensors 18(1), 267 (2018). https://doi.org/10.3390/s18010267

Gomez, C.A., Shami, A., Wang, X.: Machine learning aided scheme for load balancing in dense IoT networks. Sensors 18(11), 3779 (2018). https://doi.org/10.3390/s18113779

Gong, Z., Li, X., Liu, J., Gong, Y.: Machine learning in explaining nonprofit organisations' participation: a driving factors analysis approach. Neural Comput. Appl. 31, 8267–8277 (2019)

Hossen, M.I., Michael, G.K.O., Connie, T., Lau, S.H., Hossain, F.: Smartphone-Based context flow recognition for outdoor parking system with machine learning approaches. Electronics **8**(7), 784 (2019). https://doi.org/10.3390/electronics8070784

Huang, J., Deng, Y., Yang, Q., Sun, J.: An Energy-Efficient train control framework for smart railway transportation. IEEE Trans. Comput. **65**(5), 1407–1417 (2016)

Huang, Z., Tang, J., Shan, G., et al.: An efficient passenger-hunting recommendation framework with multitask deep learning. IEEE Internet Things J. **6**(5), 7713–7721 (2019)

Hurst, W., Montanez, C.A.C., Shone, N., Al-Jumeily, D.: An ensemble detection model using multinomial classification of stochastic gas smart meter data to improve wellbeing monitoring in smart cities. IEEE Access **8**, 7877–7898 (2020)

Hwang, S., Lee, Z., Kim, J.: Real-time pedestrian flow analysis using networked sensors for a smart subway system. Sustainability **11**(23), 6560 (2019). https://doi.org/10.3390/su11236560

Idowu, S., Saguna, S., Ahlund, C., Schelen, O.: Applied machine learning: Forecasting heat load in district heating system. Energy Build. **133**, 478–488 (2016)

Iqbal, R., Doctor, F., More, B., Mahmu, S., Yosuf, U.: Big data analytics: computational intelligence techniques and application areas. Technol. Forecast. Soc. Chang. **153**, (2020). https://doi.org/10.1016/j.techfore.2018.03.024

Jung, D., Tuan, V.T., Tran, D.Q., Park, M., Park, S.: Conceptual framework of an intelligent decision support system for smart city disaster management. Appl. Sci. **10**(2), 666 (2020). https://doi.org/10.3390/app10020666

Ke, R., Zhung, Y., Pu, Z., Wang, Y.: A smart, efficient, and reliable parking surveillance system with edge artificial intelligence on IoT devices. IEEE Trans. Intell. Transp. Syst. (2020). https://doi.org/10.1109/TITS.2020.2984197

Ke, X., Zhang, Y.: Fine-grained vehicle type detection and recognition based on dense attention network. Neurocomputing **399**, 247–257 (2020). https://doi.org/10.1016/j.neucom.2020.02.101

Khan, S., Muhammad, K., Mumtaz, S., Baik, S.W., de Albuquerque, V.H.C.: Energy-efficient deep CNN for smoke detection in foggy IoT environment. IEEE Internet Things J. **6**(6), 9237–9245 (2019)

Kim, H., Ben-Othman, J.: Toward integrated virtual emotion system with AI applicability for secure CPS-Enabled smart cities: AI-Based research challenges and security issues. IEEE Network **34**(3), 30–36 (2020). https://doi.org/10.1109/MNET.011.1900299

Kofod-Petersen, A.: How to do a Structured Literature Review in computer science (2015). https://research.idi.ntnu.no/aimasters/files/SLR_HowTo2018.pdf

Le, L.T., Nguyen, H., Dou, J., Zhou, J.: A comparative study of PSO-ANN, GA-ANN, ICA-ANN, and ABC-ANN in estimating the heating load of buildings' energy efficiency for smart city planning. Appl. Sci. **9**(13), 2630 (2019a). https://doi.org/10.3390/app9132630

Le, L.T., Nguyen, H., Zhou, J., Dou, J., Moayedi, H.: Estimating the heating load of buildings for smart city planning using a novel artificial intelligence technique PSO-XGBoost. Appl. Sci. **9**(13), 2714 (2019b). https://doi.org/10.3390/app9132714

Li, D., Deng, L., Cai, Z., Franks, B., Yao, X.: Intelligent transportation system in macao based on deep self-coding learning. IEEE Trans. Industr. Inf. **14**(7), 3253–3260 (2018)

Liang, F., Hatcher, W.G., Liao, W., Gao, W., Yu, W.: Machine learning for security and the internet of things: the good, the bad, and the Ugly. IEEE Access **7**, 158126–158147 (2019)

Lin, W., et al.: Analysing the relationship between human behavior and indoor air quality. Sens. Actuator Networks **6**(13) (2017). https://doi.org/10.3390/jsan6030013

Liu, C.H., Chen, Z., Zhan, Y.: Energy-efficient distributed mobile crowd sensing: a deep learning approach. IEEE J. Sel. Areas Commun. **37**(6), 1262–1276 (2019a)

Liu, Y., et al.: DeepStore: an interaction-aware Wide&Deep model for store site recommendation with attentional spatial embeddings. IEEE Internet Things J. **6**(4), 7319–7333 (2019b)

Liu, Y., Yang, C., Jang, L., Xie, S., Zhang, Y.: Intelligent edge computing for IoT-based energy management in smart cities. IEEE Network **33**(2), 111–117 (2019c). https://doi.org/10.1109/MNET.2019.1800254

Liu, Y., Zhang, W., Pan, S., Li, Y., Chen, Y.: Analysing the robotic behaviour in a smart city with deep enforcement and imitation learning using IoRT. Comput. Commun. **150**, 346–356 (2020)

Mannion, P., Duggan, J., Howley, E.: Parallel reinforcement learning for traffic signal control. Procedia Comput. Sci. **52**, 956–961 (2015)

Manzanilla-Salazar, O.G., Malandra, F., Mellah, H., Wette, C., Sanso, B.: A machine learning framework for sleeping cell detection in a smart-city IoT telecommunications infrastructure. IEEE Access **8**, 61213–61225 (2020)

Martinez Garcia, J., Zoeke, D., Vossiek, M.: MIMO-FMCW Radar-based parking monitoring application with a modified convolutional neural network with spatial priors. IEEE Access **6**, 41391–41398 (2018)

Medapati, P.K., Murthy, P.H.S.T., Sridhar, K.P.: LAMSTAR: for IoT-based face recognition system to manage the safety factor in smart cities. Trans Emerging Tel Tech. e3843 (2019). https://doi.org/10.1002/ett.3843

Miraftabzadeh, S.A., Rad, P., Choo, K.K.R., Jamshidi, M.: A privacy-aware architecture at the edge for autonomous real-time identity reidentification in crowds. IEEE Internet Things J. **5**(4), 2936–2946 (2018)

Mo, X., Zhang, L., Li, H., Qu, Z.: A novel air quality early-warning system based on artificial intelligence. Environ. Res. Public Health **16**(19), 3505 (2019). https://doi.org/10.3390/ijerph16193505

Mohanta, B., Das, P., Pantaik, S.: Healthcare 5.0: A paradigm shift in digital healthcare system using Artificial Intelligence, IOT and 5G Communication. In: 2019 International Conference on Applied Machine Learning (ICAML), Bhubaneswar, pp. 191–196 (2019). https://doi.org/10.1109/icaml48257.2019.00044

Niu, X., Zhu, Y., Cao, Q., Zhang, X., Xie, W., Zheng, K.: An online-traffic-prediction based route finding mechanism for smart city. Int. J. Distrib. Sens. Networks **11**(8) (2015). https://doi.org/10.1155/2015/970256

Obinikpo, A.A., Kantarci, B.: Big sensed data meets deep learning for smarter health care in smart cities. Sens. Actuator Network **6**(4), 26 (2017). https://doi.org/10.3390/jsan6040026

Palka, D., Brodny, J., Rizaoglu, T., Bagci, U., Maščeník, J.: Literature research in the field of technology assessment using a tool of a systematic literature review. Multidisciplinary Aspects Prod. Eng. **1**, 109–115 (2018)

Park, S., Lee, S., Park, S., Park, S.: AI-based physical and virtual platform with 5-layered architecture for sustainable smart energy city development. Sustainability **11**(16) (2019). https://doi.org/10.3390/su11164479

Ping, P., Xu, G., Kumala, E., Gao, J.: Smart street litter detection and classification based on faster R-CNN and edge computing. Int. J. Softw. Eng. **30**(4), 537–553 (2020)

Ponce, H., Gutierrez, S.: An indoor predicting climate conditions approach using Internet-of-Things and artificial hydrocarbon networks. Measurement **135**, 170–179 (2018). https://doi.org/10.1016/j.measurement.2018.11.043

Qin, L., Yu, N., Zhao, D.: Applying the convolutional neural network deep learning technology to behavioural recognition in intelligent video. Tech. Gazzete **25**, 528–535 (2018)

Qiu, J., Du, L., Zhang, D., Su, S., Tian, Z.: Nei-TTE: Intelligent traffic time estimation based on fine-grained time derivation of road segments for smart city. IEEE Trans. Industr. Inf. **16**(4), 2659–2666 (2020)

Rjab, A.B., Mellouli, S.: Smart cities in the era of artificial intelligence and internet of things: literature review from 1990 to 2017. In: Proceedings of the 19th Annual International Conference on Digital Government Research: Governance in the Data Age, pp. 1–10 (2018). https://doi.org/10.1145/3209281.3209380. Article No. 81

Rjab, A.B., Mellouli, S.: Artificial intelligence in smart cities: systematic literature network analysis. In: Proceedings of the 12th International Conference on Theory and Practice of Electronic Governance, pp. 259–269 (2019). https://doi.org/10.1145/3326365.3326400

Rojek, I., Studzinski, J.: Detection and localisation of water leaks in water nets supported by an ICT system with artificial intelligence methods as a way forward for smart cities. Sustainability 11(2), 518 (2019). https://doi.org/10.3390/su11020518

Roldan, J., Boubeta-Puig, J., Martinez, L.J., Ortiz, G.: Integrating complex event processing and machine learning: an intelligent architecture for detecting IoT security attacks. Expert Syst. Appl. 149, (2020). https://doi.org/10.1016/j.eswa.2020.113251

Sanam, T.F., Godrich, H.: A multi-view discriminant learning approach for indoor localization using amplitude and phase features of CSI. IEEE Access 8, 55947–55959 (2020)

Serban, A.C., Lytras, A.M.D.: Artificial intelligence for smart renewable energy sector in Europe - Smart energy infrastructures for next generation smart cities. IEEE Access 8, 77364–77377 (2020)

Shen, M., Tang, X., Zhu, L., Du, X., Guizani, M.: Privacy-preserving support vector machine training over blockchain-based encrypted IoT data in smart cities. IEEE Internet Things J. 6(5), 7702–7712 (2019)

Sholla, S., Mir, R.N., Chishti, M.A.: A neuro fuzzy system for incorporating ethics in the internet of things. J. Ambient Intell. Humaniz. Comput. (2020). https://doi.org/10.1007/s12652-020-02217-2

Shousong, C., Xiaomin, G., Xiaoguang, W., Ying, C.: Research on urban land price assessment based on artificial neural network model. IEEE Access 7, 180738–180748 (2019)

Shu, W., Cai, K.: A SVM multi-class image classification method based on DE and KNN in smart city management. IEEE Access 7, 132775–132785 (2019)

Talamo, M., Valentini, F., Dimitri, A., Allegrini, I.: Innovative technologies for cultural heritage. Tattoo sensors and AI: the new life of cultural assets. Sensors 20(7), 1909 (2020). https://doi.org/10.3390/s20071909

Vazquez-Canteli, J., Ulyanin, S., Kampf, J., Nagy, Z.: Fusing TensorFlow with building energy simulation for intelligent energy management in smart cities. Sustain. Cities Soc. 45, 243–257 (2018). https://doi.org/10.1016/j.scs.2018.11.021

Venkatesh, J., Aksanli, B., Chan, C.S., Akyurek, S.A., Simunic Rosing, T.: Modular and personalised smart health application design in a smart city environment. IEEE Internet Things J. 5(2), 614–623 (2018)

Wan, C.H., Hwang, M.C.: Value-based deep reinforcement learning for adaptive isolated intersection signal control. IET Intell. Transp. Syst. 12(9), 1005–1010 (2018)

Wang, H., Xue, Q., Cui, T., Li, Y., Zeng, H.: Cold start problem of vehicle model recognition under CrossScenario based on transfer learning. Comput. Mater. Continua. 63(1), 337–351 (2020)

Wang, S.: Smart data mining algorithm for intelligent education. J. Intell. Fuzzy Syst. 37(1), 9–16 (2019)

Webster, J., Watson, R.: Analysing the past to prepare for the future: writing a literature review. MIS Q. 26(2) (2002). http://www.jstor.org/stable/4132319

Wolfswinkel, J.F., Furtmueller, E., Wilderom, C.P.M.: Using grounded theory as a method for rigorously reviewing literature. Eur. J. Inf. Syst. 22(1), 45–55 (2013)

Wu, X., Park, Y., Li, A., Huang, X., Xiao, F., Usmani, A., Huang, X.: Smart detection of fire source in tunnel based on the numerical database and artificial intelligence. Fire Technol. (2020). https://doi.org/10.1007/s10694-020-00985-z

Xiong, M., et al.: Person re-identification with multiple similarity probabilities using deep metric learning for efficient smart security applications. J. Parallel Distrib. Comput. 132, 230–241 (2017). https://doi.org/10.1016/j.jpdc.2017.11.009

Xu, R., Cheng, Y., Liu, Z., Xie, Y., Yang, Y.: Improved long short-term memory based anomaly detection with concept drift adaptive method for supporting IoT services. Future Gener. Comput. Syst. **112**, 228–242 (2020). https://doi.org/10.1016/j.future.2020.05.035

Yao, H., Gao, P., Wang, J., Zhang, P., Jiang, C., Han, Z.: Capsule network assisted IoT traffic classification mechanism for smart cities. IEEE Internet Things J. **6**(5), 7515–7525 (2019)

Yassine, A., Singh, S., Alamri, A.A.: Mining human activity patterns from smart home big data for health care applications. IEEE Access **5**, 13131–13141 (2017)

Zhang, P., Zhao, Q., Gao, J., Li, W., Lu, J.: Mining human activity patterns from smart home big data for health care applications. IEEE Access **7**, 63550–63563 (2019)

Zhang, Q., Zhou, D., Zeng, X.: HeartID: a multiresolution convolutional neural network for ECG-based biometric human identification in smart health applications. IEEE Access **5**, 11805–11816 (2017)

Zhao, B., Teo, Y.S., Ng, W.S., Ng, H.H.: Data-driven next destination prediction and ETA improvement for urban delivery fleets. IET Intell. Transp. Syst. **13**(11), 1624–1635 (2019a)

Zhao, L., Wang, J., Liu, J., Kato, N.: Routing for crowd management in smart cities: a deep reinforcement learning perspective. IEEE Commun. Mag. **57**(4), 88–93 (2019b). https://doi.org/10.1109/MCOM.2019.1800603

Zhou, T., Shen, J., He, D., Vijayakumar, P., Kumar, N.: Human-in-the-loop-aided privacy-preserving scheme for smart healthcare. IEEE Trans. Emerg. Top. Comput. Intell. (2020). https://doi.org/10.1109/TETCI.2020.2993841

Web Librarianship

International Initiatives and Advances in Brazil for Government Web Archiving

Jonas Ferrigolo Melo$^{(\boxtimes)}$ (iD) and Moisés Rockembach (iD)

Federal University of Rio Grande do Sul, Porto Alegre, RS, Brazil
jonasferrigolo@gmail.com, moises.rockembach@gmail.com

Abstract. This study aimed to illustrate some government web archiving initiatives in several countries and stablish an overview of the Brazilian scenario with regard to the preservation of content published on government websites. In Brazil, although there is a robust set of laws that determine the State to manage, access and preserve its documents and information, there is still no policy for the preservation of web content. The result is the erasure and permanent loss of government information produced exclusively through websites. It is noticed that there are several government initiatives for web archiving around the world, which can be used as examples for the implementation of a Brazilian policy. It is concluded that the long-term maintenance of governmental information available on the web is fundamental for public debate and for monitoring governmental actions. To ensure the preservation of this content, the country must define its policy for the preservation of documents produced in a web environment.

Keywords: Digital preservation · Websites · Web archiving · Government web archiving

1 Introduction

Contemporary society is closely inserted in the technological context, from daily routines that have become computerized, to the most complex actions using artificial intelligence. As social agents, we also interact in this scenario, as Manuel Castells had already predicted, in the 90s, when the trilogy was published "The Information Age: economy, society and culture" in which it states that the Network Society is the result of the social appropriation of a set of information and communication technologies [1]. Here we use the concept of the Network Society as Castells [1] defends: reposition of the State as the front that can boost the power that digital communications and relations are promoting with technological advances.

As a result, our social memory is also being produced as a consequence of these digital devices. After all, the Internet reflects the construction of a visual narrative: social networks are as real as relationships not intermediated by the machine [2]. In this sense, while part of our social memory has migrated to the digital environment and is printed in a database, we have to guarantee mechanisms for its preservation.

One of the challenges of the Network Society linked by digital technologies is the digitalization of the method of communications in society, which are increasingly done through commercial interests and not necessarily coinciding with the public interest [3]. What must be emphasized to understand the relationship between technology and society is that the role of the State in promoting technological innovations. The technology expresses a society's ability to boost technological dominance through its social institutions, including the state [1].

In this way, it is understood that governments should promote the transparency of their actions, through continuous accountability and the strengthening of relations between the State and society as a way of guaranteeing social memory and the full maintenance of democracy itself. Government communication is one of the main sources of information about democratic governments and governed places [4]. As well as official documents, the media content produced by the Government and disseminated especially through institutional websites, allow access to official statements and information approved by the State. For this reason, governments have a fundamental role in establishing routines to preserve this heritage that is being produced exclusively through the Internet.

The National Archive of United Kingdom (UK) says that a substantial part of your government's current records are produced only in digital format and the lack of a strategy for archiving and preserving that content will inevitably lead to the disappearance of important information for the future [5]. With the development of computer networks and the Internet, the process of producing, storing, accessing, using and consuming information has transformed the way governments treat official data [6]. In any case, regardless of the way information is produced, as a way of guaranteeing the fundamental rights of citizens, it is the duty of the State to provide access to content published on the web, especially within the scope of government websites that have critical information and are under the optics and jurisprudence of the Access to Information Law that regulates, among others, item XXXIII of article 5 of the Federal Constitution of Brazil, since everyone has the right to receive information from their public bodies of their particular interest, or of collective or general interest [7].

We understand that, in addition to making content available on government portals, it is also the duty of the State to guarantee access to this information, as a way of combating possible problems that result from the lack of preservation of web pages. Possible access problems can result from the purposeful modification of content, its deletion or technical problems such as link rot, among other issues. The official websites can be understood as places of memory of governments, as they keep news, photos, videos and documents that speak about public administrations, politics and the transformations undergone by the population [4, 8].

The web archiving is widely recognized due to its use with regard to historical, cultural and intellectual preservation. Countries with a high Internet insertion rate have established archiving initiatives to crawl and store web content, which disappears quickly and needs to be accessed for long-term use [9]. The geographic distribution of the web archiving initiatives is still unequal: in Latin America, only Chile constituted its own web archiving initiative, in which it currently has five collections. In Brazil, according Rockembach [10], there is still no systematic web archiving, covering national domains, although there are other actions that are contributing to this increase, which will be discussed in detail below.

In search of solutions, we consider that through the web archiving, governments will be able to preserve and maintain the evidence of their services and actions, in order to make them accessible for future research purposes and also as records of the evolution of their own actions. Holub and Rudomiro [11] say that due to the dynamic nature of the web, its explosive growth, short lifespan, instability and similar characteristics, the importance of archiving it has become invaluable for future generations. According to Rockembach and Pavão [12] if there is no digital preservation of the content produced on the web, much of what was developed in this medium will be lost forever. This requires a web archiving solution that relies on structured technical policies and procedures. The National Archives, UK [13] declares that web archiving is a vital process to ensure that people and organizations can access and reuse knowledge in the long term and meet their information retrieval needs.

Here we aim to discuss the government web archiving, based on international initiatives and the current situation in Brazil. We expect that research in this field will contribute to achieving a sense of community, national identity and entrenchment among Brazilian citizens. In the sense, the implementation of a national web archiving policy, in a way, preserves information that shapes national identity through its development in political circles. The web is increasingly used as a tool for social communication and interactions between public authorities and civil society and, over time, the web archive may form a record of events that capture the nation's environment and accompany development Brazilian national identity. The preservation of these records provides a valuable source of documented heritage for current and future generations, creating a sense of community and belonging.

Discussing this theme based on the experiences of different countries in the world, provides an exchange of technical and scientific information, making the result based on positive experiences in government web archiving. Brazil, due to being inserted in the theme for a few years, needs this scientific contribution to assist in the development of studies.

2 Government Web Archiving

The websites of government agencies play an important role in the dissemination of government information to the general public, and have established themselves as fundamental tools for the search for information from public policies, in general. Considering the dynamic nature of the web, the contents of these institutional websites can change quickly and some information can be permanently removed, which may cause a break in the information scope and lead to the loss of valuable information for research and even for accountability of the actions of the government.

According to Lala and Joe [14], even the scenario of large content productions on websites and even with countless initiatives around the world that work with this perspective of preservation, there is still a long way to go before recognizing the value and the importance of archiving websites. One of the largest and most used web archives in the world, containing more than three billion URLs, is the UK Government Web

Archive (UKGWA), maintained by The National Archives, UK. The UKGWA's mission is to preserve government-owned web content in all its formats, even though its central content consists of material published by state departments [13]. These websites are identified by The National Archives and government organizations [15].

The initiative is part of a broad program by The National Archives that involves managing the British government's web heritage. In 2017, a guidance was provided to government digital teams in which questions about website management and maintenance are elucidated, in order to ensure that the government's web presence can be successfully archived and permanently accessible in the UKGWA. In "The UK Government Web Archive: Guidance for digital and records management teams", information is provided on the functioning of the websites' archiving process, the catch schedule, the limitations of what can be captured and made available through this preservation system and the circumstances in which content can be removed. The National Archives' approach to government web archiving involves remote and automatic collection of websites according to a schedule, using a crawler [16].

This process of selecting the web content to be preserved is one of the first steps that must be established when building a web archive. In archival science this process is known as appraisal; librarians call collection development. This process of web curation remains inevitable, even in the digital context [17]. Although there are technical barriers, such as storage capacity, for example, the curation process is fundamental, considering that maintaining everything is not a viable strategy for a number of reasons [17]. For web archiving, it is a selection process in which the websites that will be preserved are chosen, based on one or more criteria [18]. The development of the selection policy will also define the form of collection of the websites [19], which may include the description of the context, the intended users, access to mechanisms and the expected uses of the web archive [20].

In relation to the US government websites, two specific initiatives that are relevant to our analysis. One of the initiatives is the collaborative web archive End of Term Web Archive: US Government Websites (EOT), created since 2008 by a group of institutions that developed collections made up of federal government websites (.gov,.mil) in the legislative, executive and judicial spheres of the US government. Websites that were at risk of changing (for example whitehouse.gov) or disappearing completely during government transitions were captured. Currently, EOT is comprised of the collections of websites from the end of the Bush administration (2008) and the end of the two terms of the Obama administration (2012 and 2016). [21]. Another USA initiative that preserves government websites is the movement promoted by scientists to safeguard government information on climate change, anticipating the risk that data and information from.gov, such as EPA websites (Environmental Protection Agency) and NOAA (National Oceanic and Atmospheric Administration), could be lost or become unavailable with the transition to the new administration that was taking effect with the election of Donald Trump [22].

The initiative by the State of Sarawak, Malaysia, is a government web archiving initiative that also deserves mention. Jamain et al. [5], says that the objective of the government initiative is to preserve the evidence of web content published by departments and agencies of the public administration of the State of Sarawak, to contribute to facilitated access to information and to the provision of information to research, in addition to being in accordance with the legislation of the State Library of Sarawak regarding the legal deposit. The capture of websites takes place every two months and, in addition to textual documents, includes the safeguarding of static images, sound recording, films and other multimedia formats made available on government portals [5].

The legal deposit has been a normative tool used to justify the need to preserve websites in some countries. Although, until the twentieth century, the legal deposit had its scope focused only on books and printed publications. In addition to the advent and massive use of sound recording, video and, in general, digital information, the scope of the legal deposit has been extended to include a variety of document formats. This proposition came from the Charter on the Preservation of the Digital Heritage, published in 2003 by UNESCO, in which it promoted the adoption of measures for the preservation of digital information, considering, for example, the possibility of using the legal deposit for materials of the web [23].

Traditionally, national libraries have the prerogative to preserve books and printed materials produced under the country's jurisprudence, a rule established by the legislation of the legal deposit. Due to UNESCO's recommendations, part of the literature on web archiving indicates that national libraries have a significant role in preserving the web [24]. This referral was positive for these organizations, especially with regard to the legal risks assumed when preserving web content. Supported by the law of legal deposit, the organization preserves its right to store, as determined by the legislation, this means that organizations, whether National Archives or National Libraries, are unique organizations, with the prerogative of preserving memory, nationally or locally [25].

In this sense, some national libraries started to build collections of websites from the early 90s [26]. These libraries started to consider web collections as a natural development of their traditional collection and as part of their duty to preserve national culture [27]. In the early 2000s, some countries added to their legal deposit laws a requirement to deposit electronic publications in online formats [28], such as Tasmania, Switzerland, Iceland and New Zealand [24]. Finland, Iceland, Norway and Sweden, for example, decided to collect local websites that meet the broad criteria of being "national" [29].

In Croatia, the National and University Library of Zagreb, in collaboration with the University of Zagreb Computing Center (SRCE) established, in 2004, from legislation, that all websites registered in the country would have a copy deposited in the Library. In 2011, the HAW (in Croatian, Hrvatski arhiv weba) or, in English, Croatia Web Archive, collected the national domain (.hr) for the first time with the intention of expanding the scope of the national collection of websites and started a thematic collection of content from web related to national events, such as the 2013 local elections for government territories [11].

Like these initiatives, there are others that promote the preservation of government websites. With these examples, it can be seen that the governmental web archiving takes place in different ways, whether from the central government's own proposals, or through articulated civil organizations or groups, or even through specific initiatives of memory places. Anyway, it is an interdisciplinary work, which involves knowledge and routines from the information sciences and technology, in which archives and libraries converge towards digital preservation, associating with the disciplines of information technology. Each initiative has its own routines, criteria, selection processes, software and flows established according to its need to compose the web file. The responsibility of documentary collection for the public domain has been established as an important tool to preserve the official memory of several countries [30].

In addition to these areas of study that are fundamental to the web archiving, there are other disciplines and professionals that have space in the formation of knowledge related to this theme, such as the communication sciences, engineering, ethics, law and many other knowledge that can contribute to the development of the field. Universities have also understood that the Internet has increasingly been the point of origin for a large volume of information, research and scientific publications, and have turned their technological resources and their technical knowledge to the preservation of contents and scientific communications produced in this area [31].

3   Government Web Archiving in Brazil

The Brazilian Institute of Geography and Statistics (IBGE - Instituto Brasileiro de Geografia e Estatística), showed that in 2016, 116 million people were connected to the Internet in Brazil, which is equivalent to 64.7% of the Brazilian population aged over 10 yrs [32]. Likewise, CGI.br/NIC.br, Regional Center for Studies for the Development of the Information Society (Cetic.br - Centro Regional de Estudos para o Desenvolvimento da Sociedade da Informação), reported that in 2018, 67% of Brazilian households had Internet, and the percentage of network users increased to 76% of the Brazilian population [33].

The Cetic.br, in a survey in the previous year, showed that 100% of federal and state public bodies use the Internet, with 90% of these bodies having websites [34]. The Digital report "in 2020", carried out by "We Are Social" and "Hootsuite", shows a complete view of the digital landscape in the country: with 150.4 million Internet users - 71% of the population - the report shows that 66% of the Brazilian population is active in social networks. These figures show that more than half of the population has access to the internet and that public institutions at the state and federal levels are present on the world wide web, showing the production of digital documents, both in the social and government spheres. However, unfortunately, there is still no preservation of this content published on the web.

Brazilian law presents several points about the State's responsibility with regard to the management, access and preservation of its documents. Among the main laws, at the level of legal hierarchy, are the Federal Constitution and Law 8,159/1991. In the Federal Constitution, article 216, paragraph 2, determines that "the public administration, under the terms of the law, is responsible for the management of governmental documentation and the measures to open its consultation to those who need it" [7]. In turn, Federal Law 8.159/1991, which provides for the national policy on archives, states in Article 1 that it is the duty of the public power "document management and special protection of archival documents, as an instrument to support administration, to culture, scientific development and as evidence and information" [35]. The same law conceptualizes that archives are "documents produced and received by public agencies, public institutions and private entities [...] whatever the support of the information or the nature of the documents"

[35]. There are dozens of Decrees, Laws, Provisional Measures, Resolutions, Ordinances and Normative Instructions, arising from these laws, which decide on the treatment of documents within the scope of public administration. Therefore, it is a matter of understanding these laws and discussing the right to access this information from the State and how much this right is guaranteed by governments.

The National Archives of Brazil, the institution responsible for drafting national management policies, excluded documents produced on the web from the scope of accepted document formats for collection at the institution, when presenting and publishing its Digital Preservation Policy, with versions in 2012 and 2016: "In the future, other more complex types of documents in digital format, such as multimedia and web pages, should also be contemplated" [36].

Recently, government websites in Brazil were the subject of the publication of Decree number 9.756/2019, which "Establishes the single portal 'gov.br' and provides for the unification rules of the Federal Government's digital channels" [37], which establishes in its article that "[…] through which institutional information, news and public services provided by the Federal Government will be made available in a centralized manner" [37]. The GOV.BR Single Portal was officially launched on July 1, 2019. Before its launch, the first page of the Portal (see Fig. 1) stated that "[…] the digital channels of the Federal Government will be unified". The single portal "[…] will gather, in one place, services for the citizen and information on the performance of all areas of the government", and went on to say that the portal will also be "[…] the door entry of institutional pages of the federal administration, such as ministries, regulatory agencies and other bodies" [38].

Fig. 1. The first page of the GOV.BR website before its launch. 2019.

In the Brazilian academy, the web archiving subject is still recent, but it already promotes discussions on digital preservation in the national scenario. In 2017, the Federal University of Rio Grande do Sul created the Research Group on Web Archiving and Digital Preservation (NUAWEB - Núcleo de Pesquisa em Arquivamento da Web e

Preservação Digital) with the objective of investigating characteristics of web archiving through national and international initiatives, dealing with both policies and technologies involved in the process [39].

The research group studies aspects of preservation, use and access over time of digital objects made available on the web, with contributions from Archival Science, Library Science, Information Science, Communication and Computer Science. NUAWEB is developing some research projects simultaneously, such as the AWEB - Web Archiving of the Brazilian Elections; and the Brazilian Web Archiving: preservation policies and technological models. The research group also presented during the International Internet Preservation Consortium Web Archiving Conference 2019, in Zagreb, Croatia, the fundamentals to foster the discussion on Brazilian web archiving initiatives, under the website https://www.arquivo.ong.br [40].

One of these surveys investigated the possibilities of archiving Brazilian Federal Government websites, with 23 government websites as object of analysis. The research consisted of checking the resources offered by these websites; archive the selected websites, using the Heritrix as web page crawler; rebuild archived websites using the software WABAC; and compare the resources available in the live and archived versions of selected websites. It is concluded that the websites of the Brazilian Federal Government are archivable without loss of relevant information and that the country lacks a public policy to systematize the archiving of government websites [41].

However, less than a year after the end of the research, it is possible to notice systemic changes in the analyzed websites, in terms of content and layout. In the example below is the Ministry of Defense website archived in December 2019 (see Fig. 2) and the same website in October 2020 (see Fig. 3). It is noticed that there has been a change in the layout, navigation menu and image layout. As there is still no systematic preservation of the websites analyzed, we can already say that unique information produced within the same government management has been lost.

Fig. 2. Ministry of Defense's website archived in 2019 December, 16. 2019.

In the political sphere, there is an important action regarding the web archiving under development since July 2015. This is the Bill (PL - Projeto de Lei) 2.431/2015,

Fig. 3. Ministry of Defense's website a live. 2020 October, 20. 2020.

authored by Deputy Luizianne Lins (PT), which "Provides on the institutional digital public heritage inserted in the world wide web and other measures" [42]. In 2015, the Bill passed through the Science and Technology, Communication and Informatics Commission (CCTCI), of the Chamber of Deputies, and had an opinion for approval with a substitute given by Deputy Fábio Sousa (PSDB), who in his report presents the project considering that its approval "[…] aims to expand the protections given to public information, more specifically that stored on the internet" [42]. The report asks for changes in the wording of the project and the addition of a paragraph that states that "[…] guidelines should be established in each body or entity that guide the periodic backup copies of critical information from the environments of the official sites" [42].

Two years later, the Bill returned to discussions in the working committees of the Chamber of Deputies, this time with the Culture Commission (CCULT): in October 2017, the Bill had an opinion for the rejection made by Deputy Evandro Roman (PSD), which in its report justifies saying that the obligation to keep all the content hosted on the official government websites "[…] brings a great operational difficulty, implying increasing expenses in storage technologies, which can make the unviable preservation" [42], goes on to say that "[…] the preservation of all content ignores the dynamic character of the world wide web, which precisely facilitates the updating and dispersion of information as soon as possible for those interested" [42]. This demonstrates that the presentation of the Bill, as well as the rapporteur ships are devoid of any in-depth study on the relevance of the topic, evidencing the lack of qualified knowledge regarding the web archiving in these contexts.

In March 2019, a new rapporteur for the Bill Project was appointed at CCULT, Deputy David Miranda (PSOL), who in December of the same year presented an opinion for the approval of the Bill with a substitute, in which he adds that "[…] it is necessary arrangements for the digital content of official sites not to be deleted at the mercy of ideological positions of a candidate or another who wins the elections" [42]. The rapporteur added in his substitute, in addition to the institutional websites already provided, the social networks "[…] such as Youtube, Facebook, Twitter, etc. […]" [42], in addition to including the personal accounts in social networks of heads of Public Authorities and holders of maximum organs of Federal Authorities, during the exercise of their mandates, considering that "[...] these political actors are the main spokespersons of such institutions" [42].

Although the laws are clear regarding the need to preserve documents produced at the governmental level, there is an understanding on the part of the scientific community that websites are not documents that would fall within this scope. However, the lay opinion also sometimes follows this line, as can be seen in the comments of the Deputies regarding the Bill that deals with the preservation of websites. The result is the permanent erasure of digital information produced exclusively on government websites, causing an erasure of political history that could be easily accessed by a large part of the population. Have we ever wondered where is the information published on past federal government websites?

Luz and Weber [43] developed a survey of government communication available on the official website of the Presidency of the Republic of Brazil produced during the government of ex-president Michel Temer (2016–2018) and ex-president Dilma Rousseff (2010–2016), discussing the impacts of preserving and erasing government communication for the country's political memory [43]. The conclusion is that websites have been subject to constant modifications to remove content from previous governments and difficulty in accessing certain sections or themes related to governments that are already closed, especially when the political lines do not converge with the current management.

The exclusion of content produced during previous mandates makes it evident that there is no policy in Brazil to safeguard content published on official websites, even when Brazilian law determines that it is the obligation of the public power to protect and grant access to official information [43]. This prevents society from having access to these contents, such as government actions, advertising campaigns, language, images, speeches and political positioning that guided the government actions of that administration.

From this scenario, it is understood that the implementation of a policy for the preservation of governmental web pages in Brazil is urgent, considering that part of the institutional memory published exclusively on the Internet is lost with government management changes, even though this is a great challenge, both in raising awareness about the information produced on the web, and regarding technical issues associated with the activity. Luz and Weber [43] warn that the lack of a policy to safeguard these contents in Brazil alerts to the consequences that the loss of information can have on the memory of a country, a city, its public policies and its reality.

4 Concluding Remarks

This study aimed to illustrate some initiatives of web archiving in many countries and to establish an overview of the Brazilian scenario with regard to the preservation of content published on government websites. It is noticed that there are several governmental initiatives for archiving on the web all over the world, whether promoted by the central government itself, by government agencies, or through the promotion of civil agents interested in the theme. In Brazil, although there is a robust set of laws that determine the State to manage, access and preserve its documents and information, there is still no policy for the preservation of this content, especially with regard to web documents. The result is the erasure and permanent loss of government information produced exclusively through websites.

Since the insertion of government actions on the Internet, the web has become a democratic space for access to information, especially in Brazil, where the Internet penetration rate is high and reaches about 70% of the population. This space for communication, dissemination and storage of public information is the official website, as it provides government data and government actions, providing disclosure, as well as management's own accountability. Furthermore, when complying with demands, it becomes a space for preserving the memory of public policies in the country, respecting the principle of transparency, inherent in a democratic state. Still, the long-term maintenance of governmental information available on the web is fundamental for public debate and for the monitoring of government actions by society.

To guarantee the preservation of this content, the State should define its policy for the preservation of digital documents, including those produced in a web environment, along the lines of similar projects carried out around the world. Selecting the preservation and web archiving techniques, the appropriate technologies, the content that will be primarily archived are some of the ways that Brazil could follow to effectively implement a web archiving initiative. It is up to us, as a scientific community, to present studies to provide the preservation of this informational content produced exclusively on the web, offering possibilities for public access to be ensured and facilitated in respect of the constitutional right to memory.

References

1. Castells, M.: A sociedade em rede - a era da informação: economia, sociedade e cultura. fundação calouste gulbenkian, lisboa (2000)
2. Dodebei, V.: Memória e patrimônio - perspectivas de acumulação/dissolução no ciberespaço. Aurora – Rev. de Arte, Mídia e Política. 0(10), 36 (2011)
3. Castells, M.: A galáxia internet - reflexões sobre internet, negócios e sociedade. Fundação Calouste Gulbenkian, Lisboa (2001)
4. Luz, A.J., Weber, M.H.: Comunicação governamental e memória política - preservação e apagamento de informações oficiais nos sites das capitais. In: Abstracts of the Encontro Anual COMPÓS, São Paulo, Faculdade Cásper Líbero (2017)
5. Jamain, J., Yahya, A.L., Muhammad, N., Rahman, A., Ayob, M.: Web archiving issues and challenges in State Government of Sarawak (Malaysia) - Do they really need their website to be archived? In: Paper presented at IFLA WLIC 2018, Kuala Lumpur, Malaysia, Transform Libraries, Transform Societies in Session 160 - Preservation and Conservation with Information Technology (2018)
6. Pimenta, M.S., Canabarro, D.R.: Democracia e capacidade estatal na era digital. In: Pimenta, M.S., Canabarro, D.R. (Orgs): Governança Digital. UFRGS/CEGOV, Porto Alegre (2014)
7. Brasil: Constituição da República Federativa do Brasil de 1988, Brasília (1988)
8. Weber, M.: Estratégias da comunicação de Estado e a disputa por visibilidade e opinião. In: Kunsch, M. (eds.), Comunicação Pública, sociedade e cidadania. Difusão, São Caetano do Sul (2011)
9. Xie, Z., Sompel, H.V., Liu, J., Reenen, J., Jordan, R.: Archiving the relaxed consistency web. In: Proceedings of the 22nd ACM international conference on Information & Knowledge Management (CIKM '13). Association for Computing Machinery, New York, NY, USA, pp. 2119–2128 (2013). https://doi.org/10.1145/2505515.2505551

10. Rockembach, M.: Arquivamento da Web - estudos de caso internacionais e o caso brasileiro. RDBCI – Rev. Digital de Biblioteconomia e Ciência da Informação **16**(1), 7–24 (2018)
11. Holub, K., Rudomino, I.: Croatian web archive - an overview. Review of the National Center for Digitization, 25 (2014). https://urn.nsk.hr/urn:nbn:hr:203:894084. Accessed 10 Oct 2020
12. Rockembach, M., Pavão, C.M.G.: Políticas e tecnologias de preservação digital no arquivamento da web. Rev. Ibero-americana Ciência da Informação **11**(1), 168–182 (2018). https://doi.org/10.26512/rici.v11.n1.2018.8473
13. The National Archives: Operational Selection Policy OSP27UK. United Kingdom: The National Archives (2014)
14. Lala, V., Joe, S.: Web Archiving At The National Library of New Zealand. Paper presented at the LIANZA Conference 2006 Papers. https://opac.lianza.org.nz/cgi-bin/koha/opac-detail.pl?bib=113. Accessed 20 Oct 2020
15. Espley, S., Carpentier, F., Pop, R., Medjkoune, L.: Collect, preserve, access - applying the governing principles of the national archives UK government web archive to social media content. Alexandria **25**(1–2), 31–50 (2014). https://doi.org/10.7227/ALX.0019
16. The National Archives: The UK Government Web Archive - Guidance for digital and records management teams. The National Archives, United Kingdom (2017)
17. Milligan, I., Ruest N., Lin J.: Content selection and curation for web archiving - the gatekeepers vs. the masses. In: Proceedings of the 16th ACM/IEEE-CS on Joint Conference on Digital Libraries (JCDL '16). Association for Computing Machinery, New York, NY, USA, pp. 107–110 (2016). https://doi.org/10.1145/2910896.2910913
18. Niu, J.: An overview of web archiving. D-Lib Mag. **18**, 3–4 (2012)
19. Bragg, M., Hanna, K., Donovan, L., Hukill, G., Peterson, A.: The Web Archiving Life Cycle Model. Internet Archive (2013). https://ait.blog.archive.org/files/2014/04/archiveit_life_cycle_model.pdf. Accessed 01 Oct 2020
20. Khan, M., Rahman, A. Ur.: A systematic approach towards web preservation. Inf. Technol. Libr. **38**(1), 71–90 (2019). https://doi.org/10.6017/ital.v38i1.10181
21. EOT (2019). https://eotarchive.cdlib.org. Accessed 21 Sep 2020
22. VICE: https://www.vice.com/en_us/article/d7yeej/all-references-to-climate-change-have-been-deleted-from-the-white-house-website-5886b75d0b367c453f87dd14. Accessed 21 Sep 2020
23. UNESCO: Charter on the Preservation of Digital Heritage (2003)
24. Shveiky, R., Bar-Ilan, J.: National libraries' traditional collection policy facing web archiving. Alexandria **24**(3), 37–72 (2013). https://doi.org/10.7227/ALX.0001
25. Rockembach, M.: Inequalities in digital memory - ethical and geographical aspects of web archiving. Int. Rev. Inf. Ethics 26 (2017). https://informationethics.ca/index.php/irie/article/view/286
26. Day, M.: Preserving the fabric of our lives: a survey of web preservation initiatives. In: Koch, Traugott, Sølvberg, Ingeborg Torvik (eds.) ECDL 2003. LNCS, vol. 2769, pp. 461–472. Springer, Heidelberg (2003). https://doi.org/10.1007/978-3-540-45175-4_42
27. Lazinger, S.: Digital preservation and metadata - history, theory, practice. Libraries Unlimited, Englewood, Colorado (2013)
28. Hakala, J.: Archiving the web - european experiences. Program: Electron. Libr. Inf. Syst. **38**(3), 176–183 (2004). https://doi.org/10.1108/00330330410547223
29. Day, M.: Collecting and Preserving the World Wide Web - A Feasibility Study Undertaken for the JISC and Wellcome Trust. University of Bath, United Kingdon, UKOLN (2003)
30. Tosh, J.: The Pursuit of History, 3rd edn., pp. 76–83. Pearson Education, London (2002)
31. Ferreira, L.B., Martins, M.R., Rockenbach, M.: Usos do arquivamento da web na comunicação científica. Prisma.com, **36**, 78–98 (2016)

32. IBGE (2018). https://www.ibge.gov.br/estatisticas/sociais/populacao/9171-pesquisa-nac ional-por-amostra-de-domicilios-continua-mensal.html?=&t=o-que-e. Accessed 10 Oct 2020

33. CGI.br (2018). https://cetic.br/tics/domicilios/2018/individuos. Accessed 10 Oct 2020

34. CGI.br (2017). https://cetic.br/pesquisa/governo-eletronico. Accessed 10 Oct 2020

35. Brasil: Lei Federal 9.159/1991. Dispõe sobre a política nacional de arquivos públicos e privados e dá outras providências, Brasília (1991)

36. Arquivo Nacional do Brasil: Política de preservação digital, Versão 2, Rio de Janeiro (2016)

37. Brasil: Decreto nº 9.756, de 11 de abril de 2019 - Institui o portal único "gov.br" e dispõe sobre as regras de unificação dos canais digitais do Governo federal. Diário Oficial [da] República Federativa do Brasil, Poder Executivo, Brasília, DF, 11 Abr (2019)

38. Gov.br (2019). https://www.gov.br. Accessed 10 Oct 2020

39. NUAWEB (2019). https://www.ufrgs.br/nuaweb/. Accessed 20 Oct 2020

40. Rockembach, M., Melo, J.F.: Web archiving of Brazilian websites. Paper presented at the IIPC WEB ARCHIVING CONFERENCE 2019, Zagreb, Croatia. Abstracts & presentations.... Drop-in Talks & Drop-in Slides (2019)

41. Melo, J.F.: Arquivamento dos websites do Governo Federal Brasileiro - preservação do domínio GOV.BR. Dissertação. 133 f. Porto Alegre: Universidade Federal do Rio Grande do Sul (2020)

42. Brasil: Projeto de Lei 2431/2015. Dispõe sobre o patrimônio público digital institucional inserido na rede mundial de computadores e dá outras providências, Brasília (2015)

43. Luz, A.J., Weber, M.H.: A memória política do Brasil no site da presidência - acesso e desvios da comunicação dos governos de Dilma Rousseff e Michel Temer. Liinc Em Revista, **15**(1) (2019). https://doi.org/10.18617/liinc.v15i1.4571

Metadata Quality of the National Digital Repository of Science, Technology, and Innovation of Peru: A Quantitative Evaluation

Miguel Valles$^{(\boxtimes)}$ (iD), Richard Injante (iD), Victor Vallejos (iD), Juan Velasco (iD), and Lloy Pinedo (iD)

Universidad Nacional de San Martín, Tarapoto, Peru
mavalles@unsm.edu.pe

Abstract. The National Digital Repository of Science, Technology, and Innovation (ALICIA) of Peru is in charge of harvesting the scientific production of the universities from their institutional repositories. The aim is to determine and evaluate the availability of resources in the repositories, and if the quality of the metadata harvested by ALICIA has any relation to the metadata of the institutional repositories. To this end, a non-experimental, descriptive, comparative study was carried out after recovering the data from ALICIA, using its API rest and the data from the institutional repositories, and using organic techniques for validating broken links and web scraping, which are then stored in a database on which queries were made using non-SQL statements. There is 97.7% and 95% availability of resources in public and private institutional repositories respectively, and on average, the quality of the metadata describing the resource is between 54% and 57% . It is notable that despite all the documented interventions and the results found, there are still quality problems in the metadata registering process considering that universities are responsible for its publication.

Keywords: Data quality · Curation · Dublin core · Institutional repositories · Metadata

1 Introduction

The universities have an important mission since they are the main sources of knowledge generation and contribution to science, which among other things allows them to solve society's problems, from where the complaints about their insufficient and inefficient productivity are born and who gives them the benefit of their elitist academic and scientific position, as shown in [1].

However, even though most of the research funding is the result of government contributions [2], some of the results are available through restricted access with payments and/or editorial subscriptions to licenses [3]. This require investing time with uncertainty

E. Bisset Álvarez (Ed.): DIONE 2021, LNICST 378, pp. 96–105, 2021.
https://doi.org/10.1007/978-3-030-77417-2_7

in the achievement of the required content, being a nefarious factor in situations such as the current ones in which every minute counts.

In opposition, the open access movement is the immediate access, without payment, registration, or subscription to academic and scientific material, it has a wide community in the world that involves teachers, undergraduate, and graduate students [4], who, with their ingenuity, solve problems and share the results in a way that increases the scientific heritage and enriches knowledge [5]. Its aim is interoperability, compatibility among archives, long-term preservation, and universal access to information. In addition to promoting free access, it encourages its availability, distribution, and reproduction [6]. It emerged in response to the restricted access to knowledge in academic and scientific journals imposed by commercial publishers (Gideon, 2008), as cited by [7].

This same open access movement has caused universities to face today an enormous challenge related to the need and obligation to generate knowledge that acquires visibility, achieves impact factors that demonstrate its importance, influence, and contribution to the scientific community that uses the published results [2]. According to [8]:

"A growing number of universities around the world have begun to establish open access policies regarding the academic results of their researchers, requiring them to publish mainly in open access journals and/or to archive their pre-prints or post-prints in institutional archives".

In this sense, in Peru, with the approval of the University Law in 2014 [9], the National Superintendence of University Education (SUNEDU) has the function of supervising the fulfillment of the basic conditions of quality. The state as a strategy to standardize and improve the visibility of the scientific production of the institutional repositories approved according to [10] creates the National Science Repository, Open Access Technology and Innovation (ALICIA). This strategy has allowed the implementation of technological infrastructure in universities to make available the results of scientific production that meet the needs of the community [11]. Without a doubt, we can say that institutional repositories are the cornerstone for open access [2]. By facilitating the systematization and access to publications, they contribute to the concept of open science [12, 13].

In South America there are similar strategies, adapted to each context, this promotes the exchange of scientific production with the implementation of infrastructures for harvesting the resources available in the repositories of the universities in each country [14]. Even within [15] the national nodes of the countries have been progressively integrating the region into a platform that has international standards of interoperability. This provides the possibility of generating a centralized scheme of a federated search for scientific results that facilitates the citation process and reference of resources in an open-access scheme.

According to [16, 17] and [18], ALICIA collects, integrate, and storage through a constant process based on the OAI-PMH interoperability protocol [19], known as harvesting, then presents the data and metadata from all the institutional repositories that are within the scope of Law No. 30035 [10]. By law, universities that receive state funding have to guarantee the availability of their repositories and their resources must remain efficiently usable, updated, and collected by "ALICIA" [20]. According to [18] all the repositories use DSPACE as an institutional repository solution.

The current ALICIA guidelines are based on the "OpenAIRE Guidelines for Literature Repository Managers v. 3.0", the Dublin Core metadata schema, and the OAI-PMH exchange protocol. 39 metadata must be considered for information registration in the institutional repositories: 23 are mandatory, 9 recommended, and 7 optional [21].

Although indeed, the resources of the institutional repositories, known as gray literature, have not undergone a peer-review process to ensure acceptable quality levels as in the case with scientific articles published in indexed journals according to [22]. This should not be an excuse for that the registration of metadata describing the resources, which are subsequently harvested by ALICIA, not to meet the quality criteria required by ALICIA guidelines [23].

We must understand that ALICIA needs to review the harvested content and determine whether the repositories that are part of the system [24] comply with the requirements of the directives [21]. Strategies are needed to improve the quality of the process of registering metadata [25] avoiding the same deficiencies that affect the process of bibliographic review of research conducted in universities, as well as its visibility and impact factor.

Not only that, ALICIA must look for mechanisms of identification, healing at the source, and correction at the destination, in case the results of their searches identify resources that at the time of access are fallen links, due to the dynamics of institutional repositories and the deficient skills of its staff to ensure a proper healing and updating process.

There are many research works worldwide that verify the quality of repository metadata, such as the [23], which evaluates the problem from those responsible for the registry, or the [26] which correlates the quality of metadata with its academic visibility in Google Scholar. However, there are few studies to determine if the harvesting process is correct. If there is a need for updates on the harvested metadata sources. Or any one that compares the quality of the data harvested by the harvesting infrastructures existing per country [14] against the existing metadata in the repositories.

Ensuring metadata quality that describe the resources available in the repositories, has much to do with the acceptance reflected in citations and impact factor of the research available in them and which are documented by [27], In addition to the qualitative criteria of the researchers who review these resources, as [28] claim, poor quality of metadata recording can have a negative impact not only on the way scientists retrieve, share and use research datasets, but also on the way that manage and audit repositories.

Therefore, we wonder whether the quality of the metadata harvested by ALICIA has any relation with the metadata of the institutional repositories. In this research, we seek to identify if the repository's administrator record the information according to established directives based on an indicator that we call MQ similar to that of [26]; How good ALICIA manages to harvest these data according to the correct configuration of protocols in the repositories through comparison. In what extent the repositories ensure the availability of the resources that ALICIA claims are available. Finally, we evaluate the need for a process of curing the ALICIA database by applying a comparative analysis for both.

2 Materials and Methods

2.1 Analysis Unit

This is a non-experimental, comparative descriptive cross-sectional research. The study universe is the metadata available in ALICIA and the resources available in the repositories of the 51 public universities and 92 private universities in Peru.

Initially, we have to specify that we carried out the whole process described in materials and methods during May 18–22, 2020, on a virtual private server on the amazon web services.

2.2 Data Recovery

We extracted the data harvested by ALICIA, consuming its API rest located and documented in https://alicia.concytec.gob.pe/vufind/api/v1, through a program built in python based on the libraries "requests", "JSON" and "BeautifulSoup" for text analysis.

We compared this metadata with data extracted through a web scraping process from the institutional repositories of all the universities that are part of ALICIA. This process is necessary because ALICIA does not harvest all the metadata available through the OAI-PMH protocol.

In detail, we obtain the metadata of each of the records available in the ALICIA database through a loop recovery process. For each record the metadata dc.identifier. Uri (URL of the resource) is identified, which is used as a parameter to perform an HTTP query in which organic broken-link validation techniques are applied [29, 30], based on HTTP responses to validate the availability of the resource in the institutional repository with a valid response code and not "not found".

If the resource is available in the repository, we apply web-scrapping techniques on grey literature documented by [31], to structure the data of the obtained response that we insert in a non-SQL database on which we perform queries on 20 Dublin Core elements, specified according to [21] to generate the presented results.

For all the algorithms built, we have worked with python and the request, JSON, and BeautifulSoup libraries. For the processing and generation of the descriptive statistics, we used No SQL statements (Fig. 1).

The algorithm used for data recovery is:

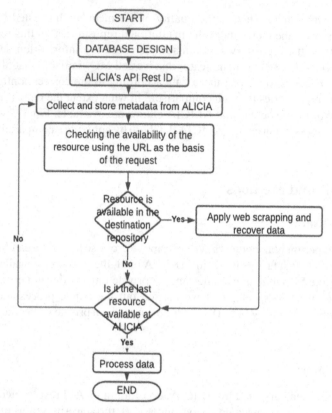

Fig. 1. Data recovery algorithm

2.3 Data Analysis and Processing

We based the evaluation criteria on the following indicators: Resource availability, Metadata quality.

Where

$$Availability = \frac{\sum_1^n resources\ that\ respond\ to\ the\ HTTP\ request}{\sum_1^m resources\ harvested\ by\ Alicia} * 100 \qquad (1)$$

$$Metadata\ quality = \frac{\sum_1^n metadata\ harvested\ by\ Alicia}{\sum_1^m metadata\ specified\ in\ the\ standard} * 100 \qquad (2)$$

3 Results

Thus, according to formula (1), In summary, we present the availability of resources in the repositories in Tables 1 and 2:

Table 1. Amount of available resources, as harvested by ALICIA and according to the web scraping we did from the repositories

Type of management	Harvested by alicia	Web scraping from the repositories	Availability
Public	111964	109393	97.7%
Private	156859	148949	95.0%

Source: own elaboration

At the date of data extraction, only 125 of 143 universities have repositories harvested in ALICIA; however, 15 of these have values in the URL metadata that do not respond to the HTTP checks performed by the algorithm. In addition, of these 87 have approved licensing, 38 have denied licensing.

According to information from SUNEDU, there are 51 state universities; however, only 39 of them have and maintain their institutional repository accessible.

Table 2. Available resources by type of university and type of resources, as harvested by ALICIA and according to the web scraping we did from the repositories

Management	Thesis type	Harvested by Alicia		Web scraping from the Repositories		Availability	
		Lic	No Lic	Lic	No Lic	Lic	No Lic
Public	Undergraduate	83.804	2.771	83.281	2.734	99.4%	98.7%
	Master's degree	16.491	4.077	16.464	2.126	99.8%	52.1%
	Ph.D.	4.616	205	4.606	182	99.8%	88.8%
Private	Undergraduate	92.563	20.300	88.732	19.043	95.9%	93.8%
	Master's degree	38.356	2.160	35.834	2.050	93.4%	94.9%
	Ph.D.	3.263	217	3.105	185	95.2%	85.3%

Source: own elaboration

Table 2 shows us an interesting characteristic, which according to our analysis, shows that universities with a denied license have problems guaranteeing the availability of the resources they have in their institutional repositories.

An important point to note is that the availability of resources harvested by ALICIA in the repositories of origin in public universities is 97.7%, in contrast to 95% in private universities. However [26], they find that 43% of public universities' resources and 60% of private universities' resources are correctly indexed in Google Scholar. The explanation can be found when we make a more detailed analysis of the content of the metadata since, apparently, according to [26] itself, the staff of private universities has more training in registering their resources in their repositories.

Continuing with the formula (2), following tables presents the quality of the metadata that describe the resources available in the repositories.

Table 3. Quantity of metadata harvested by type of university and type of resource.

Management	Thesis type	Amount of metadata per resource					
		Licensed			Licensed Denied		
		Max	Average	Min	Max	Average	Min
Public	Undergraduate	30	21.5	8	22	21	16
	Master's degree	29	22.5	14	23	21	20
	Ph.D.	29	21.8	14	22	21	20
Private	Undergraduate	34	22.1	12	26	22	13
	Master's degree	32	22.5	13	26	22	16
	Ph.D.	29	22.6	13	25	22	19

Source: own elaboration

An important fact from Table 3 is that, on average, public universities register 21.5 and private universities 22.5 metadata for undergraduate theses. In general, for both masters and doctoral theses, public universities register an average of 1 metadata less than private universities (Table 4).

Table 4. Quality of harvested metadata by type of university and type of resource.

Management	Thesis type	Amount of metadata per resource					
		Licensed			License Denied		
		Max	Average	Min	Max	Average	Min
Public	Undergraduate	77%	55%	21%	56%	54%	41%
	Master's degree	74%	58%	36%	59%	54%	51%
	Ph.D.	74%	56%	36%	56%	54%	51%
Private	Undergraduate	87%	57%	31%	67%	56%	33%
	Master's degree	82%	58%	33%	67%	58%	41%
	Ph.D.	74%	58%	33%	64%	56%	49%

Source: own elaboration

As for Table 5 (which is the summary of Table 3), on average, the public universities register 21.65 metadata and the private ones 22.29 metadata, that is, 0.64 metadata less. However, this analysis is only quantitative, and an analysis of the content of the metadata is necessary to determine the qualitative aspects of the data recording process to ensure that we are following the indications of [32] and the study of [25].

Table 5. Amount of metadata harvested by type of university.

Management	Amount of metadata per resource					
	Licensed			Not Licensed		
	Maximum	Average	Minimum	Maximum	Average	Minimum
Public	30	21,65	8	22	21	18
Private	34	22,29	12	26	22	14

Source: own elaboration

Table 6. Quality of harvested metadata by type of university.

Management	Metadata quality by resource					
	Licensed			Not Licensed		
	Maximum	Average	Minimum	Maximum	Average	Minimum
Public	77%	56%	21%	56%	54%	46%
Private	87%	57%	31%	67%	56%	36%

Source: own elaboration

Table 6 shows a very important indicator regarding the work and importance of those responsible for the repositories because according to [25], in general, they need training processes and strengthening of skills to ensure that the registration of metadata is done correctly, since as universities are responsible for the quality of the metadata recorded as mentioned [32].

4 Conclusions

After evaluating the resources harvested by ALICIA and their availability in university repositories, we concluded that public universities have 3% of unavailable resources compared to 5% of resources unavailable by private universities. The licensing process initiated 5 years ago with the enactment of the new university law could be the reason, in the process of which 25% of private universities have not achieved their licensing and have ceased to function.

Although many metadata's quality problems have been documented in the literature over the last few years (e.g., [33, 34] and even Concytec did its analysis with [25]), many of these problems are still present in the case of Peru's National Digital Repository of Science, Technology, and Innovation (ALICIA).

Finally, ALICIA must establish mechanisms for maintain available repositories of universities with denied licenses so that a significant number of researches are not left offline.

References

1. Concepción, D., Gonzáles, E., Miño, J.: Una visión actual de la ciencia como fuerza productiva directa. Rev. Univ. y Soc. **10**, 54–59 (2018)
2. Zacca-Gonzales, G.: Los repositorios en función de la ciencia abierta. Rev. Cuba. Inf. en Ciencias la Salud **30**(4), 1–3 (2019)
3. Banzato, G., Rozemblum, C.: Modelo sustentable de gestión editorial en Acceso Abierto en instituciones académicas. Principios y procedimientos. Palabra Clave (La Plata) **8**, e069 (2019). doi:https://doi.org/10.24215/18539912e069.
4. Silva-Rodríguez, A.: La sostenibilidad de la divulgación de la ciencia mediante modelos de negocios de acceso abierto. Rev. Digit. Int. Psicol. y Cienc. Soc. **2**, 21–39 (2016). https://doi.org/10.22402/j.rdipycs.unam.2.1.2016.73.21-39
5. Jacobs, N., Nixon, W.J.: Universities, Jisc and the journey to open. In: Technology, Change and the Academic Library, pp. 171–181. Elsevier (2021). https://doi.org/10.1016/b978-0-12-822807-4.00017-8
6. Cano, A., De Dios, R., García, O., Cuesta, F.: Los repositorios institucionales: situación actual a nivel internacional, latinoamericano y en Cuba. Rev. Cuba. Inf. en Ciencias la Salud **26**, 0–0 (2015)
7. Abrizah, A., Noorhidawati, A., Kiran, K.: Global visibility of Asian universities' Open Access institutional repositories. Malaysian J. Libr. Inf. Sci. **15**, 53–73 (2010)
8. Granholm, K.: Open access och spridning En kvantitativ analys av hur open access-publicerade artiklar citeras och sprids på webben. Uppsala University (2013)
9. Ley 30220: Ley Universitaria (2014). https://www.sunedu.gob.pe/wp-content/uploads/2017/04/Ley-universitaria-30220.pdf
10. Ley N° 30035: Ley que regula el repositorio nacioanl digital de ciencia, tecnología e innovación de acceso abierto 2 (2013)
11. Sandí, J., Cruz, M.: Repositorios institucionales digitales: Análisis comparativo entre Sedici (Argentina) y Kérwá (Costa Rica). e-Ciencias la Inf. **7**, 1 (2016). https://doi.org/10.15517/eci.v7i1.25264
12. Concytec: Autoridades académicas de Macrorregión Norte debaten sobre ciencia abierta y repositorios digitales (2019). https://portal.concytec.gob.pe/index.php/noticias/1791-ciencia-abierta-y-repositorios-digitales-debatidos-en-macrorregion-norte
13. Goben, A., Sandusky, R.J.: Open data repositories: current risks and opportunities. Coll. Res. Libr. News **81**, 62 (2020). https://doi.org/10.5860/crln.81.1.62
14. La Referencia. Nodos nacionales: el engranaje I LA Referencia. La Referencia (2020). https://www.lareferencia.info/legacy/nodos-nacionales-el-engranaje.html
15. La Referencia. ¿Qué es el buscador regional? I LA Referencia. La Referencia (2020). https://www.lareferencia.info/legacy/buscador-regional.html
16. Decreto Supremo N° 006-2015-PCM. Reglamento de la Ley N° 30035, Ley que regula el Repositorio Nacional de Ciencia, Tecnología e Innovación de Acceso Abierto 4 (2015)
17. Atamari Anahui, N., Díaz Vélez, C.: Repositorio Nacional Digital de Acceso Libre (ALICIA): Oportunidad para el acceso a la información científica en el Perú. An. la Fac. Med. **76**, 81 (2015). https://doi.org/10.15381/anales.v76i1.11081
18. Rivero, A.: Avances del repositorio nacional digital ALICIA: recolección del repositorio nacional digital ALICIA a las instituciones. Consejo Nacional de Ciencia, Tecnología e Innovación Tecnológica - Concytec (2018). https://repositorio.concytec.gob.pe/handle/20.500.12390/85
19. Haslhofer, B., Schandl, B.: Interweaving OAI-PMH data sources with the Linked Data cloud. Int. J. Metadata, Semant. Ontol. **5**, 17–31 (2010). https://doi.org/10.1504/IJMSO.2010.032648

20. Concytec. Reglamento RENATI (2016). https://busquedas.elperuano.pe/normaslegales/aprueban-reglamento-del-registro-nacional-de-trabajos-de-inv-resolucion-no-033-2016-sun educd-1425605-1/

21. Concytec: Directrices para el procesamiento de información en los repositorios institucionales (2020). https://portal.concytec.gob.pe/images/documentos/alicia/directrices_repositorio.pdf

22. Valderrama, J.: Literatura Gris. . Form. Univ. **4**, 1–2 (2011). https://doi.org/10.4067/S0718-50062011000600001

23. Stovold, E.: Metadata quality in institutional repositories may be improved by addressing staffing issues. Evid. Based Libr. Inf. Pract. **11**, 93–95 (2016). https://doi.org/10.18438/b81s7n

24. Concytec. Instituciones Integrantes - ALICIA (2019). https://alicia.concytec.gob.pe/vufind/

25. Francisco Talavera Chocano -Consultor, M.: Análisis de la calidad de metadatos en el Repositorio Nacional Digital de Ciencia, Tecnología e Innovación de Acceso Abierto – ALICIA. Consejo Nacional de Ciencia, Tecnología e Innovación Tecnológica - Concytec (2019). https://repositorio.concytec.gob.pe/handle/20.500.12390/399

26. Alhuay-Quispe, J., Quispe-Riveros, D., Bautista-Ynofuente, L., Pacheco-Mendoza, J.: Metadata quality and academic visibility associated with document type coverage in institutional repositories of peruvian universities. J. Web Librariansh. **11**, 241–254 (2017). https://doi.org/10.1080/19322909.2017.1382427

27. Repiso, R., Moreno-Delgado, A., Aguaded, I.: Factors affecting the frequency of citation of an article. Iberoam. J. Sci. Meas. Commun. **1**, (2020). https://doi.org/10.47909/ijsmc.08

28. Balatsoukas, P., Rousidis, D., Garoufallou, E.: A method for examining metadata quality in open research datasets using the OAI-PMH and SQL queries: the case of the Dublin Core 'Subject' element and suggestions for user-centred metadata annotation design. Int. J. Metadata, Semant. Ontol. **13**, 1–8 (2018). https://doi.org/10.1504/IJMSO.2018.096444

29. Hayat, S., Li, Y., Riaz, M.: Automatic recovery of broken links using information retrieval techniques. In: ACM International Conference Proceeding Series, pp. 32–36. Association for Computing Machinery (2018). https://doi.org/10.1145/3278293.3278296

30. Bashir, S.: Broken link repairing system for constructing contextual information portals. J. King Saud Univ. Comput. Inf. Sci. **31**, 147–160 (2019). https://doi.org/10.1016/j.jksuci.2017.12.013

31. Haddaway, N.: The use of web-scraping software in searching for grey literature. Grey J. **11**, 186–190 (2015)

32. Hrynaszkiewicz, I.: Publishers' responsibilities in promoting data quality and reproducibility. In: Bespalov, Anton, Michel, Martin C., Steckler, Thomas (eds.) Good Research Practice in Non-Clinical Pharmacology and Biomedicine. HEP, vol. 257, pp. 319–348. Springer, Cham (2019). https://doi.org/10.1007/164_2019_290

33. Swanepoel, M.: Digital repositories: all hype and no substance? New Rev. Inf. Netw. **11**, 13–25 (2005). https://doi.org/10.1080/13614570500268290

34. Palavitsinis, N., Manouselis, N., Sanchez-Alonso, S.: Metadata quality in learning object repositories: a case study. Electron. Libr. **32**, 62–82 (2014). https://doi.org/10.1108/EL-12-2011-0175

Elements for Constructing a Data Quality Policy to Aggregate Digital Cultural Collections: Cases of the Digital Public Library of America and Europeana Foundation

Joyce Siqueira[1] , Danielle do Carmo[1] , Dalton Lopes Martins[1(✉)] ,
Daniela Lucas da Silva Lemos[2] , Vinicius Nunes Medeiros[3] ,
and Luis Felipe Rosa de Oliveira[1]

[1] Universidade de Brasília, Brasilia, DF, Brazil
{joyce.siqueira,danielle.carmo,rosa.luis}@aluno.unb.br,
daltonmartins@unb.br
[2] Universidade Federal do Espírito Santo, Vitoria, ES, Brazil
daniela.l.silva@ufes.br
[3] Tainacan.org, Brasilia, DF, Brazil
viniciusmedeiros@inf.ufg.br

Abstract. Institutions around the world have sought to aggregate different cultural heritage data sources in order to provide society with comprehensive and useful services. This qualitative exploratory study presents a comparative analysis of two paradigmatic institutional aggregators, namely the Europeana Foundation in the European Union and the Digital Public Library of America in the United States. To that end, strategic aggregation documents were identified and quality policy elements analyzed and compared. As a result, nine quality-oriented data aggregation elements were selected: data providers; application process; metadata model; data exchange agreement; copyright license; call for applications; metadata use; technical criteria for data quality and data validation and publication. The elements identified and described are important in formulating processes that make it possible to aggregate digital cultural heritage collections from different Brazilian institutions and provide support for the solution currently being developed in collaboration with the Brazilian Institute of Museums (Ibram).

Keywords: Cultural data aggregation · Data quality policy · Institutional aggregator

1 Introduction

The online availability of cultural collections in the form of digital catalogs and repositories as well as their digitization and socialization on the internet (Scopigno et al. 2017; Potenziani et al. 2018; Medeiros e Sá et al. 2019) are noteworthy phenomena in contemporary times, and have intensified with the increased use of the internet as tool for work, leisure and socialization.

E. Bisset Álvarez (Ed.): DIONE 2021, LNICST 378, pp. 106–122, 2021.
https://doi.org/10.1007/978-3-030-77417-2_8

Initiatives to integrate different data sources are aimed at making better use of these technologies and providing society with comprehensive and useful services. However, it is important to consider that in order to aggregate data from different sources, they must first be organized to ensure favorable conditions for diversified high-quality analyses, integrated searches and possibilities for reuse, with a view to constructing new informational services and products.

The term "organized" refers to the minimal organization of data structure. Data structure is the set of metadata used to describe an informational object, the cataloging rules used to populate these metadata and the standards and controlled vocabularies applied to index and classify objects based on a common framework (Gilliland 2016; Zeng and Qin 2016; Abbas 2010).

Important contemporary initiatives such as Findability, Accessibility, Interoperability and Reuse (FAIR) (Wilkinson et al. 2016) are aimed at good data management practices consistent with the concept of well-structured data and establishing quality standards. To some extent, digital repository data described based on these principles can be considered quality data.

Despite the different motivations behind the search for objective criteria to analyze data quality, it is directly linked to the intention to reuse data, generating value based on the possibility of connecting it with other databases and providing an opportunity to produce knowledge.

This raises the following questions: What is the minimum metadata structure needed for aggregation? How can data quality be measured? What are the criteria that define a quality digital object?

What determines whether data meet the minimum requirements for aggregation and reuse and what can be done to ensure they reach these objectives? The responses to these and other questions should form part of the data quality policies of institutions aimed at assessing datasets from different sources in order to aggregate them.

For the purposes of the present study, a data quality policy is a set of technical recommendations to guide institutions interested in aggregating their data (DPLA 2020c; Scholz 2019). Following the policy improves user experience and helps create comprehensive coherent collections (DPLA 2020c). It also encourages partners to not only submit a minimum of metadata and content quality, but to aim for rich metadata and the highest possible data quality (Scholz 2019).

It is important to underscore that different areas of knowledge have shown interest in data aggregation techniques and the possibilities they provide, with research confirming that the issue permeates the scientific world.

Solutions were found in the areas of bioinformatics, geography, medicine, music and culture (SIQUEIRA and MARTINS 2021), among others. Examples of integrated data in different fields include genetic expression data on the fruit fly (*Drosophila melanogaster*) to select genes and validate experimental results in bioinformatics (Miles et al. 2010); geographic location data to assess water license requests in geography (Ziébelin et al. 2017); input forms and information systems in epidemiological studies to determine the cause of diseases such as obesity, depression and dementia in medical sciences (Kirsten et al. 2017); data to provide a single web portal to access a collection of historical sources on the Mediterranean from around 15 different epigraphic archives (Mannocci et al. 2014). Although only these four areas are highlighted, each of them clearly needs to manage a wide range of data in a unique way.

The aim of the present study is to find solutions in the field of culture and forms part of wider scientific initiative involving researchers from different academic institutions, who are collaborating with the Brazilian Institute of Museums (Ibram) in an important project to digitize and disseminate the collections of Brazilian museums.

Ibram is directly responsible for managing 30 museums[1] and, in 2016, adhered to the Tainacan platform, "a powerful and flexible repository platform for the WordPress tool that allows users to manage and publish digital collections" (GOV.BR 2020). It has since implemented the system in its museums as a strategy to provide online access to digital museum collections.

An important issue for the project is the ability to generate an integrated search service that allows users to access a single website in order to search for and retrieve information on all the museums that have provided their data in an open digital repository online.

As part of the intermediate phase in the scientific initiative to provide this service, the researchers involved have studied international facilities that perform similar tasks and can provide supporting frameworks and technical recommendations. The services selected were the Digital Public Library of America and Europeana Foundation, identifying the documents, procedures, criteria and fundamental stages of data aggregation, with a focus on measuring the quality of the data collected.

As such, the aim of this study is to propose essential elements for a data quality policy that contributes to aggregation procedures for online collections, with a view to: i) identifying and selecting strategic documents from the institutions investigated; and ii) analyzing and comparing the quality policy elements mapped here. In our view, these aspects are potential elements in formulating aggregation processes for collections.

The article is organized as follows: Sect. 2 describes the institutions and the methodological procedures used to collect and analyze data; Sect. 3 presents the results of the study, clarifying the elements identified and deemed relevant in a data quality policy aimed at aggregation; Sect. 4 discusses and compares how these elements are used in the context studied; and finally, Sect. 5 identifies important aspects for further research.

2 Methodology

This is a qualitative descriptive exploratory study involving two renowned institutions that aggregate cultural data, the Digital Public Library of America and Europeana Foundation, with a comparative analysis of their data quality policies in a case study design.

The Digital Public Library of America is a nonprofit organization aimed at amplifying the value of libraries and cultural organizations as Americans' most trusted sources of shared knowledge. Its integrated search portal DPLA - https://dp.la/ (DPLA 2020h) provides millions of materials from libraries, archives, museums and other cultural institutions across the country in a single point of access (DPLA 2020a).

---

[1] IBRAM, Ibram Museums, http://www.museus.gov.br/os-museus/museus-ibram/, last accessed on 10/05/2020.

The Europeana Foundation is an independent nonprofit organization that operates the Europeana Collections platform at https://www.europeana.eu/pt (EUROPEANA 2020b), with millions of items from European archives, libraries and museums, providing access to books, music, artworks and other materials via sophisticated research tools (EUROPEANA 2020a; EUROPEANA PRO 2020b).

These institutions were selected as paradigms in the field and for geopolitical reasons, since the DPLA covers the entire United States and the Europeana Foundation all of Europe. Also considered was the fact that the Europeana Foundation encompasses more than one country and several different languages while the DPLA deals with only one country and language; the relevance of the former as a data aggregation service, particularly because of its direct relationship with some of the world's most renowned cultural institutions such as the Louvre Museum and British Library; and the significant differences in the processes described for the two institutions, with the DPLA using simpler and more manual procedures than those of the Europeana Foundation.

Initial exploratory analysis also found that the DPLA model is far more compatible with the Brazilian reality than that of the Europeana Foundation.

The technical data collection and analysis methods used were a literature review and documentary analysis. The former was used to substantiate concepts and provide theoretical sustainability for the study and the latter to provide access (via the institutional websites) to documentary sources, including manuals, tutorials, video demonstrations, educational material and other content aimed specifically at supporting institutions that want to become data providers.

Documents from the two institutions selected were analyzed and interpreted using content analysis, a set of techniques aimed at obtaining indicators (quantitative or not) that enable inferences to be made regarding knowledge present in the material assessed.

The techniques involved include defining categories (shown in Table 1) to support content analysis of the selected materials. According to Bardin (2016), categories are established based on the following criteria: semantics (themes); syntax (verbs, adjectives, pronouns); lexicons (grouping synonyms or antonyms, or according to the meaning of words); and expression (grouping writing or linguistic deviations).

The criterion used in the present study was semantics, establishing themes related to the fields of information and technology. The following analysis categories (individual and comparative) were defined:

- Data providers: individual or institutional aggregators that submit data for aggregation.
- Application process: the means provided by the institutional aggregator to initiate the aggregation process.
- Metadata model: formal specification of the data model, listing the classes and properties of metadata.
- Data exchange agreement: Document governing the partnership between the aggregator and data provider.
- Copyright licenses: list of copyright licenses pre-selected by the aggregator.
- Metadata use: minimum metadata requirements for aggregation.
- Technical criteria for data quality: defining quality criteria for metadata content and/or digital media.

- Call for applications: specifications on how collections are structured to keep aggregators informed.
- Data validation and publication: prototype of aggregated data that must be validated by the provider in order to be published.

It is important to note that categorization was essential in mapping and understanding the workflow stages involved in quality assessment and data aggregation by the two institutions.

Finally, data organization and management (the final stage of the method) made it possible to obtain a conceptual theoretical corpus based on the empirical objects investigated and draw theoretical and methodological conclusions regarding the underlying quality policy of both institutions.

Table 1 presents the supporting documents for quality assessment and data aggregation that guided this study. The list of documents and their purpose is an important tool for future research.

Table 1. Documentation on metadata and data quality.

Description	DPLA	Europeana
Data model	*Metadata Application Profile* v5.0 (Gueguen et al. 2017)	*Europeana Data Model v2.3* (Clayphan et al. 2016)
Data exchange agreement	*Data Exchange Agreement* (DPLA 2017a)	Europeana, o *Europeana Data Exchange Agreement* (EUROPEANA FOUNDATION 2020)
Metadata and content quality requirements	Partnering with DPLA Metadata (DPLA 2020g); DPLA Metadata Quality Guidelines (DPLA 2020i) and DPLA Geographic and Temporal Guidelines for MAP 3.1 (DPLA 2017b)	Europeana Publishing Guide v1.8 (Scholz 2019) and Europeana Publishing Framework 2.0 (EUROPEANA FOUNDATION 2019)
Copyright licenses	DPLA Standardized Rights Statements Implementation Guidelines (DPLA 2017c)	Europeana Licensing Framework (EUROPEANA THINK CULTURE 2011)
Guidelines for content development and management	–	Europeana Content Strategy (Scholz et al. 2017)
User scenarios and their metadata requirements	–	Discovery - User scenarios and their metadata requirements v3 (Charles et al. 2015)
Guidelines on the material to be collected	Collection Development Guidelines (DPLA 2020c)	–

The DPLA lists all the useful documents and other tools for data providers at https://pro.dp.la/hubs/documentation (DPLA 2020d) and the Europeana Foundation at https://pro.europeana.eu/share-your-data/process (EUROPEANA PRO 2020c).

3   Results

The nine stages obtained are illustrated in order in Fig. 1 for each institution.

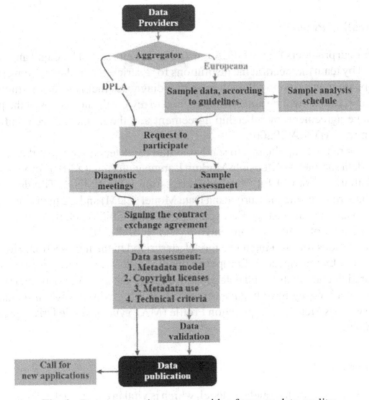

Fig. 1. Data aggregation stages, with a focus on data quality

The nine stages are described below: data providers; application process; metadata model; data exchange agreement; copyright licenses; call for applications; metadata use; technical criteria for data quality and; data validation and publication.

3.1   Data Providers

Data providers are institutions that submit their collections for aggregation. These consist of individual providers that submit their data directly to the institutional aggregator; and aggregators, who aggregate data from individual providers, typically grouped according to a specific niche such as region or theme, and submit the collections to the institutional aggregator. Both the DPLA and Europeana Foundation receive data from both types of providers based on their own specific criteria.

The DPLA currently has 43 data providers, 29 of which are aggregators, as well as around 41 million items from cultural institutions across the United States. Europeana has 38 data providers, all of which are aggregators, and over 50 million digitized items on European cultural heritage. The complete lists of data providers for both institutions are available at https://pro.dp.la/hubs/our-hubs (DPLA 2020f) and https://pro.europeana.eu/page/aggregators (EUROPEANA PRO 2020a).

3.2 Application Process

Prospective data providers for the DPLA must submit their request by email and will then be contacted by team members at the institutions for a series of mandatory conversations. Aggregators are then asked to fill out a form that diagnoses them as data partners. Once accepted, new partners will be notified and invited to review the next steps of the process: data exchange agreement, membership agreement and the metadata, tech and content information form (DPLA 2020b).

Europeana requires applicants to send an email to its team, who will then request a compacted data sample in eXtensible Markup Language format (XML)[2] or via the Open Archives Initiative Protocol for Metadata Harvesting (OAI-PMH)[3]. The data must be mapped in accordance with the Europeana Data Model (EDM) and are then automatically checked and validated against the Europeana Publishing Guide (Scholz 2019), using the Metis processing tool available from GitHub (GITHUB 2020).

Both institutions follow a metadata model, described in the next section. However, it should be noted that prospective Europeana data providers must have prior knowledge of the required documentation and data model, whereas the DPLA assists data providers in this process after they have been accepted, identifying which original metadata are compatible with its Metadata Application Profile (MAP) via a Google Drive spreadsheet (DPLA 2020e).

3.3 Metadata Model

Each institution has its own metadata model, which is vital in ensuring that an aggregation standard is in place. The Europeana Foundation developed the Europeana Data Model (EDM), a theoretical data model that allows data to be presented in accordance with the practices of different domains that contribute to the Europeana Collection (Clayphan et al. 2016).

The DPLA's Metadata Application Profile (MAP) is the basis for data structuring and validation and guides how they are stored, serialized and published. The MAP was designed based on the EDM (Gueguen et al. 2017).

---

[2] Markup language to facilitate online information sharing.

[3] Protocol for collecting metadata records in repositories.

3.4 Data Exchange Agreement

Once accepted, data providers must officially enter into a partnership with the institutional aggregator. An important part of this process is signing a data exchange agreement, which establishes the rules governing the partnership,

namely the DPLA's Data Exchange Agreement (DPLA 2017a) and Europeana Foundation's Europeana Data Exchange Agreement (EUROPEANA FOUNDATION 2020).

The most noteworthy element of these agreements is the need for the data provider to guarantee that their aggregated metadata and media are covered by a copyright license.

3.5 Copyright Licenses

As mentioned above, copyright licenses are a fundamental part of the data aggregation process because they govern how the end user will access and reuse the digital object. The institutions studied here adopt a similar stance in relation to copyrights.

In regard to metadata, both institutions require that these be submitted under a CC0 1.0 Universal Public Domain Dedication license[4]. For media, the DPLA lists 12 standardized rights statements and recommendations for copyrights (DPLA 2017a) and the Europeana Foundation 14 (EUROPEANA PRO 2015) .

The licenses are briefly analyzed to determine whether the metadata license used allows unrestricted reuse, while access and reuse of media s are graded on different levels.

3.6 Call for Applications

The DPLA does not have a well-defined schedule for including new data providers, but provides an online calendar (DPLA 2021) with new collection dates.

Europeana advises that initial data collection is performed on a first-come-first-serve basis; however, thematic collections, metadata and content quality, user demands and business priorities define which collections will receive more attention. The data are processed and published continuously throughout the year.

3.7 Metadata Use

The metadata schemas used by the two institutional aggregators cover a wide range of fields, but the present study focuses on the following categories: mandatory, mandatory when present, mandatory based on a predefined metadata group and; recommended (Table 2).

Metadata descriptions can be found in documents on the MAP (Gueguen et al. 2017) and EDM metadata models (Clayphan et al. 2016).

The DPLA lists 11 metadata elements, four **required**, four **required when present** and three **recommended**, while Europeana lists 14, six **required** and four **required**

---

[4] CREATIVE COMMONS. CC0 1.0 Universal (CC0 1.0) Public Domain Dedication (2020), https://creativecommons.org/publicdomain/zero/1.0/, last accessed on 10/16/2020.

Table 2. Metadata submission categories

No.	Metadata	DPLA	Europeana
1	Subject	SourceResource.subject[d]	dc:subject[c]
2	Location	SourceResource.spatial[d]	dcterms:spatial[c]
3	Type or genre	–	dc:type[c]
4	Time	–	dcterms:temporal[c]
5	Date created	SourceResource.date[d]	–
6	Copyright license	SourceResource.rights[a]	edm:rights e sua URI
7	Unique identifier	–	edm:ProvidedCHO, edm:WebResource, ore:Aggregation[a]
8	Thumbnail link	Aggregation.object[b]	–
9	Link to the original record	Aggregation.isShownAt[a]	edm:isShownBy ou edm:isShownAt[a]
10	Metadata populated	–	edm:ugc = true[b]
11	Name of the institution	Aggregation.dataProvider[a]	edm:dataProvider[a]
12	Name of the aggregator	Aggregation.provider[b]	edm:provider[b]
13	Textual objects	–	dc:language[b]
14	Type of resource	SourceResource.type[b]	edm:type[b]
15	Collection title	Collection.title[b]	–
16	Title	SourceResource.title[a]	dc:title[c]
17	Description	–	dc:description[c]

[a]Required
[b]Required when present
[c]Required, based on a predefined metadata group
[d]Recommended

when present. The difference in numbers is because the Europeana Foundation organizes metadata submissions by group, only one of which is mandatory, as is the case for *dc:title* and *dc:description*.

The next section addresses the quality of metadata and media content.

3.8 Technical Criteria for Data Quality

In additional to minimum metadata requirements, it is vital that the data submitted exhibit quality content, including digital media. Thus, in order to advise data providers on the quality of their own metadata and media and how to improve them when necessary, both the DPLA and Europeana provide technical guidelines for institutions interested in aggregating their data.

DPLA guidelines are developed around three tiers of recommendations for each property in the MAP, whereas the more complex Europeana framework involves two sets of tiers, one for metadata quality and the other for media quality.

The three tiers of the DPLA Metadata Quality Guidelines (DPLA 2020i) are minimal quality requirements, recommendations for improved quality and recommendations for best quality.

Guidelines under the "minimal quality requirements" tier are considered necessary for data aggregation and typically relate to correct use of the property or granularity issues. Although each property may or may not have minimal requirements, properties included in the record must follow the guidelines. For example, while there is no minimal quality requirement for the mandatory metadata property *title*, the optional *alternative title* property has a minimal requirement of not being a copy of the main title, although translated titles are acceptable.

The "recommendations for improved quality" tier is a set of guidelines aimed at improving data within the aggregation context, creating better overall consistency through actions such as the use of authorities or content standards. For example, for the optional metadata property *creator,* the use of name authority files such as LCNAF (Library of Congress Names[5]), VIAF (The Virtual International Authority File[6]) and ULAN (*Union List of Artist Names*[7]) is recommended. Another example is the optional *extent* property, with the recommended use of content standards such as CCO (Cataloging Cultural Objects[8]), RDA (Resource Description and Access[9]) and DCRM (Descriptive Cataloging of Rare Materials[10]).

Finally, the "recommendations for best quality" guidelines are related to the use of Uniform Resource Identifiers (URIs), a unique identifier that enables the use of Linked Open Data (LOD)[11]. Since linked data is still a relatively new concept in libraries, archives and museums, it is reserved for the highest quality tier (DPLA 2020b, p. 1).

Data quality and requirements for the Europeana Foundation are presented in two main documents, the Europeana Publishing Guide (Scholz 2019), which provides technical guidelines for data aggregation, and the Europeana Publishing Framework (EUROPEANA FOUNDATION 2019), whose primary objective is to inform data providers of the benefits of submitting quality metadata and media.

---

[5] LIBRARY OFCONGRESS. LC Name Authority File (LCNAF), https://id.loc.gov/authorities/names.html, last accessed on 10/19/2020.

[6] VIAF. The Virtual International Authority File, http://viaf.org/, last accessed on 10/19/2020.

[7] GETTY. Union List of Artist Names, https://www.getty.edu/research/tools/vocabularies/ulan/index.html, last accessed on 10/19/2020.

[8] VRA. Cataloging Cultural Objects (CCO), https://vraweb.org/resources/cataloging-cultural-objects/, last accessed on 10/19/2020.

[9] RDA. Resource Description and Access, https://www.rdatoolkit.org/about, last accessed on 10/19/2020.

[10] RBMS. Descriptive Cataloging of Rare Materials, http://rbms.info/dcrm/dcrmmss/, last accessed on 10/19/2020.

[11] LOD is a set of best practices for publishing and interlinking structured data on the web, establishing links between items from different sources to form a global data space (HEATH and BIZER, 2011 apud Santarém Segundo 2015).

Europeana provides guidelines for metadata and media. How data are published by data providers is classified into tiers A, B and C, with A the lowest tier and C the highest. Metadata tier A represents Europeana Collections as a basic search platform; B as an exploration platform and C as a knowledge platform.

When a data provider's metadata fall under tier A, it can use Europeana Collections as a basic search platform, that is, users can find specific objects if they have the precise information they need, with the aid of search filters. In order to meet this requirement, at least 25% of the metadata elements provided and considered relevant must have at least one language tag. Additionally, the metadata should include at least one of the enabling elements taken from the discovery and user scenarios.

The discovery and user scenarios were developed by the Data Quality Committee (DQC) and reflect user needs when accessing information, listing problems and proposing actions. The goal of these scenarios is to demonstrate specific situations to users and encourage them to provide increasingly comprehensive metadata by including enabling elements (optional metadata elements) that support specific user scenarios, with a view to ensuring greater information retrieval and better services in the collections (Charles et al. 2015).

Metadata provided under tier B allow the providing institutions to use Europeana Collections as an exploration platform, whereby users can find objects by searching for both general and specific information. This allows collections to be contextualized and discovered in thematic collections, with findability enhanced by the possibility of multilingual searches, and presented on the resource pages of the entities featured on the platform. To that end, at least 50% of the relevant metadata elements provided must have at least one language tag. The metadata should also include at least three enabling elements covering at least two distinct discovery and user scenarios and one contextual class with all the minimum required elements or a link to LOD vocabulary.

Contextual classes give meaning to objects and make them easier to retrieve. The EDM includes four contextual classes, edm:Agent, edm:Place, edm:TimeSpan and skos:Concept, which can be used to capture distinct entities that are related to the cultural heritage objects. The use of references to multilingual vocabularies and LOD is recommended, including the Getty Art & Architecture Thesaurus (AAT), Wikidata or Geonames (Charles et al. 2015).

Tier C metadata allow the data provider to use Europeana Collections as a knowledge platform. This enables users to find the object they are looking for more accurately without needing specific information on it because the contextual metadata provided generate a linked network of knowledge so that connections can be discovered through the relationships between them and their identities. Metadata can therefore be used in partner projects such as Historiana[12] (for the educational sector) and CLARIN[13] (research infrastructure project) and others involving creative industries. This contextualized and enriched information can also be used in classes, studies and applications, ensuring greater reach for the collections within and beyond the Europeana Collections platform. To make this possible, at least 75% of the relevant metadata elements provided

---

[12] HISTORIANA. Historiana, https://historiana.eu/, last accessed on 10/19/2020.

[13] CLARIN. European Research Infrastructure for Language Resources and Technology, https://www.clarin.eu/, last accessed on 10/19/2020.

must have at least one language tag. The metadata should include at least three enabling elements covering a minimum of two distinct discovery and user scenarios and two different contextual classes with all the minimum required elements or links to LOD vocabularies.

The above information clearly shows that both aggregators (DPLA and Europeana) stipulate the lowest tiers as minimum network entry requirements, but also guide institutions on how to progressively manage their data to ensure better quality, thereby exploiting search resources more efficiently and discovering information provided by the platforms.

The publication of media submitted by data providers is classified into tier 1 for Europeana Collections as a search engine, 2 as a showcase; 3 as a distribution platform for reuse and 4 as a free reuse platform.

The tiers rate the possibility of reusing digital content, with tier 1 representing the lowest possible use and 4 the highest.

Digital objects compliant with tier 1 are sufficient to allow data providers to use Europeana Collections as a search engine so that users can find their collections. The digital object is visualized via the website of the original institution and not the Europeana Collections environment.

Tier 2 data means the Europeana Collections platform can be used as a showcase for the data provider's collections, enabling users to visualize high-quality versions of the collections. The objects can also be included in one of Europeana's thematic collections, providing greater exposure.

Data that fall under quality tier 3 allow providers to use Europeana Collections as a distribution platform for noncommercial reuse, meaning users can view high-quality versions of digital objects directly on the platform and incorporate their collections in existing projects such as Historiana or CLARIN. This is made possible by rights statements that authorize some forms of reuse, allowing collections to be used on websites, applications and other noncommercial services.

Tier 4 enables data providers to use Europeana Collections as a free reuse platform, whereby data related to collections can be incorporated in projects and partnerships with creative industries; shared on social media under the hashtag #OpenCollections; used on open platforms such as those of the Wikimedia Foundation and in hackathons. To that end, the collection must be published under rights statements that allow free reuse, enabling it to be used on websites, applications and commercial and noncommercial services.

The specific requirements for each type of media (image, audio, video and 3D) are described in detail in the Europeana Publishing Guide (Scholz 2019).

3.9 Data Validation and Publication

At the end of the process, Europeana provides applicants with a link to view their data beforehand and authorize its publication or determine whether changes are needed (Scholz 2019). The DPLA does not offer this type of service.

4 Discussion

According to the data obtained, the DPLA's aggregation processes are simpler and more manual than Europeana's complex procedures. An example is that Europeana requires sample data to be submitted in line with its available documentation and initially automatically validates the data via Metis software, whereas the DPLA maps metadata manually using an electronic spreadsheet after the institution has been accepted as a data provider.

Europeana requirements are stringent and their stages complex, which are positive points when considering the pedagogical effects for the field because they establish paths to follow, but also generate added difficulty for interested data providers. The process appears to require a certain level of technical and administrative maturity on the part of providers interested in publishing their data online.

The DPLA's data model (MAP) focuses on a common interface to describe media by reflecting information on content, whereas Europeana's EDM model differs in the description underlying the content (or describing knowledge about the resource) through semantic technologies, particularly semantic vocabularies available in the LOD environment, with a view to the formal and explicit publication of knowledge on the internet.

This allows data providers that use aggregation services such as Europeana Collections to expand search opportunities for their users on knowledge networks semantically linked to web data. This includes multimedia research laboratories, creative industries, the cultural and education sectors, and social media such as Wikidata, enabling information within and beyond the aggregation platform to be explored in greater depth.

Europeana operates in a scenario that assumes greater organizational maturity in its partners' information management processes. It is important to note that the institution essentially relies on national aggregators in the respective countries to collect data, as well as some specific thematic projects, such as in the field of cinema. As a result, the data structure and quality parameters governing its initial submission processes are more stringent, whereas the DPLA differs significantly in that it helps providers map their data according to the required model, providing technical and institutional support on how to manage and submit their data in a suitable format for aggregation. This pedagogical and supportive effect is particularly important in the present study.

Both institutions provide recommendations on metadata quality within an aggregation context and a general metadata framework for application in machine-readable media content, but their frameworks differ in terms of modelling multimedia information. Despite their similar vision on metadata, there are obvious differences in their semantic proposals, with Europeana clearly expressing its goal of collaborating and becoming a data provider in open linked data semantic networks, while the DPLA does not seem to explicitly share this objective.

It is also important to underscore that Europeana indicates possible uses for aggregated data in more complex scenarios, developing and establishing partnerships with thematic data reuse projects and relationships with knowledge bases such a Wikidata. It can undoubtedly be inferred that this greater complexity in managing and aggregating data can provide important benefits, resulting in higher quality data with better potential for reuse in future applications and services.

In regard to the minimum metadata requirements for aggregation in the different mandatory categories, the DPLA also exhibited a small dataset, required only 11 fields compared to Europeana's 14. One of the fundamental differences between the two institutions was the language requirement, heavily emphasized by Europeana in a number of points, but less so by the DPLA. The fact that Europeana covers different countries and official languages makes this factor more complex, generating demands that will translate into elements of the data model.

5   Final Considerations

The results obtained indicate advances in data aggregation research in terms of providing metadata models aimed at describing digital objects and multimedia online. As observed in the present study, the two aggregator institutions studied view metadata as the product of cataloging rules and consider it vital in describing informational resources in order to expand access points and improve the management, organization and retrieval of digital objects online.

The study shows the different stages and processes involved in compiling a data aggregation policy to develop a search and retrieval service in the area of culture. The different stages were analyzed, highlighting their technical characteristics and the policies involved, and clearly demonstrating that the process includes phases of dialogue, agreements, meetings, validation and necessary adjustments to achieve the aggregation result. It is important to note that this process could not be fully automated, with human intervention essential in certain stages when decisions and adjustments are needed. The important role of documentation experts such as librarians, museologists and archivists in projects of this type is also evident.

One of the most noteworthy points observed throughout the study is the pedagogical role of these initiatives in establishing levels of quality and data assessment. For institutions without the necessary training to produce data that meet these quality criteria, these initiatives establish objective parameters that serve as an inspiration and benchmark to improve the quality of their data. This is considered a significant shortcoming in the field of museology in Brazil and possibly even in other information-related fields.

It is hoped that this type of research will provide the necessary direction and initiatives to qualify documentation efforts and establish better criteria for institutions that want to publish their data on digital repositories for subsequent collection and participation in different information networks. These efforts not only improve the data quality of providing institutions, but the quality of cultural institution websites themselves.

The objectives of the study were achieved, identifying the relevant documents and referencing those that are essential in explaining the rules, guidelines and standards adopted by the institutions in their efforts at aggregation. It is important to note that these documents provide significant guidance for researchers and students interested in the issue as well as a framework that presents them systematically in order of their importance and role in the services established. The secondary objective of analyzing and comparing the elements of the two institutions' quality policies was also met, with the difference and similarities clearly identified. In general, the DPLA's process is simpler and more collaborative, providing more direct support for data providers in adjusting and preparing their data for aggregation. The Europeana Foundation gives providers greater autonomy in the submission process and has stricter, more complex requirements, but also offers broader possibilities for data reuse and greater integration with knowledge bases generated based on semantic web principles.

In terms of gathering subsidies and information that contribute to compiling a data quality policy for the aggregation of digital repositories from museums managed by the Brazilian Institute of Museums, the study conducted a detailed analysis of the processes of both institutions and identified elements to serve as inspiration and a basis for reflection in the Brazilian initiative. It is hoped that part of the results can be used in future research in the field of Brazilian museums.

References

Abbas, J.: Structures for Organizing Knowledge: Exploring Taxonomies, Ontologies, and Other Schema. Neal-Schuman Publishers, New York (2010)

Bardin, L.: Análise de Conteúdo. 3ª Reimpressão da 1, vol. 70. Edições, São Paulo (2016)

Charles, V., Isaac, A., Hill, T. (ed.): Discovery - User scenarios and their metadata requirements v3. Europeana Foundation (2015). https://pro.europeana.eu/files/Europeana_Professio nal/EuropeanaTech/EuropeanaTech_WG/DataQualityCommittee/DQC_DiscoveryUserScena rios_v3.pdf. Accessed 23 Sept 2020

Clayphan, R., Charles, V., Isaac, A.: Europeana Data Model – Mapping Guidelines v2.3. Europeana Think Culture (2016). https://pro.europeana.eu/files/Europeana_Professional/Share_ your_data/Technical_requirements/EDM_Documentation/EDM_Mapping_Guidelines_v2. 3_112016.pdf. Accessed 23 Sept 2020

DPLA: about us (2020a). https://dp.la/about. Accessed 28 Sept 2020

DPLA: Becoming a Service Hub (2020b). https://pro.dp.la/prospective-hubs/becoming-a-servic e-hub. Accessed 05 Oct 2020.

DPLA: Collection Development Guidelines (2020c). https://pro.dp.la/hubs/collection-develo pment-guidelines. Accessed 23 Sept 2020

DPLA: Documentation and Tools (2020d). https://pro.dp.la/hubs/documentation. Accessed 27 Sept 2020

DPLA: DPLA Data Exchange Agreement (2017a). https://digitalpubliclibraryofamerica.atlass ian.net/wiki/download/attachments/85406202/DataExchangeAgreement2017March.docx? version=1&modificationDate=1494427927703&cacheVersion=1&api=v2. Accessed 16 Oct 2020

DPLA: DPLA Geographic and Temporal Guidelinesfor MAP 3.1 (2017b). https://docs.google. com/document/d/1b2iJI90I24hUp-8kCfnZhAQcefBt0vPUjMzhDn9HOZ0/edit#heading=h. s8e27gcgg4cl. Accessed 23 Sept 2020

DPLA: DPLA Metadata Quality Guidelines. Digital Public Library of America (2020i). https:// docs.google.com/document/d/1dITqEYEWsMX1a2pLPmkL78k1LN2b4im03spn8_QFscY/ edit#heading=h.8lontuvk72d8. Accessed 23 Sept 2020

DPLA: DPLA Partners Crosswalk, MAP v4.0 (2020e). https://docs.google.com/spreadsheets/ d/1BzZvDOf4fgas3TD21xF40lu2pk2XW0k2pTGJKIt6438/edit#gid=1453046017. Accessed 12 Oct 2020

DPLA: DPLA Standardized Rights Statements Implementation Guidelines (2017c). https://docs. google.com/document/d/1aInokOIIsgf-B4iMTXU33qYN5B2jA3s91KgWoh7DZ7Q/edit# heading=h.ma4e226diad0. Accessed 23 Sept 2020

DPLA: Our Hubs (2020f). https://pro.dp.la/hubs/our-hubs. Accessed 16 Oct 2020

DPLA: Partnering with DPLA Metadata (2020g). https://docs.google.com/document/d/1gshwQ 0Oj84l5q-_JxHo7wjyLeXln9lCg6SUtIMkPu_g/edit. Accessed 23 Sept 2020

DPLA: Portal Digital Public Library of America (2020h). https://dp.la/. Accessed 16 Oct 2020

DPLA. Hub Re-ingest Schedule (2021). https://digitalpubliclibraryofamerica.atlassian.net/wiki/ spaces/CT/pages/84969744/Hub+Re-ingest+Schedule. Accessed 27 May 2021

Europeana: about us (2020a). https://www.europeana.eu/en/about-us. Accessed 28 Oct 2020

Europeana: Europeana Collections (2020b). https://www.europeana.eu/pt. Accessed 16 Oct 2020

Europeana Foundation: Europeana Data Exchange Agreement (2020). https://pro.europeana. eu/files/Europeana_Professional/Publications/Europeana%20Data%20Exchange%20Agre ement.pdf. Accessed 27 Sept 2020

Europeana Foundation: the more you give, the more you get. Europeana Publishing Framework 2.0 (2019). https://pro.europeana.eu/files/Europeana_Professional/Publications/Publishing_F ramework/Europeana%20Publishing%20Framework%20V2.0%20English.pdf. Accessed 27 Sept 2020

Europeana Pro: available rights statements (2015). https://pro.europeana.eu/page/available-rights-statements. Accessed 27 Sept 2020

Europeana Pro: Europeana Aggregators (2020a). https://pro.europeana.eu/page/aggregators. Accessed 16 Oct 2020

Europeana Pro: Europeana Foundation, About us (2020b). https://pro.europeana.eu/about-us/fou ndation#europeana-foundation. Accessed 28 Sept 2020

Europeana Pro: Share your data (2020c). https://pro.europeana.eu/share-your-data/process. Accessed 27 Sept 2020

Europeana Think Culture: Europeana Licensing Framework. Institute for Information Law, the Bibliothèque nationale de Luxembourg and Kennisland in cooperation with Europeana (2011). https://pro.europeana.eu/files/Europeana_Professional/Publications/Europeana%20L icensing%20Framework.pdf. Accessed 23 Sept 2020

GITHUB: Europeana - Metis Framework (2020). https://github.com/europeana/metis-framework. Accessed 12 Oct 2020

Gilliland, A.J.: Setting the stage. In: Baca, M. (ed.) Introduction to Metadata, 3 edn. Getty Research Institute, Los Angeles (2016)

GOV.BR: Projeto Tainacan. Ministério do Turismo - Instituto Brasileiro de Museus (Ibram) (2020). https://www.museus.gov.br/acoes-e-programas/projeto-tainacan/. Accessed 12 Oct 2020

Gueguen, G., et al.: Metadata Application Profile, version 5.0. DPLA - Digital Public Library of America (2017). https://drive.google.com/file/d/1fJEWhnYy5Ch7_ef_-V48-FAViA72OieG/ view. Accessed 23 Sept 2020

Kirsten, T., Kiel, A., Rühle, M., Wagner, J.: Metadata Management for Data Integration in Medical Sciences. Datenbanksysteme für Business, Technologie und Web (2017). https://dl.gi.de/han dle/20.500.12116/627. Accessed 05 Oct 2020

Mannocci, A., Casarosa, V., Manghi, P., Zoppi, F.: The Europeana network of ancient Greek and Latin epigraphy data infrastructure. In: Closs, S., Studer, R., Garoufallou, E., Sicilia, M.-A. (eds.) MTSR 2014. CCIS, vol. 478, pp. 286–300. Springer, Cham (2014). https://doi.org/10. 1007/978-3-319-13674-5_27

Medeiros e Sá, A., Ibañez Vila, A.B., Rodriguez Echavarria, K., Marroquim, R., Luiz Fonseca, V.: Accessible digitisation and visualisation of open cultural heritage assets. In: Selma, R., Karina, R.E. (eds.) Eurographics Workshop on Graphics and Cultural Heritage. The Eurographics Association (2019). https://doi.org/10.2312/gch.20191349

Miles, A., Zhao, J., Klyne, G., White-Cooper, H., Shotton, D.: OpenFlyData: an exemplar data web integrating gene expression data on the fruit fly Drosophila melanogaster. J. Biomed. Inform. **43**(5), 752–761 (2010). https://doi.org/10.1016/j.jbi.2010.04.004

Potenziani, M., Callieri, M., Dellepiane, M., Scopigno, R.: Publishing and Consuming 3D Content on the Web: A Survey, vol. 10, no. 4. Foundations, Trends R in Computer Graphics, and Vision (2018). https://doi.org/10.1016/10.1561/0600000083

Santarem Segundo, J.E.: Web Semântica, Dados Ligados e Dados Abertos: Uma Visão dos Desafios do Brasil Frente às Iniciativas Internacionais. XVI Encontro Nacional de Pesquisa em Ciência da Informação (XVI ENANCIB) (2015). http://www.brapci.inf.br/index.php/art icle/download/43838. Accessed 17 Oct 2020

Scholz, H.: Europeana Publishing Guide v1.8. Europeana Foundation (2019). https://pro.eur opeana.eu/files/Europeana_Professional/Publications/Europeana%20Publishing%20Guide% 20v1.8.pdf. Accessed 23 Oct 2020

Sczholz, H., et al.: Europeana Content Strategy. Europeana Foundation (2017). https://pro.europe ana.eu/files/Europeana_Professional/Publications/Europeana%20Content%20Strategy.pdf. Accessed 23 Sept 2020

Scopigno, R., Callieri, M., Dellepiane, M., Ponchio, F., Potenziani, M.: Delivering and using 3D models on the web: are we ready? Virtual Archaeol. Rev. **8**(17), 1–9 (2017). https://doi.org/10. 4995/var.2017.6405

Siqueira, J., & Martins, D. L.: Workflow models for aggregating cultural heritage data on the web: A systematic literature review. J. A. Inform. Sci. Tech. 1–21 (2021). https://doi.org/10.1002/ asi.24498

Wilkinson, M.D., et al.: The FAIR guiding principles for scientific data management and stewardship. Sci. Data **3** (2016). Article number: 160018. https://doi.org/10.1038/sdata.201 6.18

Zeng, M., Qin, J.: Metadata, 2nd edn. ALA Neal-Schuman, Atlanta (2016)

Ziébelin, D., Hobus, K., Genoud, P., Bouveret, S.: Heterogeneous data integration using web of data technologies. In: Brosset, D., Claramunt, C., Li, X., Wang, T. (eds.) W2GIS 2017. LNCS, vol. 10181, pp. 35–47. Springer, Cham (2017). https://doi.org/10.1007/978-3-319-55998-8_3

Scholarly Publishing and Online Communication

A Strategy for the Identification of Articles in Open Access Journals in Scientific Data Repositories

Patrícia Mascarenhas Dias⊙, Thiago Magela Rodrigues Dias$^{(\boxtimes)}$ ⊙, and Gray Farias Moita⊙

Federal Center for Technological Education of Minas Gerais, Belo Horizonte, Brazil

Abstract. This work aims to identify articles published in open access journals registered in the Lattes Platform curricula. Currently, the curricular data of the Lattes Platform has been the source of several studies that adopt bibliometric metrics to understand scientific evolution in Brazil. However, when registering a publication in a curriculum, only basic information of the journal is informed. Therefore, in order to quantify the publications that were made in open access journals, a strategy that uses DOAJ data is proposed, validating the publications and thus obtaining a process that allows identifying which publications were made in this format of communication. As a result, it was possible to quantify in an unprecedented way the set of publications by Brazilians in open access journals.

Keywords: DOAJ · Lattes platform · Data link · Open access

1 Introduction

The traditional printed format of science communication is gradually giving way to new electronic formats, due to the rise of information and communication technology. In the context of bibliometric research and studies, scientific communication emerges today as a central element at various levels of discussion. Therefore, the scientific journal appears as an important mechanism for communicating research results.

[7] states that the scientific journal performs at least four essential functions: certification of science with the support of the scientific community; communication channel between scientists and wider dissemination of science; scientific file or memory and record of authorship of scientific discovery.

According to several studies, journals, mainly those available in electronic format - have been growing since the last decade. It can be said that journals, in all areas of knowledge, have the role of being a filter for the recognition of works that have been accepted. For [11], publication in a magazine recognized by the area is the most accepted way to register the originality of the work and to confirm that the works were reliable enough to overcome the skepticism of the scientific community.

In this context, in the early years of the 21st century the Open Access Movement, whose definition is "to make available to any internet user to read, download, copy,

E. Bisset Álvarez (Ed.): DIONE 2021, LNICST 378, pp. 125–135, 2021.
https://doi.org/10.1007/978-3-030-77417-2_9

distribute, print, search or reference the full text of articles or use them for other purposes without any barriers, as long as the work is properly recognized and cited", encouraged the appearance of journals in this format [5].

Despite the numerous benefits that open access journals provide, there is a need for a joint effort so that the main element of the whole process, scientific information, is accessible to all interested parties. To this end, some initiatives have already been undertaken, such as the creation of digital repositories to store and organize scientific literature in accordance with international interoperability standards and the search for awareness of the main actors involved in the process of production, publishing and evaluation of scientific information, to make such content available in digital environments open to the general public.

Neubert et al. [8] state that open access assumes an important role in the entire context of scientific activity, as it allows the researcher to have access to the results of other studies without the cost barriers and difficulties of access, in addition to promoting visibility and dissemination the results of the scientific activities of each researcher and each university.

Open access scientific publication is part of a broader scenario in favor of opening knowledge in general (open access, open data, open educational resources, free software, open licenses) and is essentially a movement towards the design of information and knowledge as public goods [4].

It is worth mentioning that there is generally a limited amount of resources to promote research and a large number of researchers or institutions interested in these resources. Therefore, the broader and more accurate this understanding of scientific production, the greater the possibility of determining resources correctly. However, this type of assessment is an extremely complex task, as it involves the analysis of different characteristics, both quantitative and qualitative. In addition, there is no consensus on which measures or characteristics should be considered for the assessment of scientific productivity [3].

Bearing in mind that a large part of scientific research in the country is financed with public resources, usually in public educational institutions or research centers, it is expected that the results of such studies will be disseminated without any type of barrier, mainly financial. In this context, coupled with the advantages that open access publications have, such as availability, visibility and accessibility, several efforts are being made to ensure that more and more scientific articles are published in open access journals.

Therefore, understanding how the publications of a certain group of researchers have been carried out in open access journals, makes it possible to identify an overview of the current stage of this type of communication in Brazil. It also allows to verify if in certain areas of knowledge this type of publication tends to be more frequent.

This type of study is characterized as an important mechanism to evaluate the evolution of publications in open access journals by Brazilian researchers, allowing to verify whether the incentive policies for the publication of research in this communication format have achieved satisfactory results.

2 Related Works

In the work of [12], the authors analyzed the policies of open access to scientific information and the proposals for action, with an emphasis on government initiatives in different countries. It was identified that the movement of free access to scientific information was already a concern officially registered in several countries, although with different degrees of development. Among these differences are the policy determinations themselves, as some oblige public institutions and researchers to make their research results available in open access, while others only suggest the involvement and participation of these researchers and institutions in the movement.

[1], the main open access scientific communication channels used by researchers are identified, and the factors involved in adhering to the self-archiving of their scientific production are analyzed. The objective of the work was to identify the main channels of scientific communication in open access used by researchers from public universities in the State of Rio de Janeiro. The list of 47 CNPq Advisory Committees for Research Productivity Scholarships was used and a stratified probabilistic sampling by knowledge area was carried out, following the division by Advisory Committee (Agrarian Sciences, Biological Sciences, Exact and Earth Sciences, Science of the Health, Human Sciences, Applied Social Sciences, Engineering, and Linguistics, Letters and Arts). From the selection of researchers contemplated by the CNPq Research Productivity Scholarship program in 2010, whose list is available on the website of this federal agency, those linked to public universities in the State of Rio de Janeiro with post-graduate courses were identified. stricto sensu graduation. After identifying the e-mail addresses of those selected, correspondence was sent containing the form with closed and open questions attached to the following categories: informational behavior, open access publication and adherence to institutional repository.

In general, the results of the research point to a change in the attitude of these researchers in relation to the publication of research results in open access channels. Some areas present publications in formal channels of scientific communication, such as electronic journals, and self-archiving in institutional or thematic repositories. Others are more part of individual or group research initiatives, often anticipating institutional policies. The researchers were unanimous regarding the advantages of open access publishing, and the democratization of knowledge was pointed out by the majority as the main advantage of this adhesion. In addition to this aspect, the benefit of communication between peers – "exchanges", "partnerships" and "dialogues" - also appears in the speeches of the researchers in the knowledge production process. It is also signaled the importance of using this open communication channel at two different times: for the researcher to access the information for their research and to make their results available, allowing them greater visibility and impact.

In order to explore the national and international scenario and thus present an investigation that seeks a technological solution to effect open access to research data, [10] propose a methodology divided into five stages: a) identification of practices of open access to research data in Brazilian institutions; b) mapping your users and their needs; c) proposal for a web portal to bring together the national community; d) survey of services and technological solutions existing in the international scenario for the sharing of research data; e) proposing recommendations to support the creation of research data repositories in national institutions and their aggregation to a research network with open access to research data. As a result, international initiatives and strategies are proposed for the creation of a research data repository and for the creation of communities of practice around the subject.

For [6], the spread of the open access movement in Latin American and Caribbean countries, driven by the growth of regional and national initiatives such as the creation of digital magazine libraries in open access and the establishment of government policies of support, has provided evidence of the significant role of open access for the participation of these countries in global scientific production. In the work, open access publications from Latin American and Caribbean countries are mapped, through a bibliometric analysis of the publications indexed by WoS and SciELO during the period from 2005 to 2017. It is found that the publications have intensified significantly in the period examined, and that although there is an increase in the number of publications in this format, in some countries its magnitude does not translate into a relative weight of open access in the total number of publications.

In the work of [9], the authors analyze documents published in open access between the years 2012 and 2016 by authors with Brazilian affiliation and identify the profile of these publications. For this, data from 930 journals and 63,847 documents were collected from WoS. It is also noteworthy that the Brazilian scientific production in open access is characterized by an endogenous profile, and that policies are still necessary to encourage the publication of articles in open access, mainly in international journals.

Considering the works that analyze publications in open access journals, it is clear that most of them analyze small sets of individuals, in addition to using international data repositories, thus neglecting publications from some areas of knowledge and, therefore, not representing significantly the Brazilian production in open access as a whole.

The vast majority of studies that evaluate the open access movement do not have publications in this type of format as their main object of study, but repositories or journals in open access. Therefore, although the works presented in this section are important to understand the existing initiatives and opinions of Brazilian researchers, as well as the main open access repositories in Brazil. A comprehensive study of Brazilian researchers who have published widely published papers in open access journals is necessary.

3 Development

For the process of data extraction for the analyzes to be carried out in the context of this work, curricular data from the CNPq Lattes Platform were used. A large part of the funding notices for research projects, carried out by various funding agencies, use data registered in the applicants' curricula as one of the forms of evaluation of the proposals. Therefore, there is a great incentive for researchers to keep their curriculum information up to date. This makes Lattes Platform curricula an excellent source of data for analysis. For this same reason, several works have used the Lattes Platform as a data source for several studies on different topics, such as networks of scientific collaborations, analyzes of productivity, academic genealogy, among others [2, 3].

Considering that the majority of related works analyzed only specific groups of individuals, and considering that the manipulation of large amounts of curricula from the Lattes Platform is not a trivial task, since there are problems involving information retrieval and efficient algorithms for handling large volumes of data, LattesDataXplorer [2], a framework for data extraction and treatment, developed by the research group of this work was used.

As already explained, a curriculum registered in the Lattes Platform can contain various information capable of helping to understand the evolution of Brazilian science from different perspectives. However, to serve the purposes of this work, only data from publication of articles in open access journals were considered. In view of this, an extension of LattesDataXplorer was proposed with the inclusion of non-existent a priori components, which would evaluate for each article published in a journal (6,985,179), of each of the individuals (5,901,161) (data collections in October 2018), if the journal in which that article had been published was open access (Fig. 1). Therefore, with the proposal for this extension, only authors and publications in open access journals could be analyzed.

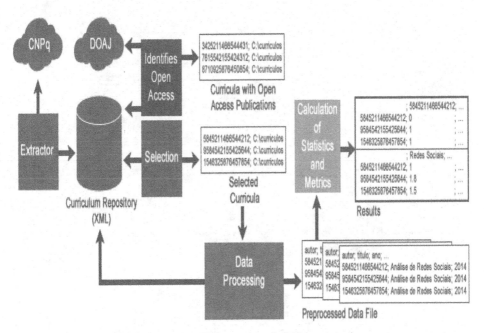

Fig. 1. LattesDataXplorer extended. Source: Authors.

Initially, using LattesDataXplorer, all resumes registered on the Lattes Platform in October 2018 were collected and stored in the local repository. Then, the component developed and called "Identifies open access" was used to retrieve all open access journals registered on the Directory of Open Access Journals (DOAJ) portal, an online directory that indexes and provides access to open access journals. In February 2019, the DOAJ indexed 12,324 journals and 3,513,782 articles. DOAJ has been a source of data and reference on open access journals for several studies.

Collected the data of the journals on the DOAJ portal in October 2018, the same period of collection of the curricula for the analyzes presented in the present work, 12,171 open access periodical titles were retrieved, containing data such as title, ISSN and eISSN, among other information.

In order to optimize the computational processing of curricula as much as possible, whenever a publication whose ISSN or eISSN of the journal was contained in the list of open access journals extracted from DOAJ, immediately the identifier of the curriculum under analysis was inserted in the list of curricula in access. open, and the next curriculum of the set under analysis was evaluated.

After analyzing all the resumes that make up the local repository, a list containing all resumes with open access publications is generated, and it becomes the basis for the "Data processing" component, which now incorporates the methods proposed in this work (Fig. 2).

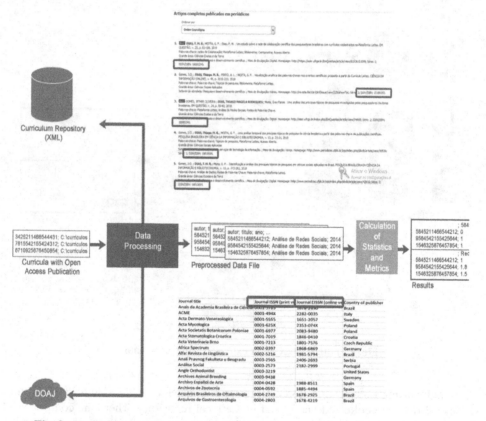

Fig. 2. Method for identifying publications in open access journals. Source: Authors.

With the list of curricula that have articles in open access, the identification of publications in this format is performed with the processing of the curricula, using the "Data Processing" module of LattesDataXplorer, in order to generate the pre-processed data files that summarize information of interest and that will serve as the basis for calculating the metrics.

In addition to general data on researchers with open access publications that will compose some of the archives, such as data on academic training, areas of expertise, guidelines and professional practice, each of the articles recorded in the section "Complete articles published in journals" was analyzed. Of each curriculum contained in the "List of curricula with publications in open access". For each article in each curriculum, it was verified and analyzed whether the ISSN or eISSN of the publication was present in the list of journals recovered from DOAJ. Thus, it was possible to identify the entire number of articles in open access journals (Fig. 3).

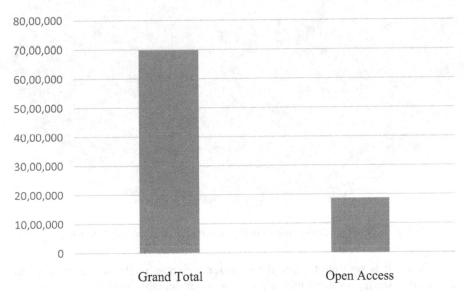

Fig. 3. Number of publications in journals registered in the curricula. Source: Authors.

As can be seen, of the total set of articles published in journals, considering the entire history of publications recorded in all resumes registered in the Lattes Platform (6,985,179 publications), a percentage of 26.76% (1,869,585) was published in open access journals, taking into account the list of journals recovered from DOAJ. This percentage of publications in open access is relevant, above all, for considering the entire publication history of each researcher. It is noticed that publications in open access journals have been receiving attention and adherence by researchers year after year, presenting themselves as a trend in dissemination and scientific communication, especially in recent years. A temporal evaluation was carried out in order to assess the growth of publications year by year.

4 Results

Using the extension proposed in this work for LattesDataXplorer, all authors who published at least one article in an open access journal (370,431) were identified. These

authors, despite being a small number of individuals in relation to the whole set registered in the Lattes Platform (6.27%), have a great representativeness when considering
the total number of articles published in journals (approximately 76%) (see Fig. 4).

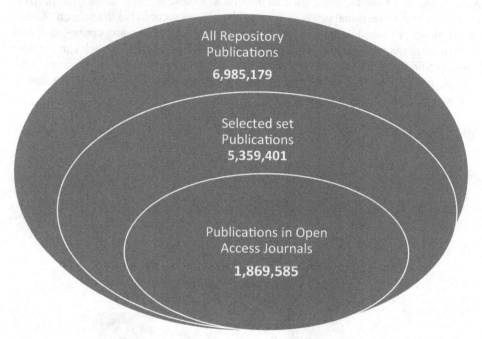

Fig. 4. Number of publications in registered journals. Source: Authors.

Therefore, it is possible to note the representativeness of the set to be analyzed
in this work. Bearing in mind that it includes a considerable portion of the authors
who have published articles in journals in Brazil, the results presented may provide an
unprecedented view on the evolution of articles in open access, as well as serve as a
basis for several other works.

It is possible to inform in the curricula the areas, subareas and specialties in which a
given individual operates. When analyzing the areas of activity of the group of individuals, it is possible to notice great diversity in the distribution of curricula in each major
area, as well as an irregular distribution in the number of areas that each major area has.
Therefore, an analysis based on the areas of activity is important (Fig. 5).

As can be seen, there is no uniform distribution of the number of areas in each large
area. The large area of Linguistics, Letters and Arts has only three areas, while the large
areas of Biological Sciences and Engineering have 14 areas each. When registering the
large areas of activity, the individual may not inform the "area" field. In these cases,
individuals were also categorized as "Not Informed". Due to the small number of individuals (0.81%) who reported "Others" as a large area, the analysis of their areas was
not considered.

In order to verify the most representative areas of knowledge of the analyzed group,
the area of Medicine (33,966) stands out, composing the large area of Health Sciences,

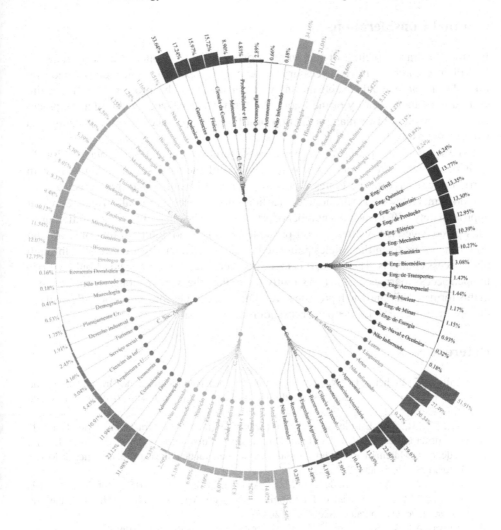

Fig. 5. Distribution of authors by their areas of expertise. Source: Authors.

the most representative, as already presented. The representativeness of the field of Medicine is so considerable that it alone has practically the same number of individuals as the total sum of the major areas of Linguistics, Letters and Arts and Engineering combined. Following, the areas of Education (17,066), Agronomy (15,705), Nursing (12,734), Administration (11,717), Chemistry (10,490) and Psychology (10,511) stand out. These seven areas alone are responsible for housing approximately 30% of the total set of individuals analyzed.

5   Final Considerations

In order to draw a picture of the publication of articles in open access journals by Brazilian researchers, it was necessary to develop components that, incorporated into LattesDataXplorer, could enable the analyzes carried out in this project. Thus, the entire curriculum data repository of the Lattes Platform was analyzed, enabling an unprecedented study on the Brazilian production of articles in open access journals using data from DOAJ as well.

The set of articles published in open access journals has as authors a total of 370,431 individuals, which represents approximately 6% of the total set of individuals with curricula registered in the Lattes Platform. It should be noted that this percentage of authors is much lower than the number of articles in open access journals, which represent approximately 27% of the total number of articles published in journals of all individuals. This percentage is very close to that presented by [4], who point out that only around 30% of the total scientific articles published in the world annually are available through open access channels. Therefore, it is identified here that the percentage of publications in open access journals in Brazil is slightly lower than the world average of publications in this format. This study will provide several new researches that aim to broadly analyze the production of articles in open access journals in Brazil.

References

1. Chalhub, T., Pinheiro, L.V.R.: Acesso aberto à informação científica no Brasil: Um estudo das universidades públicas do estado do Rio de Janeiro. Rio de Janeiro (Relatório Final de Atividades) (2011)
2. Dias, T.M.R.: Um Estudo Sobre a Produção Científica Brasileira a partir de dados da Plataforma Lattes. 2016. 181 f. Tese (Doutorado) - Curso de Programa de Pós-graduação, Modelagem Matemática e Computacional, Centro Federal de Educação Tecnológica de Minas Gerais, Belo Horizonte (2016)
3. Digiampietri, L.A.: Análise da Rede Social Acadêmica Brasileira. 2015. 160 f. Tese (LIVRE DOCÊNCIA) - Informação e Tecnologia, Escola de Artes Ciências e Humanidades da Universidade de São Paulo, São Paulo (2015)
4. Furnival, A.C.M., Silva-Jerez, N.S.: Percepções de pesquisadores brasileiros sobre o acesso aberto à literatura científica. Percepções de Pesquisadores Brasileiros Sobre O Acesso Aberto à Literatura Científica, João Pessoa, vol. 27, no. 2, pp. 153–166 (2017
5. Leta, J., Costa, E.H.S., Mena-Chalco, J.P.: Artigos em Periódicos de Acesso Aberto: um Estudo com Pesquisadores Bolsistas de Produtividade do CNPq. Revista Eletrônica de Comunicação, Informação e Inovação em Saúde [s.l.], vol. 11, pp. 1–6 (2017)
6. Minniti, S., Santoro, V., Belli, S.: Mapping the development of Open Access in Latin America and Caribbean countries. An analysis of Web of Science Core Collection and SciELO Citation Index (2005–2017). Scientometrics 117(3), 1905–1930 (2018). https://doi.org/10.1007/s11 192-018-2950-0
7. Mueller, S.P.M.: O círculo vícios o que prende os periódicos nacionais. Datagramazero, Brasília, vol. 0, no. 4, pp. 1–8 (1999)
8. Neubert, P.S., Rodrigues, R.S., Goulart, L.H.: Periódicos da Ciência da Informação em acesso aberto: uma análise dos títulos listados no DOAJ e indexados na Scopus I Open access journals in information Science. Liinc em Revista, [s.l.], vol. 8, no. 2, pp. 389–401. Liinc em Revista (2012). https://doi.org/10.18617/liinc.v8i2.497

9. Pavan, C., Barbosa, M.: Article processing charge (APC) for publishing open access articles: The Brazilian scenario. Scientometrics **117**(2), 805–823 (2018). https://doi.org/10.1007/s11 192-018-2896-2
10. Pavão, C.G., Rocha, R.P., Gabriel Junior, R.F.: Proposta de criação de uma rede de dados abertos da pesquisa brasileira. Rdbci: Revista Digital de Biblioteconomia e Ciência da Informação, Campinas, vol. 16, no. 2, pp. 329–343 (2018)
11. Rodrigues, R.S., Oliveira, A.B.: Periódicos Científicos na America Latina: títulos em Acesso Aberto indexados no ISI e SCOPUS. Perspectivas em Ciência da Informação, Belo Horizonte **17**(4), 76–99 (2012)
12. Silva, T.E., Alcará, A.R.: Políticas de acesso aberto à informação científica: iniciativas governamentais. In: IX Encontro nacional de pesquisa em ciência da informação, 9, São Paulo. Anais. São Paulo, pp. 1–14 (2008)

How to Spot Fake Journal: 10 Steps to Identify Predatory Journals

Adilson Luiz Pinto[1]([⊠]) [iD], Thiago Magela Rodrigues Dias[2],
and Alexandre Ribas Semeler[3] [iD]

[1] PGCIN, Federal University of Santa Catarina, Florianópolis, Brazil
adilson.pinto@ufsc.br
[2] Federal Center for Technological Education of Minas Gerais, Belo Horizonte, Brazil
[3] Geosciences Institute, Federal University of Rio Grande do Sul, Porto Alegre, Brazil

Abstract. Nowadays, with overload information on the web, it is common to come across information and questionable content. This scenario is called fake news (for news), fake conference (for events), and predatory journals (for suspicious academic journals). Predatory journals explore the model of academic productivism, meeting the need for speed on the part of researchers in publishing, publish, or perish. A device created by academic immediacy. This study focused on analyzing the theme through methods and tools for the identification of predatory journals. A model is proposed for identifying these journals, using already established websites, databases, and repertoires. After monitoring somes emails between January and April 2020, we identified a series of patterns in the forms of communication of these fake editorial boards. As a result of this research we suggest a ten practical recommendations steps to identify predatory journals: 1) ISSN; 2) inclusion in predatory journal lists; 3) web page and domain; 4) editorial information (call for papers, previous issues; indexing, plagiarism identification, editorial board members); 5) standards of published papers; 6) DOIs, and ORCID identification; 7) indexing status; 8) Article Processing Charge; 9) spelling and typographical errors on the web page and the papers; 10) If none of these actions have any effect and you still have doubted about the seriousness of the journal, consult a specialist on the subject.

Keywords: Fake journals · Predatory journals · Journals verification

1 Introduction

Nowadays, with overload information on the web, it is common to come across information and questionable content. This scenario is called fake news (for news), fake conference (for events), and fake and predatory journals (for suspicious academic journals). By definition, predatory journals mean a publication that misleads researchers into publishing in exchange for some form of payment (in submission or publication). Publishing under an Article Processing Charge (APC) requirement, without attention to publication ethics, peer review, or databases indexing [3].

© ICST Institute for Computer Sciences, Social Informatics and Telecommunications Engineering 2021
Published by Springer Nature Switzerland AG 2021. All Rights Reserved
E. Bisset Álvarez (Ed.): DIONE 2021, LNICST 378, pp. 136–144, 2021.
https://doi.org/10.1007/978-3-030-77417-2_10

In this context, as the information can be understood as correct because a predatory journal is something scientific, it is also necessary to be concerned with the ethical question. For example, it would be the same as saying that Hydroxychloroquine or chloroquine are 100% reliable to fight the infected of SARS-Cov-2 (Covid-19), but in health, you can't publish a study that indicates a medication and it can be used as truth [4], and in fact, we know that recent studies indicate that it is not, as it can cause irregular heartbeat and even lead to death.

An additional case is a fake editorial board, like OLLIE dog, who published articles in scientific media and currently is part of seven scientific predatory editorial advice and even he can review work [5]. Ollie dog owner Mike Daube created his fake curriculum and since then the dog has received numerous invitations to publish and be part of groups of scientific journals. This is a predatory journals strategy, occurs because it aims to recruit young researchers as scientific editors, generally at the beginning of their careers and who at low rates can be part of the editorial board [6], supposedly helping in the course of academic careers. The marketing of predatory journals is acuteness designed, which guarantees publication at an acceptable price, and guaranteed representation in large databases.

Another, predatory journals strategy is fake metrics, especially since researchers are constantly evaluated, for a variety of reasons, they need to justify the output of their research as a form of validation. Some of the fake metrics are (a) Scientific Journal Impact Factor available at http://sjifactor.com/; (b) Global Impact Factor - available at http://globalimpactfactor.com; and (c) Universal Impact Factor available at http://uif actor.blogspot.in.

A third strategy relates topper review, predatory journals never do peer review, these journals only make minimal editions of a foreign language, if this evaluation process applies. These journals do not verify the veracity of the data used, they promise peer review within 24 or 48 h, and a paper published in a few weeks. Predatory journals promise agility in publication and not quality.

Predatory journals explore the model of academic productivism, meeting the need for speed on the part of researchers in publishing, publish, or perish. A device created by academic immediacy.

A predatory journals also act as open access journals. This is clear in return for payment. The values are between 60 and 3.000 dollars to publish a paper. In this context, they facilitate the stages of the scientific process, promise speed review, publication in a few days, and indexing on databases of international prestige. Careless researchers, unaware of the negative consequences of publishing in predatory journals, become the victims of academic productivism that leads them to abuse quick solutions to publicize their publications, in search of better salary conditions, promotion of research, and increase their scientific indicators [7]. But it's not just young researchers who fall into the traps of predatory journals.

Interestingly, even the prestigious international databases indexing predatory journals in their databases. This shows that the process, in some cases, is meticulous to deceive researchers and even journals qualification systems. For example, SelcukBesir Demir [8] conducted an intense search of Bells' list of predatory journals (n = 2708) and identified that these journals were being indexed in several databases, such as the power of infiltration of these types of content in the medium academic. In the databases indexed in the Web of Science, 3 journals were found; while in Scopus database, 53 journals were found.

According to the scenario presented on predatory journals, this study will focus on analyzing the theme through methods and tools for the identification of fake journals and predatory. Thus, this paper proposes a model for identifying these journals, using already established repertoires, such as the Bell's list [9], Kscien's list [10], UlrichsWeb proposing a checklist pattern to identify predatory journals.

2   Methodology

The purpose of analyzing the predatory journals arises from the constant receiving emails with invitations to publish papers, after we received numerous emails about the possibility of publishing articles in international journals, with speed and ease, upon payment fees. After monitoring 35 emails between January and April 2020, we identified a series of patterns in the forms of communication of these fake editorial boards. Thus, we used the parameters proposed by Tove Faber Frandsen [11] and Jeffrey Beall [12] to identify the veracity of these (35) emails, and we applied it to 10 random journals (Table 1):

Table 1. Predatory journals list

Journal title	ISSN	Email editor	Web site
International Multilingual Journal of Science and Technology (IMJST)	2528-9810	editor@imjst.org	http://imjst.org/
Revista Iberoamericana de Ciências	2334-2501	editor@reibci.org	http://www.reibci.org
Journal of Multidisciplinary Engineering Science and Technology (JMEST)	2458-9403	editor@jmest.org	https://www.jmest.org/
Journal of Travel Tourism and Recreation	2642-908X	jttr@sryahwapublications.com	https://www.sryahwapublications.com/journal-of-travel-tourism-and-recreation
Columban Journal of Life Sciences	0972-0847	cjlseditor@yahoo.com	http://cjlscience.org
Research Journal of Library and Information Science	2637-5915	rjlis@sryahwapublications.com	https://www.sryahwapublications.com/research-journal-of-library-and-information-science
International Interdisciplinary Journal of Scientific Research	2200-9833	editor.ijsrnet@gmail.com	https://iijsr.org/;
World Journal of Pharmaceutical and Life Sciences	2454-2229	editor@wjpls.org	https://www.wjpls.orges
Kashmir Economic Review	2706-9516	managing.editor@ker.org.pk	
Turkish Journal of Scientific Research (TJSR)	2148-5135	editor@tjsr.com	

3 Results and Discussions

We firstly identified the ISSN existence and validity since this item is the journal's universal identification. A journal without an ISSN number can be discarded without major concerns. The ISSN Portal was used for this number identification. According to the ISSN Center, the quality of a journal can be validated by means of the UlrichsWeb Global Serials Directory (http://ulrichsweb.serialssolutions.com), a quality control system for serial publications managed by the ProQuest Company. Secondly, some lists for predatory journals were consulted, including:

1) https://beallslist.weebly.com/;
2) https://predatoryjournals.com/journals/;
3) https://archive.fo/9MAAD#selection-283.0-293.73;
4) http://kscien.org/predatory.php

These four lists are the most extensive lists to locate a predatory journals (Table 2).

Table 2. Representation and visibility of journals (Items 1, 2, and 3)

ISSN	Portal ISSN	Ulrichs web	Bell's list	Predatory journal	Scholarly OA	KScien List	Own website
2528-9810	Yes	No	No	No	No	No	Yes
2334-2501	Yes	No	No	No	No	No	Yes
2458-9403	Yes	No	No	No	Yes	Yes	Yes
2642-908X	Yes	No	No	No	No	No	Yes
0972-0847	Yes	No	No	Yes	Yes	Yes	No
2637-5915	Yes	No	No	No	No	No	No
2200-9833	No	No	No	Yes	Yes	Yes	Yes
2454-2229	Yes	No	No	Yes	Yes	Yes	Yes
2706-9516	No	No	No	Yes	Yes	Yes	Yes
2148-5135	Yes	No	No	Yes	Yes	No	No

The journals 2528-9810 and 2458-9403 claim that their publishers Naci Kalkan of Bitlis University, but the institution and the name of the publisher do not even appear on alleged pages of the journals. The journals 2334-2501, 2642-908X, 0972-0847, 2637-5915, 2148-5135 do not present any information beyond the name, country of publication, and means of communication (online). The journals 2200-9833 and 2706-9516 do not actually exist, as can be seen from the lack of registration on the ISSN Portal.

We did not identify any periodical in the Bell´s list; five journals are on the Stop Predatory Journals list and the KScien List. The list of journals from Scholarly Open Access: Critical analysis of scholarly open-access publishing (List of Standalone Journals) was the most efficient, with 6 titles identified.

Concerning the web domains of these journals, only 3 did not have their own names. Most have their own names, but just browse a little to identify that the editorial is very vague and the existence of an exorbitant number of editors, It is evident that part of this editorial board is either fake or was recruited by young researchers (Table 3).

Table 3. Indexing of journals (Items 4, 5, and 6)

ISSN	DOI	Call open	Number	2019 n = articles	General editor	Plagiarism	Google scholar	ORCID
2528-9810	No	Yes	Uneven	40	No	Yes	No	No
2334-2501	No	Yes	Even	48	No	No	Most	No
2458-9403	No	Yes	Uneven	327	No	Yes	Most	No
2642-908X	No	Yes	Even	19	Yes	No	Most	No
0972-0847	No	No	Uneven	0	Yes	No	No	No
2637-5915	No	Yes	Even	20	No	Yes	Minority	No
2200-9833	No	No	Uneven	0	No	No	No	No
2454-2229	No	Yes	Uneven	41	Yes	No	Most	No
2706-9516	No	Yes	Uneven	5	Yes	No	Yes	No
2148-5135	No	No	No	0	No	No	No	No

Source: Survey data, 2020.

Other checks in the journal's webpage are necessary, such as:

1) check in the previous volumes and issues if the articles presented were not published in another journal;
2) check the number of articles published per volume, issue, and year;
3) the journal submissions webpage;
4) identify if the journal has a Digital Object Identifier (http://www.doi.org), and check if it belongs to the journal in CROSSREF;
5) check the durability of the magazine on the web.

According to these criteria, we will analyze the websites of the journals that have sent us an e-mail (Table 4).

The payment of fees for submission and publication of articles is one of the main characteristics of a predatory journal. In our research we found that these values varied between $25 and $300).

Regarding the indexing status of the journals, none were identified in mainstream databases for the academic community (see Table 4, 5).

Because they are not in databases, some predatory journals use a fake metrics system to validate their impact. In this scenario, some non-existent scientific indicators were found, such as the Universal Impact Factor.

Table 4. Analysis of the journals' website (Items 1, 2, and 3)

ISSN	Website characteristics
2528-9810	Journal created in 11/2019 editions were missing, the volumes published in the journal have no temporal regularity, and the number of published articles varies from (1) to (7) studies in a single volume. Some articles consulted revealed that part of its contents had already been published in other journals. The editor of the journal was not on the editorial board of the journal's website, what was included were the members of the editorial board, authors with few publications in databases like Google Scholar, WoS and Scopus. The authors of the articles have generic names and without ORCID
2334-2501	Journal created in 2014, without DOI, submissions open to April 2020, volumes published regularly, but on different areas of knowledge; without editorial advice; without plagiarism checker information, most of the articles were indexed by Google Scholar, authors without ORCID
2458-9403	Journal created in 2014; without DOI and ORCID, publications have irregularities in the number of articles (high number of publications, equal to (327) papers); without editor-chief; indexed by Google Scholar, but not all articles. This is an important feature of - predatory journals, as they reveal the APC, thus the number of publications shows that the journal has already obtained financial returns
2642-908X	Journal created in 2019, a few published volumes, without temporal regularity; it has an editor, but a simple check on the editor's curriculum at his institution revealed that he does not mention the journal, without DOI and ORCID; without plagiarism; most papers are indexed by Google Scholar
0972-0847	The journal presents little information; does not offer direct access to papers, previous numbers, DOI, ORCID, plagiarism checker, and content indexed in Google Scholar. It has an editor-chief, but he is a politician and not a scientist, according to information from Wikipedia
2637-5915	Journal created in 2017, located in India, regularity in published volumes (5); without DOI and ORCID; little indexed is in Google Scholar; plagiarism checker; submissions always open; without information about the general editor, only information from the editorial board
2200-9833	Journal created in 2014, volumes have temporal irregularity, closed in 2017, requesting submissions by email, without plagiarism checker, indexed in Google Scholar, no numbers published between 2018 and 2019
2454-2229	The webpage contains general editors, ISSN validated by the ISSN portal; created in 2016; irregularity in the number and papers of published without plagiarism, partially indexed by Google Scholar; without DOI and ORCID
2706-9516	Journal created in 1984, information on the editorial board until 2010, submissions for the second issue of 2019; only 5 articles published in 2019, without DOI and ORCID; papers without the same format style; edition is open, indexed by Google Scholar and plagiarism has not been identified

(continued)

Table 4. (*continued*)

ISSN	Website characteristics
2148-5135	Journal indexed in WorldCat, it does not present any information about its volumes and issues, editorial board, indexing, and identification metadata of papers and authors

Table 5. Payments, spelling errors, fake metrics, and indexing in international databases (items 7, 8, and 9)

ISSN	APC	WoS indexing	Scopus indexing	Misspelling	Fake factor impact
2528-9810	US$ 25	No	No	Yes	Yes
2334-2501	No	No	No	No	Yes
2458-9403	US$ 45	No	No	No	Yes
2642-908X	US$ 300	No	No	Yes	No
0972-0847	No	No	No	No	Yes
2637-5915	US$ 300	No	No	Yes	No
2200-9833	No	No	No	No	No
2454-2229	US$ 50	No	No	Yes	Yes
2706-9516	No	No	No	No	No
2148-5135	No	No	No	No	No

Source: Survey data, 2020

4　Practical Recommendations and Concluding Remarks

The main characteristics of predatory journals are fast publication promises, irregularities in the publication schedule and frequency, and no editorial guidelines, and editors lacking academic. In this context, to help young researchers to publish in quality journals, we suggest a ten practical recommendations steps to identify predatory journals:

1) Check the ISSN at (ISSN Portal), UlrichsWeb Global Serials Directory, these web portals provide registration and history of the journals. Both systems are quite efficient;
2) Check the lists of fake journals and predatory;
3) Check the domain of the pages of the journal: Every coherent journal has its own page or is linked to some scientific institution, be it a publisher, university, associations, or research institutions. Always doubt when it is linked to blogs or pages of institutions unknown to the academic world;
4) Check the update of the web page: A quality journal always has called for submission, or at least indicates when it will be receiving scientific papers; informs previous editions, with basic information and in many cases access to articles; it has information from the editor, the editorial board and reviewers; it is essential to

consult some of these characters to see if they are real; also check if the articles are on Google Scholar, most are not enough, but the entire collection; if you can, use a plagiarism sniffer in recent articles, because that says a lot about the journal;

5) Check the constancy of the journal: quality journals publish based on international standards, and follow recommendations from big databases;

6) Check if the Journal has unique digital identifiers. the journal must have a Digital Object Identifier (DOI), as well as the authors the ORCID; these two identifiers are factors of informative quality for the standardization of papers and names of authors;

7) Check if the Journal is indexed on the bases academy prestige: Indexing on platforms brings confidence to the journal; just being in the Directory of Open Access Journals (DOAJ) does not guarantee that the journal is good, you need to obtain more information, such as having a list of where the journal is represented; another detail is whether it is in serious databases, directories, and repositories such as Web of Science, Scopus, and others;

8) Check if there are publication fees: That the journal is paid is not a problem, but that to submit an article you have to make payments does not seem very logical; another thing, with the payment system, the journal guarantees that it will be evaluated in 48 h, it also seems very dangerous, since this process is the most important of the whole process and requires time; ensuring publication in weeks also seem me very worrying, after all the process of managing and editing a journal is costly and requires a good group of people to be of quality (editors, reviewers, librarians, among others);

9) Check that the journal is well-edited: In this case, revisions come in and in many cases in different languages, as many journals guarantee multilingual publications in their calls; check what the standards are if any manager is used in the journal;

10) Consult an expert on the subject of the journal: Predatory journals are made to deceive the researcher, easily and quickly, but we believe that a specialist can be consulted; look for someone from your institution, with papers on the subject of the alleged journal, and ask for his opinion, it can make all the difference in recognizing something unreal.

As a proposal for the continuity of this work, a tool is being developed that includes a set of automatic methods for the identification of fake journals and predatory. Such methods will compose a framework that gives information about a certain journal, or a list of journals, such as title, ISSN, e-ISSN, or website, in an automatic way to class if they are and predatory journals.

References

1. IFLA: How To Spot Fake News (2017). https://www.ifla.org/publications/node/11174
2. Kiely, E., Robertson, L.: How To Spot Fake News. Factcheck Posts, posted on November 18 (2016). https://www.factcheck.org/2016/11/how-to-spot-fake-news/
3. Kahan, S., Kushner, R.F.: New Year's resolution: say no to fake journals and conferences. Obesity 25(1), 11–12 (2017). https://doi.org/10.1002/oby.21738

144   A. L. Pinto et al.

4. Hutson, S.: Publication of fake journals raises ethical questions. Nat. Med. **15**(6), 598 (2009). https://doi.org/10.1038/nm0609-598a
5. Wilcken, H.: Dog of a dilemma: The rise of the predatory journal. MJA Insight **75**(9), 1795–1796 (2017). https://insightplus.mja.com.au/2017/19/dog-of-a-dilemma-the-rise-of-the-pre datory-journal/
6. Sorokowski, P., Kulczycki, E., Sorokowska, A., Pisanski, K.: Predatory journals recruit fake editor. Nature **543**(7646), 482–483 (2017). https://www.nature.com/news/polopoly_fs/1.21662!/menu/main/topColumns/topLeftColumn/pdf/543481a.pdf
7. Samuel, A.J., Aranha, V.: Valuable research in fake journals and self-boasting with fake metrics. J. Pediatr. Neurosci. **13**(4), 517–518 (2018). https://doi.org/10.4103/JPN.JPN_66_18
8. Demir, S.: Scholarly databases under scrutiny. J. Libra. Inform. Sci. **52**(1), 150–160 (2020). https://doi.org/10.1177/0961000618784159
9. Beall, Jeffrey: Best practices for scholarly authors in the age of predatory journals. Ann. R. Coll. Surg. Engl. **98**(2), 77–79 (2016). https://doi.org/10.1308/rcsann.2016.0056
10. Kakamad, F.H., et al.: Kscien's list; a new strategy to hoist predatory journals and publishers. Int. J. Surg. Open **17**, 5–7 (2019). https://doi.org/10.1016/j.ijso.2019.01.002
11. Frandsen, T.F.: How can a questionable journal be identified: frameworks and checklists. Learn. Publish. **32**(3), 221–226 (2019). https://doi.org/10.1002/leap.1230
12. Beall, J.: Criteria for determining predatory open-access publishers (2012). https://schola rlyoa.files.wordpress.com/2012/08/criteria-2012.pdf
13. Harzing, A.-W.: Publish or Perish (2007). https://harzing.com/resources/publish-or-perish

Dissemination Strategies for Scientific Journals on YouTube and Instagram

Mayara Cabral Cosmo[1] , Priscila Machado Borges Sena[2](✉) ,
and Enrique Muriel-Torrado[3]

[1] Federal University Rio de Janeiro (UFRJ), Rio Janeiro, Brazil
mayaraccosmo@hotmail.com
[2] Brazilian Federation of Librarians Associations, Information Scientists, and Institutions
(FEBAB), São Paulo, Brazil
priscilasena.pesquisa@gmail.com
[3] Information Science Department (UFSC), Federal University Santa Catarina, Florianópolis,
Brazil
enrique.muriel@ufsc.br

Abstract. Strategies for dissemination to scientific journals on YouTube and Instagram are proposed. It is characterized as exploratory and descriptive, with a qualitative approach to the data collected. First, searches were made in international databases to find scientific production that would cover the theme of dissemination in scientific journals through social media. Second, strategies for scientific dissemination in social media were established. As a result, six fundamental guidelines for the use of social media were described, covering both strategic and operational aspects: 1) Basic purpose of the use of social media; 2) Properties of the content; 3) Definition of the target audience; 4) Strategic and operational aspects of the use of social media and production of content; 5) Ethical and legal aspects involving the use of social media and production of content and; 6) Crisis management through a complaint or negative criticism about the posted content. The fourth guideline also outlined eight steps for scientific journals. Also, guidelines on the strategic use of YouTube and Instagram for the dissemination of science by scientific journals were listed. It is concluded that there is an incipient number of investigations that provide strategies for the use of social media by scientific journals and that the use of these for scientific dissemination requires a well-defined strategy to achieve the objectives proposed by the scientific journal.

Keywords: Social media · Scientific journal · Scientific communication

1 Introduction

In scientific communication, the dissemination of information occurs both formally, with the publication of books, articles in journals, abstracts, and complete works in annals of events, thematic or institutional repositories, among others. As much informally in an informal way, without the evaluation of a commission of experts, with the dissemination

E. Bisset Álvarez (Ed.): DIONE 2021, LNICST 378, pp. 145–153, 2021.
https://doi.org/10.1007/978-3-030-77417-2_11

through informal means like blogs, personal websites, academic social media (Research-Gate or Academia.edu) or popular social media like Twitter, Facebook, Instagram, etc. [1]. The latter being conceptualized [2] as scientific dissemination.

The profile of the target audience differs in the two communication processes [2]. In scientific communication, the public generally has a specific background that makes them familiar with the themes, concepts, and scientific process. In scientific dissemination, the public does not necessarily have a technical-scientific background to understand the entire production process in science and technology (S&T).

Both processes contribute to the expansion of the dissemination of science, as they are not mutually exclusive. Even with the insertion of new means of scientific communication, the importance of journals and articles has not been shaken [1]. Thus, this research is guided by the questioning of how to stimulate the dissemination of scientific journals in social media strategically?

Given the above, the goal is to propose dissemination strategies for scientific journals on YouTube and Instagram. For this purpose, the following section describes the methodological procedures adopted in the composition of the bibliographic survey, as well as the criteria used for the compilation of strategies.

Having said that, it approaches how the theme is worked in the scientific literature, the suggested strategies following the type of social media, and the practical and theoretical contributions.

2 Methodologic Procedures

This research can be characterized as exploratory and descriptive, with a qualitative approach to the data collected. First, searches were made in international databases to find scientific production that would cover the theme of dissemination in scientific journals through social media. Second, strategies for scientific dissemination in social media were established.

Scopus and Web of Science (WoS) were used for offering worldwide coverage of scientific production. According to systematic searches carried out on July 27, 2020, without period delimitation and with language specification in Spanish, English, and Portuguese, the total of documents recovered in the combination of the terms **scholarly communication AND scientific journal AND social media**, was in 54 documents, and for the terms **scientific communication AND scholarly communication AND social media,** in 121 documents (Table 1).

This was followed by the reading of the titles, abstracts, and keywords of 175 documents, to select those that came closest to the objective of the research. In this process, it was also sought to eliminate documents that did not come from scientific journals, since this delimitation was not applied at the time of searches.

With the total number of documents recovered by both combinations of terms, a new duplicity check was performed, which resulted in 50 not duplicated documents to be analyzed in their entirety on the main theme focused on strategies for dissemination of scientific journals. After reading, the final portfolio consisted of 34 documents. Of these, the most relevant were used to highlight the gaps that supported the reflection proposed and sought in this work, strategies for dissemination of scientific journals on YouTube and Instagram. This resulted in 13 research described in the following section.

Table 1. Quantity of documents retrieved per database related to the terms scholarly communication AND scientific journal AND social media and scientific communication AND scholarly communication AND social media.

Terms	Scopus	WoS	Total	Not duplicated
Scholarly communication AND scientific journal AND social media	40	32	72	54
Scientific communication AND scholarly communication AND social media	99	67	166	121

3   Scientific Communication in Social Media

The use of social media in scientific communication has received attention from researchers from different parts of the world, according to articles analyzed from the systematic searches described in the section on methodological procedures.

There is research on: the use of Twitter to communicate and conserve scientific event information, with instruction for the incorporation of hashtags and relevant user names [3]; about automated Twitter accounts that distribute links to scientific articles deposited in the arXiv preprint repository [4]; the use of Twitter and other social media to foster research ideas, collaboration and academic activity [5]; the broader involvement with scientific literature from tweets containing links to scientific articles [6]; the publication of scientific research on Facebook and Twitter, specifically related to the Zika virus in 2016, with emphasis on the predominance of the English language [7]; the potential or actual social impact of shared research on Twitter and Facebook social media [8]; the analysis of journal articles with high citation counts but low Twitter mentions and vice versa, with the aim of obtaining an overview of the differences between citation counts and Twitter mentions of academic articles [9]; and the methodology approach to measure and visualize the diffusion patterns of scientific research literature, with the aim of detecting when research articles spread beyond the academic community or lead to exchanges with the general public [10].

The research focused on the use of social media by scientific journals was also identified, the focus of this article. Social media with their function in scientific communication by complementing the scientific journals, and their active use as a stimulus to the demand for them. From research on 100 scientific and academic journal sites, they found that 19 used social media such as Facebook, Twitter, and blogs. However, the social media strategies of the sites were diverse about the content generated, including in some cases, with most coming from outside the journal [11]. An exploratory study on profiles of 30 scientific journals on Twitter, to analyze the management of profiles, identify the characteristics of the most interactive contents and proposing effective practices that motivate strategic management. The results revealed outstanding practices and certain deficiencies in the strategic management of social profiles. They mentioned ongoing research on the generation of effective social media management standards for academic journals based on ongoing analysis and regular quantification of results, to identify fields for improvement that can promote growing strategies and contribute to the creation of a brand journal [12].

There is research on Altmetrics [13–15], with reflections on the journals, but with a strong point in analyses of external mentions. They explored the use of Twitter as an Altmetrics indicator to evaluate the social impact of a scientific journal [13]. They described the presence and visibility online of selected journals from Korea, indexed in the Web of Science in the Journal Citation Reports of the Science Citation Index 2016, with the greatest online attention being located on Twitter, followed by Facebook [14, 15] analyzed the Altmetrics as to the limitations that hinder the visibility of science carried out in Brazil and concluded with an emphasis on the need to valorize the activity of scientific dissemination, especially in the democratization of society's access to scientific results, the construction of scientific policies that value "these activities in career progression and evaluation of research projects.

In view of the panorama presented, in order to add to the research and practices realized, it is presented in the next section suggestions of strategies of dissemination for scientific journals in YouTube and Instagram, as well as the justification of delimitation of these social media.

4 Strategies for Scientific Dissemination in Social Media

To guide the use of social media, some considerations are necessary, which can serve as guidelines for the strategic and operational planning of the use of these platforms. The **first guideline** deals with the basic purpose of the use of social media: the production and sharing of content. According to [16, p. 61] "Social media is a group of Internet-based applications, which are guided by the ideological and technological foundations of Web 2.0 and allow the creation and exchange of user-generated content". Therefore, the use of social media platforms presupposes the creation and sharing of content, as well as the interaction between people who are part of the network. For this, the content needs to have certain features, which refers to the **second guideline**.

There are some properties of content that need to be observed for it to fulfill its purpose of disseminating science. It is possible to admit the effectiveness of good content in face of the usefulness and relevance that leads to the intended audience. For this, both the use of social media and the production of the content itself needs to be done in a strategic, planned, and assertive way. In this sense, it is possible to use Content Marketing, which according to [17, 147], "[…] is an approach that involves creating, selecting, distributing, and expanding content, which is interesting, relevant, and useful for a clearly defined audience with the objective of generating conversations about this content […]". The **third guideline** arrives, defining the target audience.

The use of social media, as well as the production of content, will only make sense when it is clear what the profile of the audience you want to serve is. Based on this perspective, we recommend the studies based on ethnography, a research method that consists of participating observation and acting in the online environment, where different "computer-mediated forms of communication are used as a source of data for the understanding and ethnographic representation of cultural and communal phenomena" [18, p. 3].

In addition, it is necessary to pay attention to accessibility issues, as most social media platforms provide specific resources to make content more accessible. And finally, it is necessary to avoid academic jargon and have special attention to the use of simple language, clear and appropriate to the target audience.

The **fourth guideline** relates to the strategic and operational aspects of the use of social media and content production. In this regard, the eight major steps (Table 2) defined by [17, p. 151], which guide Content Marketing, can be adapted:

Table 2. Eight steps for content marketing in scientific journals based on [17].

Steps	Description
1) Setting goals	The focus and scope of the journal must be aligned and be measurable through key metrics;
2) Mapping the audience	The audience to be reached should be determined, from whom are the people to be served by the content;
3) Content design and planning	Raise and select the content ideas, as well as perform the appropriate planning, paying attention to the relevance of the themes, appropriate formats, and genuine narratives;
4) Content creation	As the most important step, it requires you to invest time and energy in the production of high-quality, original, and rich content. In addition to establishing a schedule and defining the people who will perform such activities, to ensure the sustainability of content creation;
5) Content distribution	Distribution through own channels (websites, blogs, e-mails) or conquered media (YouTube, Instagram, Facebook, Twitter, etc.) is required. It should be noted that content produced on a given platform can be distributed on other channels, provided that the necessary adaptations are made regarding the formats;
6) Content expansion	Identify relevant actions to leverage the content produced, generate relevant conversations, and interact with the audience;
7) Content performance evaluation	It covers the assessment of strategic and tactical performance indicators. The strategic scope is related to the goals set in the first step, how close to achieving them, or what still needs to be done to make this happen. Tactics refer to the assessment of key metrics, with the aid of social listening tools and data analysis;
8) Performance improvement	From the tracking and performance evaluation of the content, it is possible to define periodic improvements in the approach of the themes, as well as in the creation, distribution, and expansion of the content to be produced.

The **fifth guideline** refers to the ethical and legal aspects involving the use of social media and the production of content. In this item, [19, p. 175], they stress that it is necessary to observe local legislation, for example, the Brazilian legislation on the use of the Internet, the terms of use of each platform and copyright, and the recent General Law on Personal Data Protection (Lei Geral de Proteção de Dados Pessoais - LGPD), which deals with the issue of privacy of personal data available on the Internet.

Finally, the **sixth guideline**, also raised by [19, p. 177], deals with the guidelines for crisis management through a complaint or negative criticism about the posted content. In this context, it is necessary to value urbanity and diplomacy, but at the same time maintain a firm stance regarding the message to be passed on.

Thus, in outlining the six fundamental guidelines for the use of social media about scientific dissemination in journals, it seeks to point out that the production of authentic, relevant, and high-value content is essential to achieve the established objectives and obtain concrete results. This only happens when efforts are directed to serve a specific audience, defined from validated methodologies, as well as appreciate assertive communication.

Thus, it is necessary to attend to the operational factors of content creation and pay attention to the ethical and legal aspects, as well as to the crisis management orientation. Only after observing these guidelines, it is possible to define which platforms are important for the strategy outlined by the scientific journal.

The selection of the social media to be used goes through a primordial question, which refers to the presence of the target audience in the chosen platforms. According to the Digital 2020 report July Global Statshot, there is an increase of more than 10% in the number of users in social media worldwide if we compare with last year's statistics. "With more than 376 million new users since July 2019, this translates into almost 12 new users every second [...]" [20]. However, the demographic distribution of this usage is still not uniform across the globe.

Regarding the most used social media, the ranking is led by Facebook, followed by Youtube, Whatsapp, already Twitter appears in the last place in a list that has 16 social media. However, the positioning of this ranking may vary according to the country analyzed, as in the Digital in 2020 report, presented in January 2020, in which in the Brazilian scenario the ranking of the most used social media has another configuration, with Youtube in the first place, followed by Facebook, Whatsapp and Instagram - this time Twitter appears in 6th place [21].

Thus, based on the information mentioned, it seeks to guide the strategic use of YouTube and Instagram for the dissemination of science by scientific journals, to contribute to existing practices, as well as bring new possibilities for action.

4.1 YouTube

With over 2 billion users worldwide, YouTube offers a platform focused on creating audiovisual content, which, according to the Digital 2020 survey July Global Statshot, remains the audience's preferred format [20]. The interesting thing is to try to develop a diverse editorial line, which pays attention to variations of exhibition time, themes, and types of content.

In Brazil there is Science Vlogs Brasil, a quality seal for scientific publishers, which ensures that a video with this seal it is conveying serious scientific information, with recognized sources that are representative of the current scientific and academic consensus, being constantly analyzed by peers in a favorable network of mutual help and constant communication [22]. It has the purpose of giving reliability to the dissemination of scientific knowledge, as well as helping in the fight against misinformation. By joining as a member, the channel can participate in a support network that has important collaborators such as doctor Dráuzio Varella, paleontologist Pirula, and E-farsas.

Regard the types of content that can be developed, some examples are listed: Interviews; Webinars; Videos; Tips and tutorials; Reviews; Questions and Answers; Dissemination, coverage, and transmission of in-person events; Online events; Live events; Calls for publication, among others.

4.2 Instagram

In a social media strategy, the platform can be used to make the journal better known to the public, since it has a significant potential to attract new followers. Ideally, the account should be registered as a content creator or business, because only then is it possible to access and follow the profile metrics, such as the growth of the follower base, public profile, interactions with content, among others.

Another relevant point is to pay attention to the four formats of native content of the platform: stories, feed, reels, and IGTV. The stories have the purpose of sharing short content, which is available for up to 24 h. The types of content can be a record of the backstage work of the magazine and its collaborators, disclosure of events and calls for publication, interactive content such as question boxes, tests and polls, lives and/or live interviews.

The feed gathers the main publications of the account and supports three types of content: images, videos of up to 1 min, and carousel posts, which work as a kind of album with a limit of 10 images per post. With creativity, it is possible to develop in-depth and useful content on a given theme, provided it does not exceed the limits of 2,200 characters in the legend.

IGTV works as an exclusive profile video channel. With a different format than the feed, the videos can be produced in the vertical position (with the proportion of 9:16) or horizontal (with the proportion of 16:9), with display time that must vary between 60 s and 1 h. The types of content for YouTube (item 4.1) also apply to IGTV and there is the possibility of organizing them in series.

Finally, the latest feature, reels, has the purpose of distributing short audiovisual content (from 15 to 30 s) in which followers can interact and recreate the content from the audio replication. The challenge imposed by this format is to have the creativity to produce relevant content, even with this time limitation. Short tutorials, quick tips, events disclosure, calls for publication, and interview excerpts are some examples of the types of posts that can be produced in reels.

5 Final Considerations

Proposing dissemination strategies for scientific journals on YouTube and Instagram, was the goal that guided the research addressed in this work. In which it is evident that most of the scientific articles published bring research with the use of social media by the researchers themselves, which indirectly reflects in access to scientific journals through disseminated articles. However, an incipient number of investigations provides strategies for the use of social media by scientific journals, since among the analyzed articles we found only the research of [12] with direct suggestions of strategies.

It is important to emphasize that the use of social media for the purpose of scientific dissemination requires a well-defined strategy to achieve the objectives proposed by the scientific journal. Thus, six fundamental guidelines for the use of social media have been described, covering both strategic and operational aspects: 1) Basic purpose of the use of social media; 2) Properties of the content; 3) Definition of the target audience; 4) Strategic and operational aspects of the use of social media and production of content; 5) Ethical and legal aspects involving the use of social media and production of content and; 6) Crisis management through a complaint or negative criticism about the posted content.

Also, in the fourth guideline, eight steps were outlined for scientific journals: 1) Setting goals; 2) Mapping the audience; 3) Content design and planning; 4) Content creation; 5) Content distribution; 6) Content expansion; 7) Content performance assessment and 8) Performance improvement.

The sequence of this research will consist of investigating which scientific journals are using YouTube and Instagram as a dissemination channel and whether the practices are aligned with the guidelines described for the strategic and operational planning of these platforms.

References

1. Vanz, S.AS., Silva Filho, R.C.: O protagonismo das revistas na comunicação científica: histórico e evolução. In: Carnerio, F.F. B., Ferreira Neto, A., Santos, W. (eds.) A comunicação científica em periódicos. Appris, Curitiba (2019)
2. Bueno, W.C.: Comunicação cientifica e divulgação científica: aproximações e rupturas conceituais. Informação & Informação 15(1), 1–12 (2010)
3. Bombaci, S.P., et al.: Using Twitter to communicate conservation science from a professional conference. Conserv. Biol. 30(1), 216–225 (2016)
4. Haustein, S., et al.: Tweets as impact indicators: examining the implications of automated "bot" accounts on Twitter. J. Assoc. Inform. Sci. Technol. 67(1), 232–238 (2016)
5. Gerds, A.T., Chan, T.: Social media in hematology in 2017: dystopia, utopia, or somewhere in-between? Curr. Hematol. Malignancy Rep. 12(6), 582–591 (2017)
6. Robinson-García, N., et al.: The unbearable emptiness of tweeting—about journal articles. PLoS ONE 12(8), (2017)
7. Barata, G., Shores, K., Alperin, J.P.: Local chatter or international buzz? Language differences on posts about Zika research on Twitter and Facebook. PLoS ONE 13(1), (2018)
8. Pulido, C.M., et al.: Social impact in social media: A new method to evaluate the social impact of research. PLoS ONE 13(8), (2018)

9. Ye, Y.E., Na, J.-C.: To get cited or get tweeted: a study of psychological academic articles. Online Information Review (2018)
10. Alperin, J.P., Gomez, C.J., Haustein, S.: Identifying diffusion patterns of research articles on Twitter: a case study of online engagement with open access articles. Publ. Underst. Sci. 28(1), 2–18 (2019)
11. Kortelainen, T., Katvala, M.: "Everything is plentiful—Except attention". Attention data of scientific journals on social web tools. J. Inform. 6(4), p. 661–668 (2012)
12. Tur-Viñes, V., Segarra-Saavedra, J., Hidalgo-Marí, T.: Use of Twitter in Spanish communication journals. Publications 6(3), 34 (2018)
13. Zhao, Y., Wolfrm, D.: Assessing the popularity of the top-tier journals in the LIS field on Twitter. Proc. Assoc. Inform. Sci. Technol. 52(1), 1–4 (2015)
14. Holmberg, K., Park, H.W.: An altmetric investigation of the online visibility of South Korea-based scientific journals. Scientometrics 117(1), 603–613 (2018)
15. Barata, G.: More relevant alternative metrics for Latin America. Transinformação 31 (2019)
16. Kaplan, A.M., Haenlein, M.: Users of the world, unite! The challenges and opportunities of Social Media. Bus. Horiz. 53(1), 59–68 (2010)
17. Kotler, P., Kartajaya, H., Setiawan, I.: Marketing 4.0: do tradicional ao digital. Sextante, Rio de Janeiro (2017)
18. Corrêa, M.V., Rozados, H.B.F.: A netnografia como método de pesquisa em Ciência da Informação. Encontros Bibli: revista eletrônica de biblioteconomia e ciência da informação 22(49), 1–18 (2017)
19. Prado, J.M.K., Correa, E.C.D.: Bibliotecas universitárias e presença digital: estabelecimento de diretrizes para o uso de mídias sociais. Perspectivas em Ciência da Informação 21(3), 165–181 (2016)
20. Digital 2020 July Global Statshot. https://wearesocial.com/blog/2020/07/digital-use-around-the-world-in-july-2020. Accessed 21 Oct 2020
21. Digital in 2020. https://wearesocial.com/digital-2020. Accessed 21 Oct 2020
22. Ayrolla, D.: O projeto. 1 fev. 2016. https://www.blogs.unicamp.br/sciencevlogs/2016/02/01/o-projeto/. Accessed 21 Oct 2020

Digital Humanities and Open Science: Initial Aspects

Fabiane Führ$^{(\boxtimes)}$ and Edgar Bisset Alvarez

Universidade Federal de Santa Catarina, Florianópolis, Brazil
`fabiane.fuhr@ufpr.br, edgar.bisset@ufsc.br`

Abstract. The digital humanities have acquired more and more visibility and their field of action has expanded due to the increasing digitalization and the large volume of data arising from these processes. Collaborative researches on the impact that is in line with the dimensions of open science impact the scientific production chain. It aims to identify which aspects of open science are approached in the publication regarded to digital humanities. To achieve the general objective it indicates some specific objectives: it identifies the scientific production about open science and digital humanities indexed in the databases Web of Science (WoS), Scopus, and Scientific Electronic Library Online (SciELO) and; it describes how open science is approached in each paper from the corpus and how it is related to digital humanities. It uses the bibliographic manager Zotero to organize the bibliographic data and it uses the software Atlas.ti to the qualitative analysis and applies the data mining tool Sobek and Voyant Tools in the data. From the 13 papers analyzed, only 3 do not use projects or programs related to digital humanities to present the discussion. The data mining tools do no show the relation between digital humanities and open science. It shows the importance of data management and the necessity to have guiding documents. It also points to the relevance of metadata pattern, to work to make the data suitable to FAIR principles, to train researchers and citizens to promote collaboration among different institutions and people that have a diverse background to value open science.

Keywords: Digital humanities · Open science · Data mining · Bibliometric · Scientific production

1 Introduction

Digital Humanities (HD) is a comprehensive field of research, which has been considered by some as a set of practices, by others as a new field of study or a new discipline, or just a new look for research in the Humanities [1].

Digital Humanities gained visibility with the publication of the book *A companion to Digital Humanities*, edited by Susan Schreibman, Ray Siemens, and John Unsworth, released in 2004. However, its origin dates back to research developed in the 1940s and 1950s by Father Roberto A. Busa, who created the Index Thomisticus, which is

E. Bisset Álvarez (Ed.): DIONE 2021, LNICST 378, pp. 154–173, 2021.
https://doi.org/10.1007/978-3-030-77417-2_12

considered the first application of computing to linguistic studies in the works of São Tomás de Aquino [2].

According to Dalbello [3], the HD timeline begins with the introduction of computational methods in the literary, philological, and philosophical fields in the 1950s and continues with a focus on searchable multimedia files and structured texts in the 1980s and 1990s, in addition to editing project texts and electronic collections from the past decades [2000s].

The field of research in Digital Humanities has grown exponentially due to the increasing digitization of documents, which has generated a large mass of data, expanded the number of collaborative research, and changed the scientific production chain. However, Digital Humanities projects are not restricted only to accessibility and dissemination of knowledge. Still, they also concern about the ways of creating and disseminating them [4], the convergences between Digital Humanities and Information Sciences [5–8]. Bibliometric studies on the subject are examples of studies that have been developed [9–11].

This same chain of scientific production has changed due to the movement in favor of Open Science. This movement has been consolidated in many countries since the movement for Open Access, whose decisive landmark was the Budapest Declaration published in 2002. Open Science aims to share and access publications and research data, promote the scientific process's opening, assist in the transfer of knowledge, expand the social and economic impacts of science, and reinforce the social responsibility of science [12].

Thus, Open Science is a movement that proposes new forms of collaborative, interactive, and shared production of information, knowledge, and culture [13]. This movement has gained more and more visibility and has been extended to the most diverse investigation areas.

In this context, it is important to observe in publications on Digital Humanities how the theme of open science has developed. Therefore, this article's general objective is to identify which aspects of Open Science are being addressed in publications referring to Digital Humanities, in the Web of Science (WoS), Scopus, and Scientific Electronic Library Online (SciELO) databases.

The specific objectives are:

a) Identify the scientific production on Open Science and Digital Humanities indexed in the Web of Science, Scopus, and SciELO databases from 2014 to 2020;
b) Describe how Open Science is approached in each of the articles of the research corpus and how they relate to the Digital Humanities.

One of the motivators for the growth of research and publications on Open Science is believed to have been the development of the movement for open access to knowledge, which had its starting point in the Budapest Declaration of 2002. However, all changes require a period for their maturation, which can be seen in this research that aims to analyze when and how open science came to be among the proposals and projects related to Digital Humanities.

2 Methodology

To identify scientific production in the area of Digital Humanities and relate it to Open Science, the expression [("digital humanities" AND "open science")] was defined as a search strategy, but as the databases also have articles published in Portuguese and Spanish, the strategy for the search expression [("humanidades digitais" OR "humanidades digitales" OR "digital humanities") AND ("ciência aberta" OR "ciencia abierta" OR "open science")].

The SciELO Network was opened in 1998, with the creation of the SciELO Brazil Collection. The project was an initiative of the Fundação de Amparo à Pesquisa do Estado de São Paulo (FAPESP) and the Latin American and Caribbean Center on Health Sciences Information (BIREME/PAHO/WHO). In 2002, the project started to be financially supported by the Conselho Nacional de Desenvolvimento Científico e Tecnológico (CNPq) [14]. Currently, the SciELO Network has national collections from the following countries: Argentina, Bolivia, Brazil, Chile, Colombia, Costa Rica, Cuba, Spain, Mexico, Paraguay, Peru, Portugal, Uruguay, South Africa. They are also under development three new collections: Ecuador, Venezuela, West Indies. As it is an open-access program for international cooperation in scientific and academic communication, it is believed that its coverage is representative of the present study.

The Scopus and WoS databases are international databases maintained by the major information companies Elsevier and Clarivate Analytics. Both strategies were applied in the Scopus, Web of Science (WoS), and Scientific Electronic Library Online (SciELO) databases on July 20, 2020. The search was carried out in the title, keyword, and summary fields. There was no application of a filter concerning the publication date, as it is a recent topic. Thus, it would be possible to obtain a greater number of records for analysis.

Both strategies returned the same results since the database's index not only articles in English but also in Spanish and Portuguese. A filter was applied to these results so that only scientific articles were retrieved. Thus, 21 articles were retrieved, 9 articles from the Scopus database and 10 from the Web of Science. The year of publication of these articles corresponds to the years 2014 to 2020. The 2 articles retrieved from the SciELO database were published in 2020.

The bibliographic manager Zotero was used to collect, store, and organize the articles retrieved in the bibliographic survey. After the organization, 8 duplicate files were excluded. Thus, 13 articles remained for analysis.

Of the 13 articles that make up the research corpus, 30.77% are found only in the Scopus database, and another 30.77% only in the WoS. 23.08% are found in Scopus, and WoS databases, and another 15.38% are found in Scopus, WoS, and SciELO databases. Table 1 shows the distribution of articles by database.

Table 1. Number of articles per database

Database	Number of articles
SciELO, Scopus, WoS	2
Scopus, WoS	3
Scopus	4
WoS	4

Source: The authors (2020)

Regarding the publication date, 2019 was the year with the highest number of publications (5), 2017, 2018, and 2020 have 2 publications, and 2014 and 2015 have 1 publication per year.

For qualitative analysis of this research's corpus, the software ATLAS.ti[1] was used, a software for qualitative data analysis, which allows analyzing texts, audios, photos, videos, and working with documents in different file formats [15]. The *Sobek*,[2] text mining tool, was developed from an algorithm initially created by Schenker in 2003 and modified by the Research Group Gtech.Edu at the Federal University of Rio Grande do Sul to make the more accessible tool to educational practice, analyzing relevant words in a text and representing them graphically [16]. Canadians Stéfan Sinclair at McGill University and Geoffrey Rockwell at the University of Alberta developed the *Voyant Tools*[3] app. The application allows you to work with text or text collections and perform basic mining functions. One advantage is that it allows you to work with documents in different formats, supports large volumes of texts, allows the interaction between tools that facilitate navigation and exploration of different scales, among other options useful to the researcher [17].

3 Presentation and Analysis of Results

The research was carried out in 13 articles, of which 5 were written in English, and another 5 in German. 2 articles were written in Portuguese and 1 in Spanish. Regarding the country of publication of the journal, it was observed that Austria is the most productive country with 3 publications. Germany and Brazil have 2 publications, and the other countries only 1 publication (Sweden, USA, Spain, Switzerland, England, and one International[4] publication), as can be seen in Graph 1.

---

[1] https://atlasti.com/.

[2] http://sobek.ufrgs.br/#/.

[3] https://voyant-tools.org/.

[4] ERCIM News magazine is registered with the ISSN as an international publication, as it is a publication of the European Research Consortium for Informatics and Mathematics, which includes research establishments from different European countries [18].

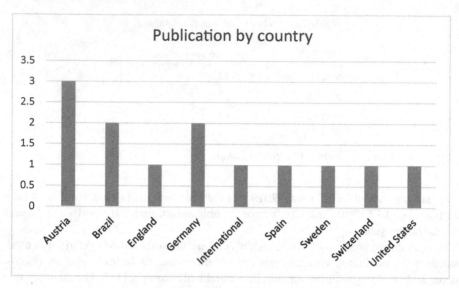

Graph 1. Publication by country

The journals were analyzed concerning Qualis Capes (Quadriênio 2013–2016). And about the impact factor (Journal Impact Factor) provided by the Journal Citation Reports of the Web of Science Group and the journals that were not identified in the Journal Impact Factor, were analyzed to verify if they were on the list of the Emerging Source Citation Report, of Web of Science Group, both Clarivate Analytics [19]. Regarding CiteScore, which is a metric that indicates the impact of the survey. CiteScore is developed by Scopus, Elsevier's database [20].

Table 2. Analysis of qualis, impact factor, emerging source and citescore of journals

Periodicals	Number of articles	Qualis[a]	Journal impact factor	Emerging source citation report	CiteScore
Bibliothek - Forschung und Praxis	1	*	*	Yes	*
Cataloging and classification quarterly	1	A2	*	Yes	1.2
ERCIM news	1	*	*	Yes	*
Estudos históricos	2	B1	*	Yes	0.1
Information research	1	A1	0.763_2019		1.7

(*continued*)

Table 2. (*continued*)

Periodicals	Number of articles	Qualis[a]	Journal impact factor	Emerging source citation report	CiteScore
Literary and linguistic computing	1	A1	1.125_2016		N/A
Mitteilungen der Vereinigung Österreichischer Bibliothekarinnen & Bibliothekare (VÖB)	3	*	*	*	0.3
Profesional de la Informacion	1	A1	1.580_2019	*	2.1
Publications	1	B5	*	Yes	1.8
Zeitschrift für Germanistik	1	*	*	**	N/A

[a]Qualis from Literary and Linguistic Computing refers to the area of Linguistic and Literature assessment. The others belong to the Communication and Information area.
* There was no information about the journals in the sources consulted
** This journal is indexed in the collection: Art & Humanities Citation Index

It was observed that the German-language journals do not present information in Qualis Capes, which can be justified by the linguistic barrier, since Qualis Capes reflects where teachers and researchers in the area have published the results of their research [21].

To analyze the corpus of this research, the software ATLAS.ti was used, a software for qualitative data analysis, which allows the analysis of texts, graphics, audios, and videos. After reading the articles in advance, 25 categories were established for marking and coding during the reading. The categories were grouped into Families, as assigned by the software.

For the Digital Humanities Family, 16 categories were established (Fig. 1) for marking and analysis. Figure 1 shows 17 nodes (1 representing the Digital Humanities Family and the others representing the 16 categories).

Open science was also the subject of research in the articles. Thus, the Open_Science Family was designated, divided into 9 categories (Fig. 2) for marking and further analysis.

All articles were analyzed for their insertion in the area of Digital Humanities and Open Science. Thus, Table 3 presents the 13 selected articles, their authors, and the projects and programs related to Digital Humanities that are described and addressed in the publication.

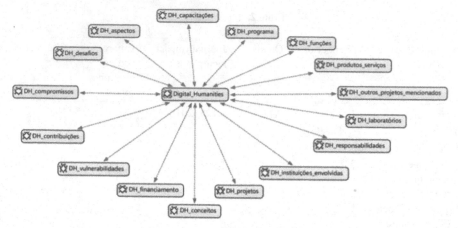

Fig. 1. Digital_Humanities family

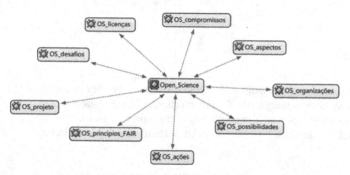

Fig. 2. Open_Science family

Regarding the type of authorship, it was observed that 07 articles are of unique authorship. The other articles have shared authorship distributed as follows: 3 articles have 3 authors and only one article published by 2 authors, 4 authors, and 8 authors.

Concerning projects and programs, only 03 authors (Knöchelman 2019; Steinerova 2018; Baum 2017) do not use projects or programs in Digital Humanities for their discussions. However, they present relevant questions regarding the themes of digital humanities and open science.

3.1 Text Mining

The texts were gathered by language. In this way, 4 sets of texts were obtained in .txt files: German (brought together in a single document the 5 texts in German); Spanish (only a text in Spanish); English (brought together the 5 texts in English) and Portuguese (brought together the 2 texts in Portuguese).

Table 3. Authorship and projects related to digital humanities

Article	Year	Authors	Project/Program
01	2020	Rollo, MF	Program Memória de Todos
02	2020	Ferla, LAC; Lima, LFS; Feitler, B	Projeto Implementação da tecnologia de Sistemas de Informações geográficas (SIG) em investigações histórícas (2012-2013); Projeto Pauliceia 2.0; Projeto de avaliação e edição de verbetes da Enciclopédia Virtual Wikipédia
03	2019	Lahti, L; Marjanen, J; Roivainen, H; Tolonen, M	Bibliographic data science
04	2019	Knöchelmann, M	
05	2019	Anglada, LM	Collection of Museu do Prado; Transcribe Betham; eBird e Mapa Literali Catalã
06	2019	Stigler, J; Klug, HW	Project Konde (acronym for Kompetenznetzwerk Digitale Edition)
07	2018	Hagmann, D	PHAIDRA repository and excavation "Molino San Vincenzo"
08	2015	Blumesberger, S	Projekts e-Infrastructures Austria
09	2014	Wells, JJ; Kansa, EC; Kansa, SW; Yerka, SJ; Anderson, DG; Bissett, TG; Myers, KN; Demuth, RC	The Digital Index of North American Archaeology (DINAA)
10	2019	Wuttke, U; Spiecker, C; Neuroth, H	Project PARTHENOS (acrônimo de Pooling Activities, Resources and Tools for Heritage E-research Networking, Optimization and Synergies)
11	2018	Steinerova, J	
12	2017	Bassett, S; Di Giorgio, S; Ronzino, P	Project PARTHENOS
13	2017	Baum, C	

To identify relevant information in the texts, it was decided to analyze them using the text miner Sobek. However, the system is adapted only to the English and Portuguese languages. When applying the German and Spanish texts, the results were inconsistent since the most frequent words considered by the system were connective.

For this reason, we sought another text mining tool that was open access and that allowed text mining in other languages. Thus, the tool chosen was Voyant Tools.

When applying the English texts in the Sobek software, the following result was obtained:

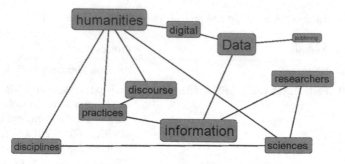

Fig. 3. Sobek mining result English – 15 results

It is possible to observe that the most prominent words are Data, Humanities, and Information. The combination of terms emphasizes the discourse and practices between these elements and the other agents involved, such as disciplines, sciences, researchers, digital (which can refer to the object, the environment, among others), and publication.

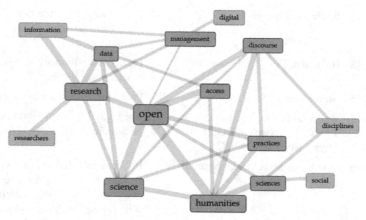

Fig. 4. *Voyant Tools* English mining result – 15 results

In Voyant Tools, it is observed that the keywords are represented in blue. The most relevant word is open (with 255 occurrences), and it has a strong connection (represented by the thickest line) with the terms science (182) and humanities (170). Between the term research (156) and information (101), there is a stronger link, and with the term researchers (43), the link is a little weaker. It is observed that the terms information, researchers, digital, disciplines, social are presented in orange; that is, they are close words, which the system calls co-occurrence.

An important piece of information evident in Voyant Tools and that did not appear in Sobek was the link between the terms Open and Science, which is one of the objects of this research.

Fig. 5. Sobek Portuguese mining result – 15 results

In Fig. 5, which represents the mining of the Portuguese text in the Sobek software, the prevalence of the term digital humanities and its connection with the terms heritage and digital and the terms knowledge and digital, are observed. The term knowledge also joins science and society. Another link that can be observed is between the digital humanities, the humanities, and the community. The terms project and history have isolated links to the term digital humanities.

The mining result on Voyant Tools (Fig. 6) showed two groups of results. The keyword knowledge is linked to similar terms, such as production, sharing, creation, and scientific. The link between human and digital keywords is strong, and this link is surrounded by the words close together: are, axes, field, scientific, and community. Thus, the system's efficiency is questioned since co-occurrences are and axes are not relevant to the results.

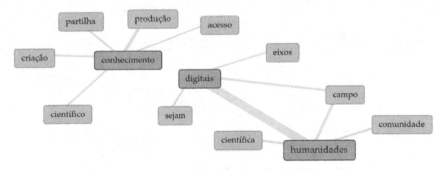

Fig. 6. Voyant Tools mining result Portuguese – 13 results

Figure 7 results in mining in German-language texts. Again, there is a certain fragility since the keywords digitalen and digitale represent the term "digital" and have a connection with digital co-occurrence, which has the same definition. However, most of the other co-occurrences related to digital are tecnologien (technologies), zeitalter (time), geisteswissenchaften (humanities), verfahren (process/procedure), forschungsinfrastrukturen (research infrastructures) and literaturwissenschaft (literary studies) are relevant to the studies.

Fig. 7. Resultado mineração *Voyant Tools* Alemão – 14 resultados

Another keyword is daten, which means data. This is linked to digitaler (digital), archivierung (archiving), Phaidra (an acronym for Permanent Hosting, Archiving, and Indexing of Digital Resources and Assets), repositorium (repository) and are aligned with the studies in question.

The German text was also applied to the Sobek miner, but the result was even more inconsistent than found on Voyant Tools since the miner is only enabled for English and Portuguese.

Figure 8 shows the mining result in Voyant Tools for the text in Spanish. However, just as in Sobek, the result using this tool is inconsistent because the three keywords two - "no" and "a" - do not bring a relevant meaning, as well as the co-occurrences linked to them, which were sólo (only, unique), más (more) and hay (has).

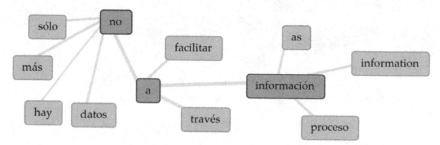

Fig. 8. Spanich Voyant Tools mining result – 12 results

For this reason, it was decided to use the word cloud, which despite including the articles and terms that would normally be excluded, presents the most relevant terms in the text, such as: información (49 ocorrências), ciencia (24), bibliotecas (18), cambios (18) ou cambio (16), digitales (12), abierta (11), investigación (11), centros (10), humanidades (9), among others with less relevance.

Fig. 9. Spanish Voyant Tools word cloud

The results using the Sobek software for the English and Portuguese languages were consistent; however, the same consistency was not maintained for the other languages of the research corpus. The Voyant Tools software, chosen for mining in other languages, also showed inconsistencies, mainly in Spanish.

3.2 Digital Humanities and Actions for Open Science

Among the results found, it was observed that the first researches that address digital humanities and open science are the studies by Wells et al. (2014), in which the project The Digital North American Archeology (DINAA) is presented, which aims to develop models and approaches to publishing and index archaeological data inventoried in historical and archaeological sites in large areas of North America [22].

In 2015, Blumesberger addressed the project to expand repository infrastructure in Austria (Projekts e-infrastructures Austria). The author emphasizes the team's work in selecting objects for scanning, presenting selection reports, defining the necessary metadata, uploading to the repository, and managing the objects until viewing. With the information collected, the team created the Data Management Plan, a document that brings considerations about all phases of the project, from creation to transfer of the data package, also addressing data management in the repository, as well as the reuse and visualization of objects. The group also discusses the use of Linked Open Data (LOD) [23].

Basset, Di Giorgio, and Ronzino (2017) present the PARTHENOS project as a Horizon 2020 project that aims to reinforce research in linguistic studies, cultural heritage, history, archeology, and related areas. To make this possible, PARTHENOS is building an interdisciplinary environment that allows humanities researchers to have access to data, tools, and services based on common policies, guidelines, and standards [24].

In 2019, Wuttke, Spiecker, and Neuroth also brought actions on the Projeto PARTHENOS. According to the authors, the most important areas for the project are to define how long-term archiving will take place, the management of intellectual property, its implementation, the development of common standards, services and methods for interdisciplinary and subsequent use, data use, as well as additional training and education. Throughout the article, the actions developed in work packages are presented, and the most relevant packages are packages 4, 7, and 8. Package 4 has developed a standardization kit consisting of documentation on the relevant human and cultural sciences standards in digital media. This package can also be used for teaching, and in partnership with packages 7 and 8, develops communication and training materials on the importance of standardization for the research process [25].

In 2018, Hagmann presented the PHAIDRA repository from the University of Vienna, which stores long-term digital data from scientific research in an environment with sustainable standards. Thus, the author considers PHAIDRA ideal for storing archaeological data, such as the data collected at Sítio Molino San Vincenzo in Tuscany, Italy. The University of Vienna has been carrying out archaeological investigations on the site since 2012. There is data collected in digital and analog format. These are scanned for further processing. The author emphasizes the use of information and communication technologies (ICT) in the archelogy, which is a prerequisite for generating digital data in archeology [26].

Returning to the year 2019, Lathi et al. present the project Bibliographic data science. They analyze the science of bibliographic data as a study that derives from the area of data science. The Finnish and Swedish National Bibliographies (FBN and SNB), the English Short-Title Catalogue (ESTC), and the Heritage of the Printed Book database (HPBD) are used for the analysis. These bibliographies cover more than 6 million entries for printed products in Europe and elsewhere. According to the authors of this study, it is possible to demonstrate, through a qualitative approach, the history of the book [27].

According to the authors, efforts for large-scale automated harmonization can improve overall reliability and commensurability between metadata collections, complementing LOD and other technologies that focus on data management and distribution. Bibliographic data science aims to fill an important gap in the area since many bibliographic metadata present inaccurate entries, collection bias, and missing information.

For Anglada (2019), Digital Humanities are a movement that affects information centers' life and evolution. According to the author, society is approaching when what is not on the network is no longer important, so libraries and documentation centers must shift their attention to the document so that it has easy access. Besides, the author suggests promoting citizen participation, as is the case with the following initiatives: Prado Museum Collection, the Transcribe Betham project, the eBird project and the Literali Català Map [28].

The author suggests that it is possible to clearly observe the effects of open science on the Digital Humanities since researchers need digital objects to be accompanied by information about the processes that made them readable by the computer to be shared and reused by different groups later.

Stiegler and Klug (2019) exhibit the Konde Project (an acronym for *Kompetenznetzwerk Digitale Edition*), which defines the prerequisites for establishing a platform digital editions, which aims to develop and preserve cultural heritage. The authors claim that digital editions are a product of the digital discipline of Humanities. Computer-aided methods are used to create, research, and disseminate publications from reliable scientific sources [29].

With the Konde project, Austria intends to become a leader in innovation in digital publishing. To this end, it relies on the participation of researchers and professionals from libraries, archives and museums to strengthen research centers and effect collaboration between participants.

Memória de Todos is the program presented by Rollo (2020) and aims to promote heritage education, literacy, digital skills development, and the democratization of access to historical research tools and promote the collection, sharing, and preserving of memories and testimonies [30].

For the author, the Digital Humanities comprise the use of tools, information research, organization, content storage, or even the programming and use of databases or computing tools in their entirety. However, these skills are scarce in the humanities community itself. Also, according to the author, digital has brought a series of opportunities to expand the contents, especially the immaterial contents that can be registered and preserved for future generations.

The authors Ferla, Lima and Feitler present the Projeto Implementação da tecnologia de Sistemas de Informações Geográficas (SIG) technology in historical investigations (2012–2013) and the Pauliceia 2.0 Project that explore the possibilities of geo-technologies in historical investigations. These studies were developed by Grupo de Pesquisas Hímaco – História, Mapas e Computadores and had the participation and support of the Núcleo de Acervo Cartográfico do Arquivo Público do Estado de São Paulo (Apesp) [31].

In addition to the previous projects, there was also the Project's development for the evaluation and editing of entries in the Wikipedia Virtual Encyclopedia, carried out during the disciplines of Modern History I and II of UNIFESP, in which it was proposed to change the entries in Wikipedia. This project supported the Brazilian team linked to the Wikipedia Foundation, which offered lectures and provided guidance on the first steps for editing the content.

To include the Pauliceia 2.0 project within the scope of open science, the project was presented in 2017 in the auditorium of the Archive of the State of São Paulo and proposed to discuss, receive criticisms and suggestions. Besides, this moment was used to request empirical material to support the computational codes' tests. In October 2018, a new presentation was made to present the platform's beta version, in which the community was invited to help with the tests.

The authors Baum (2017), Steinerova (2018), and Knöchelmann (2019) that address Digital Humanities and Open Science but do not use projects for the foundation will be addressed from this point on.

Baum (2017), in your article *Digital gap or Digital turn?* It addresses literary studies and the digital age. For the author, digital has brought a series of advantages since, in digital environments, codes, scripts, and annotation decisions can be disseminated and even discussed and revised more easily through versioning. However, the author also brings up crucial issues such as the gap between first and third-world countries caused by technologies and socioeconomic issues, such as ethnicity, gender, nationality, and education, which include access to digital literacy [32].

In the context of open science, Baum (2017) states that the ability to connect research itself does not prove that everything presupposes digital processes or that everything is subordinate. Mainly because there are some issues imposed by digital that are difficult to reflect in the analog environment due to its high technical complexity, which requires digital expertise and collaborative work in this sector. According to the author, the term Open Science groups strategies and processes that aim at the digitalization opportunity to make the components of the scientific process accessible and reusable, that is, to bring new opportunities for science, society, and business.

Baum (2017) notes that literary studies are a major challenge for open science. It is a vast and heterogeneous field with various authors and several individual researchers, working groups, institutions, funding agencies, publishers, internet providers, and the networked public. Thus, open access to specialized scientific literature should be allowed, ensuring that it can be referenced and cited, in addition to promoting new forms of specialized communications and the recognition of collaborative research and publications.

In the article *Perceptions of the information environment by researchers: a qualitative study* (2018), Jela Steinerova seeks to understand how to research and academic practices represent a challenge for improving information services and information infrastructures. In her research, the author sought to identify the perceptions and uses of open access and open science resources by researchers. Steinerova identified that researchers consider open access advantageous, mainly due to the increase in citations and publication speed. Still, they also express concerns about commercial influences and evaluations of digital publications [33].

Steinerova also noted that many researchers agree with European open access policies, while others fear the lower quality of digital publishing. Transparency and open access were identified as factors of open science and participation, collaboration, peer networking, and information sharing. Technological determination has been identified in the big data sciences, such as astrophysics, physics, genetics, and others. In the Humanities, the development of digital libraries and archives in cultural heritage was observed. Besides, other open science factors were mentioned, such as policies, evaluation of results, access to data, and publication.

According to Steinerova's research, researchers' social networks share data, information, and publications. Open science was perceived as an advantage, especially open access sources, in addition to interdisciplinary cooperation and advertising. Still, gaps were also observed in the coordination of open science and access to publications and data and concerns about commercial influences and access to finance.

Marcel Knöchelmann, in his article, *Open science in the humanities, or: Open humanities?* (2019) states that open science has deficiencies in addressing the humanities, so it is necessary to think about a discourse on open humanities. For the author, the arguments for the need for this discourse are: the humanities are a by-product of open science, as they do not have their own discourse; the fragmentation of discourses about open practices in humankind requires a unification of these discourses and, mainly, the inadequacies of current scientific communication practices, since there are differences in the communication practices of scientific and human disciplines [34].

According to Knöchelmann the term open humanities has been used previously, but that does not mean an open humanities discourse. For the author, the importance of an open humanities discourse brings together thinking that the humanities need to be open to the transfer of interdisciplinary knowledge, especially concerning digital humanities; Humanities also need a transdisciplinary space in which to shape their digital and open future, working to open up their practices, problems, and implementations. Another issue pointed out by Knöchelmann concerns the use of copyright licenses since they are important for the authorship's progressive understanding.

3.3 Emphasis on Open Science

Some articles in the corpus emphasize more than others some characteristic aspects of open science. These aspects will be addressed in this session.

Basset, Di Giorgio, and Ronzino (2017) emphasize that because it is a project of Program Horizon 2020, data management should be concerned with meeting the FAIR principles (Findable, Accessible, Interoperable, Reusable); that is, the data must be traceable, accessible, interoperable and reusable. Thus, the projects financed by the Program must implement a Data Management Plan (PGD) to improve and maximize access and reuse of the research data generated by the project. What reinforces the actions of the European Union for the circulation of knowledge.

The authors state that the PGD PARTHENOS model is divided into three levels: the first level includes a set of essential general requirements, regardless of the discipline; the second level includes specific requirements, and the third level is project-based. To assist in completing the PGD, a set of instructions for specific disciplines will be provided. According to the authors, the PGD PARTHENOS model allows researchers to freely access, mine, explore, reproduce and disseminate their data and identify the tools needed to use raw data and validate research data, or to provide their own tools, taking a significant step towards the realization of open science.

Wuttke, Spiecker, and Neuroth (2019) emphasize the FAIR principles and the values of Open Science, increasing the cooperation between the existing research infrastructures, stimulating the exchange of both technical and semantic standards, and expanding the disciplinary boundaries, making that the discussions come out of the walls of the scientific communities and start to involve representatives of data centers, memory institutions, and research associations. According to the authors, these actions are an important contribution to the European Open Science Cloud (EOSC), which aims to be an environment in which researchers can store, analyze and reuse data for research, education, among other purposes. others.

According to Hagmann (2018), PHAIDRA is based on the Fedora Commons Repository and presents information on authorship, licensing, historical research framework, among others. In addition to including more detailed versioning, classification and categorization information. All objects and collections have persistent identifiers (permanent link, handle, and DOI), making it easier to quote. All of these actions allow data to be reused. Besides, data records receive a Creative Commons Attribution 4.0 International (CC BY 4.0) license concerning free and open access whenever possible.

For Lathi et al. (2019), LOD represents a crucial step in making the most of digital resources by integrating web sources and open and reusable metadata.

According to Anglada (2019), the Open Science movement makes science faster, more accurate, and reusable. For the author, there are three criteria for making Open Science, it must be open, collaborative, and made with and for society. According to the author, what is observed is that the Open Science and Digital Humanities movement are a reflection of the profound changes in the investigation and research processes. Thus, it is not a matter of making more scientific dissemination than establishing new relationships between people and science. Citizens can not only be interested in science, but they can also contribute with their participation and contributions.

Stiegler and Klug (2019) state that the KONDE project has a consistent orientation towards Open Science (Open Access, Open Data, and Open Source). The project also guarantees long-term and free access to research data and allows laypeople to be actively involved, if necessary, and scientifically, if indicated, in the sense of addressing citizen science.

According to Rollo (2019), open science has enormous potential for transformation, especially concerning the Digital Humanities work, since it widens the interconnections between the humanities and society. According to the author, open science focuses not only on open access, open innovation, citizen science, but also on the challenges of archiving and storage, preservation and curation of data, and digital information produced on a large scale society.

3.4 Challenges for Open Science

In 2014, Wells et al. emphasized that the financial issue is a source of concern, as the financing of public goods, such as open data, usually requires public or philanthropic support. The authors also point out that the project's financial sustainability requires lawmakers to understand that databases are America's national heritage.

Another challenge mentioned by Wuttke, Spiecker, and Neuroth (2019) is the large volume of data, tools, and digital methods that have not yet been adequately addressed by scientific communities. The authors also emphasized that communities and project participants have deficits regarding the FAIR principles and the EOSC concept.

Hagmann (2018) describes as a challenge the great diversity of data and obsolete supports. Thus, the challenge is to make these archaeological data available sustainable through effective long-term archiving. For this, the data must be saved in formats suitable for archiving, which allows the development of future research and the preservation of this data.

Lathi et al. (2019) emphasize that the lack of open data availability is a major bottleneck for bibliographic data science's transparent and collaborative development. Furthermore, they criticize, in a way, the term open science because, for the authors, this terminology is not concerned with the humanities but with a grouping of scientific disciplines.

4 Conclusion

Although the Budapest Declaration dates back to 2002 and the open science movement was driven by it [35], publications were found that relate Digital Humanities to open science only from 2014. For the volume of publications only in 2019, greater expressiveness was observed in the number of publications.

The languages with the largest number of publications are English and German, with 5 publications, and Austria is the most productive country with 3 publications. Regarding Qualis Capes, of the 6 journals that appear on the Qualis list, 4 fall into stratum A.

It was hoped that the use of the text miner would show the relationship between digital humanities and open science, but this was not the case. In the analyzed corpus (groupings of texts in German, Spanish, English and Portuguese) many times not even the relationship between the humanities and the digital was evident. This may have occurred because the number of keywords retrieved was small (around 15 results) or due to some other unidentified error in the tool.

Regarding open science findings, the importance of the Data Management Plan (PGD) became evident. For a PGD to be well prepared, there must be documents that guide the researcher, in the different phases of each project, regarding the preparation of a PGD for each area of knowledge, since each area has different specificities.

The standardization of metadata, the use of software that allows versioning of research data, the adoption of permanent links, the attribution of Creative Commons licenses, and, above all, thinking that the data generated must meet the FAIR principles, or that is, data needs to be localizable, accessible, interoperable and reusable, facilitating and making open science possible.

Besides, the creation and development of communication materials, the provision of training aimed at training researchers and citizens, and the collaboration of professionals from different institutions with different backgrounds combined with citizens' participation create a favorable environment for the valorization of open science.

The valorization of open science extends to the spaces that this physical and digital heritage occupies. It expands the importance and gives visibility to the spaces for guarding and preserving memory, such as libraries, archives, and museums. Spaces for education, property protection, democratization, collection, sharing, and memory preservation. Essential spaces for the humanities and the advancement of science.

References

1. de Sousa, M.C.P.: As humanidades digitais globais: Anotações. https://humanidadesdigitais.org/hd2015/anotacoes/. Accessed 13 abr 2019
2. Guerreiro, D., Padre Roberto Busa, S.J.: Bibliotecas e Humanidades Digitais. https://bdh.hypotheses.org/tag/index-thomisticus. Accessed 01 June 2020
3. Dalbello, M.: A genealogy of digital humanities. J. Doc. 67(3), 480–506 (2011)
4. de Almeida, M.A., Damian, I.P.M.: Humanidades digitais: um campo praxiológico para mediações e políticas culturais. In: XVI Encontro Nacional de Pesquisa em Ciência da Informação. João Pessoa, PB (2015)
5. Sula, C.A.: Digital humanities and libraries: a conceptual model. J. Library Admin. 53, 10–26 (2013)
6. Burghardt, M., Wolff, C., Womser-Hacker, C.: Informationswissenschaft und digital humanities. Inf. Wissenschaft Praxis 66(5–6), 287–294 (2015)
7. Koltay, T.: Library and information science and the digital humanities: perceived and real strengths and weaknesses. J. Doc. 72(4), 781–792 (2016)
8. Pimenta, R.M.: Por que humanidades digitais na ciência da informação? Perspectivas pregressas e futuras de uma prática transdisciplinar comum. Informação e Sociedade: Estudos 30(2), 1–20 (2020)
9. Puerta Diaz, M., Bisset Alvarez, E., Marín-Arraiza, P., Dutra, M.L.: Análisis de redes sociales en laboratorios de Humanidades Digitales en Brasil: desarrollo y desafíos de la producción científica. In: X Encuentro Internacional de Investigadores y Estudiosos de la Información y la Comunicación, Cuba (2019)

10. Puerta Diaz, M., Bisset Alvarez, E., Vidotti, S.A.B.G.: Humanidades digitais: visualização da produção científica. In: II Workshop de Informação, Dados e Tecnologia (WIDaT 2018), João Pessoa (2018)
11. Tang, M., Cheng, Y.J., Chen, K.H.: A longitudinal study of intellectual cohesion in digital humanities using bibliometric analyses. Scientometrics 113(2), 985–1008 (2017)
12. dos Santos, P.X., de Almeida, B.A., Henning, P.: Livro verde - ciência aberta e dados abertos: mapeamento e análise de políticas, infraestruturas e estratégias em perspectiva nacional e internacional. Fiocruz, Rio de Janeiro (2017)
13. Albagi, S., Maciel, M.L., ABDO, A.H.: Ciência aberta em questão. In: Ciência aberta, questões abertas. Brasília, IBICT (2015)
14. Packer, A.L., Cop, N., Santos, S.M.: A rede SciELO em Perspectiva. In: Packer, A.L., Cop, N., Luccisano, A, Ramalho, A., Spinak, E. (eds.) SciELO – 15 anos de acesso aberto: um estudo analítico sobre acesso aberto e comunicação científica. UNESCO, Paris (2014)
15. Atlas, T.I.: Huge Variety Of Media Types. https://atlasti.com/product/what-is-atlas-ti/. Accessed 29 Jan 2021
16. GTECH.EDU. Mineração de texto educacional, http://sobek.ufrgs.br/uploads/sobek_quick_reference_guide_pt.pdf. Accessed 30 Sept 2020
17. LARHUD.   Voyant   Tools.   http://www.larhud.ibict.br/index.php?title=Voyant_Tools. Accessed 27 Sept 2020
18. ERCIM.EU.   ERCIM   Flyer.   https://www.ercim.eu/download/ERCIM-flyer-web.pdf. Accessed 04 Sept 2020
19. WEB OF SCIENCE GROUP. Master journal list. https://mjl.clarivate.com/home. Accessed 02 Oct 2020
20. SCOPUS.   Sources.   https://www-scopus-com.ez22.periodicos.capes.gov.br/sources.uri. Accessed 19 Aug 2020
21. CAPES. Critérios de classificação Qualis – Ensino. https://www.capes.gov.br/images/stories/download/avaliacaotrienal/Docs_de_area/qualis/ensino.pdf. Accessed 21 Aug 2020
22. Wells, J.J., et al.: Web-based discovery and integration of archaeological historic properties inventory data: the Digital Index of North American Archaeology (DINAA). Literary Linguist. Comput. 29(3), 349–360 (2014)
23. Blumesberger, S.: Der welt der metaddaten im universum von repositorien. Mitteilungen der VÖB 68(3/4), 515–528 (2015)
24. Bassett, S., Di Giorgio, S., Ronzino, P.A.: Data management plan for digital humanities: the PARTHENOS Model. ERCIM, Special Theme (111) (2017)
25. Wuttke, U., Spiecker, C, Neuroth, H.: PARTHENOS – eine digitale Forschunsinfrastruktur für die Geistes- und Kulturwissenschaften. Bibliothek – Forschung und Praxis 43(1), 11–20 (2019)
26. Hagmann, D.: Überberlegungen zur nutzung von PHAIDRA als repositorium fü r digitale archäologische daten. Mitteilungen der VÖB 71(1), 53–69 (2018)
27. Lathi, L., et al.: Bibliographic data science and the history of tha book (c, 1500–1800). Cataloging Classif. Q. 57(1), 5–23 (2019)
28. Anglada, L.M.: Muchos cambios y algunas certezas para las bibliotecas de investigación, especializadas y centros. El profesional de la informacion 28(1) (2019)
29. Stiegler, J., Klug, H.W.: KONDE – Ein netzwerk bringt forschuns – und GLAM-institutionen zusammen. Ein projektbericht. Mitteilungen der VÖB 72(2), 431–439 (2019)
30. Rollo, M.F.: Desafios e responsabilidades das humanidades digitais: preservar a memória, valorizar o patrimônio, promover e disseminar o conhecimento. Programa Memória para todos. Estudos Históricos 33(69), 19–44 (2020)
31. Ferla, L.A.C., Lima, L.F.S., Feitler, B.: Novidades no front: experiências com humanidades digitais em um curso de história na periferia da grande São Paulo. Estudos Históricos 33(69), 111–132 (2020)

32. Baum, C.: Digital gap oder digital turn? Literaturwissenschaft und das digitale Zeitalter. Zeitschrift für Germanistik **27**, 316–328 (2017)
33. Steinerova, J.: Perceptions of the information environment by researchers: a qualitative study. Inf. Res. **23**(4), (2018)
34. Knöchelmann, M.: Open science in the humanities, or: open humanities? Publications **7**(65), 1–17 (2019)
35. Dos Santos, P.X., de Almeida, B.A.: HENNING, p. 11 (2017)

Research Data Sharing: Framework of Factors that Can Influence Researchers

Elizabete Cristina de Souza de Aguiar Monteiro$^{(\boxtimes)}$ (ID)
and Ricardo César Gonçalves Sant'Ana (ID)

São Paulo State University, Av. Hygino Muzzi Filho, 737, Marília 17525-900, Brazil
ricardo.santana@unesp.br
http://www.marilia.unesp.br

Abstract. Health emergencies contribute to a greater sharing of research data among the scientific community, however, there are other factors that can influence researchers to share or retain their data. The objective is to identify the factors that can influence the perception and attitude of researchers in sharing data sets. The specific objective was to present a framework with the identified factors. Documentary and exploratory research using the method of content analysis and application of the institutional theory, the theory of planned behavior and the research model for data sharing behaviors. The data revealed that indicators of the Cognitive, Normative, Career, Resources and Social pillars influence the perception and attitude of sharing data. It is hoped that the results will allow a perception of the factors that influence researchers to share their data.

Keywords: Research data sharing · Data management · Open Science

1 Introduction

The pandemic caused by Covid-19 motivated the collaboration between agents of the scientific community to share data and research results for the diagnosis, treatment, and development of vaccines with the efficiency that the situation requires. These resources were published in preprints, journals, institutional pages, and open access data repositories.

In the health area, two dimensions of interest can be perceived: social and commercial. Health is a sensitive and strategic area for society living with intense market potential, profitable and competitive, supported by a structure marked by patrimonial law (patents; publication in high impact journals) as well as by moral law (quotation, recognition, and prestige among peers) [1, 2].

In this scenario, premises such as open science and open access to research data contributed to the acceleration of discussions and collaborations by the scientific community. Open Science is an international movement of scientific practice based on open publication, open metadata, open data, open source, open educational resources, open peer review, impact and open metrics, open repositories, including data repositories,

E. Bisset Álvarez (Ed.): DIONE 2021, LNICST 378, pp. 174–180, 2021.
https://doi.org/10.1007/978-3-030-77417-2_13

under FAIR Principles (Findable, Accessible, Interoperable, Reusable). This movement provides the enhancement of the reproducibility of science, greater transparency of public funding, reuse of data in line with society's needs, considering the ethical and legal aspects [1, 3–5].

Open Science seeks to promote the transformation, openness and democratization of science and research. It is a complex concept that requires the adoption of new practices, data sharing and research results, which requires a change in the culture of the actors involved and the behavior of researchers in relation to the development of research, thus enabling the expansion of social and innovation impact [6, 7].

The presented context reinforces the relevance of the collaboration between agents of the scientific community, but also the conflicts and benefits that can emerge with the sharing of data and research results.

It is necessary to identify the different factors that can influence researchers to share or retain their data sets, which include, but are not limited to, institutional, cultural, ethical, social and personal. Data sharing contributes to increasing the transparency of research and, consequently, the credibility of its results, as well as maximizing the replication of data, accelerating scientific collaboration, testing different hypotheses, reducing costs, increasing citations, among other benefits. Once the data is retained, the benefits are not enjoyed.

The general objective of this article was to identify factors that can influence the perception and attitude of researchers to the act of sharing data sets, also proposing a framework with the identified factors.

To achieve the objective and support this discussion, the e-cienciaDatos repository was analyzed. It is a data repository resulting from the cooperation agreement between libraries in the Community of Madrid composed of six universities: Universidad de Alcalá, Universidad Autónoma de Madrid, Universidad Carlos III de Madrid, Universidad Politécnica de Madrid, Universidad Rey Juan Carlos and UNED, members of the Madroño Consortium.

This research presents an initial view of the institutional, cultural and individual factors that can influence the researchers' perception of data sharing and their attitude between sharing or retaining their data sets.

From the results of this study, it is expected to contribute to the enrichment of theoretical discussions about factors such as institutional demands and individual perceptions that can instigate or inhibit the attitude of sharing data sets. The results can also be considered in the training of researchers and in the dissemination of products and services on data management.

2 Methodological Procedures

This article uses documentary research of a descriptive nature, in which laws, regulations, manuals, guidelines, audiovisuals, infographics, statements and reports made available by the Madroño Consortium and its libraries, as well as questionnaires and interviews applied to four librarians who work in the repositories. The questionnaire was composed of 30 questions, 20 of which were closed and 10 open, the last ones following as a script for the interview. The questions addressed aspects of the activities performed with data management and its life cycle, as well as the competencies and skills required to work in the repositories, both in technical services and in user support and training. The documentary analysis was performed between the months of January to May 2020.

With the application of the Content Analysis method and the Categorical Analysis technique, six categories were defined: Cognitive, Normative, Regulatory, Career, Resources and Social. Then, for each category, thirteen indicators were identified that made up a framework [8–10]. The definition of the indicators was defined a posteriori. In order to have reliable categories and indicators for measuring and benchmarking, the search for them was conducted on themes, concepts and characteristics of the research universes. The selection and delimitation of the indicators and categories was supported by the Institutional Theory [11] Theory of Planned Behavior [12] and the Model for data sharing behavior.

3   Framework for Data Sharing Perspectives and Attitudes

Institutional Theory comes from sociological and organizational studies and helps to connect different factors intertwined with institutions, infrastructure and people, in addition to pointing out how actors perform acceptable behaviors to be legitimized in the face of institutional pressures [11, 13].

The Institutional Theory contemplates three pillars [11] that served to define the categories and organize the indicators selected for analysis. The three pillars are:

- Regulator: establishes rules, inspects them and establishes rewards or punishments in an attempt to influence future behaviors.
- Normative: introduction of the prescriptive, evaluative and mandatory dimension in social life, specify how its basic elements should be realized, defining legitimate meanings to the values and rules adopted.
- Cognitive: presence and interaction between the actors, internalized understanding of each actor based on his interpretation of the social reality in which he participates together with the active cultural system.

The Theory of Planned Behavior (TPB) is "[…] a theory designed to predict and explain human behavior in specific contexts […]" pointing as a factor to center that there is "[…] the intention of the individual to perform a certain behavior." [12].

The Theory of Planned Behavior presents a clear relationship of how an individual's attitudes, subjective norms and perceived behavioral controls influence his behaviors, mediated by behavioral intentions [12, 14]. A central factor in the Theory of Planned Behavior is the individual's intention to perform a certain behavior [12]. It is noteworthy that the performance of a behavior is a joint function of intentions and perceived behavioral control.

The Data Sharing Behaviors research model for scientists is a model proposed by Kim (2017) which, according to the author, presents an extensive map of data sharing behaviors of scientists [13]. To support his model, the author used as a basis Scott's Institutional Theory and Ajzen's Theory of Planned Behavior. The model explains how scientists decide to share their data and how the factors that involve sharing differ in different forms of behavior.

The elaboration of the framework presented in this research results from the contribution of the combination of these three theories.

4 Results and Discussions

Data analysis, using the Categorical Analysis of the Content Analysis method and based on the theories presented, resulted in the identification of six categories: Cognitive, Normative, Regulatory, Career, Resources and Social, and 13 indicators: Cultural, Data Management Plan (DMP), Open Science, FAIR Principles, Funding agencies, Repository, Rules and Laws, Benefits, Risks, Efforts, Support from librarians, Benefits to society and Ethics, which made up the framework presented in Fig. 1.

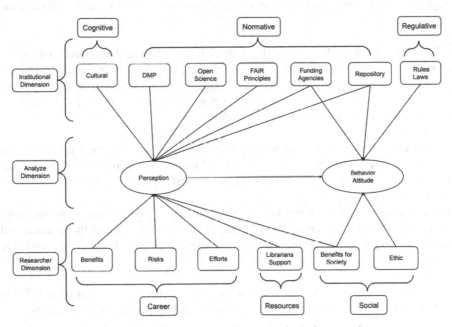

Fig. 1. Data sharing perspective and attitude framework

As illustrated in Fig. 1, the framework is made up of three dimensions: Institutional Dimension; Analysis Dimension; Researcher Dimension.

The Institutional Dimension is based on Scott's Institutional Theory (1995) being composed by the Cognitive, Normative and Regulatory categories, forming the pillars of this dimension and include the indicators:

- Cultural: points out the relationship and interaction between the actors at institutional, national and international level, highlighting the organizational system and social life with the rights, rules and responsibilities;
- Data Management Plan (DMP): document requested by funding agencies and which articulates aspects of data management by researchers;

- Open science: encompasses open scientific practices and changing the behavior of researchers related, among other aspects, to open data sharing;
- FAIR Principles: guiding principles in which they establish and regulate the procedures so that the meta (data) are Findable, Accessible, Interoperable and Reusable (FAIR), in addition to requiring changes in terms of culture and research practice;
- Development Agencies: their guidelines are being articulated according to institutional and social needs, involving the establishment of rules and how they should be applied;
- Repository: formulates policies and guidelines related to data management, conducting data sharing behavior and articulating the demand for other elements of the proposed theoretical model;
- Laws and Norms: related to the researchers' understanding of the legislation and regulations of international management spheres, their country, the university and the repository.

The focus of the analysis of the framework is the Analysis Dimension, where the researcher is. The Analysis Dimension is based on Ajzen's Theory of Planned Behavior (1991) with the perception and attitude of sharing data from researchers in which they are structured according to their context of performance, interweaving all 13 identified indicators [12]. It has two dimensions of analysis:

- Perception: corresponds to the resources and opportunities available to the researcher that impact his intentions and actions and will dictate the probability of sharing the data sets; researchers' perception of the ease or difficulty in carrying out the behavior of interest;
- Data sharing attitude: corresponds to behavior, the act of sharing your data sets. After assessing and perceiving the advantages and disadvantages of the scenario and the social and institutional pressures, the researcher decides between the available alternatives.

The Researcher Dimension is based on Kim's Data Sharing Behavior model (2017) [13] with the Career, Resources and Social categories composed of the indicators:

- Career Benefits: are the benefits perceived and expected from the sharing of data resulting from the researcher's work;
- Career Risks: are the risks perceived in the process of sharing your data;
- Career efforts: are the activities, time and energy spent on managing the data and preparing it for sharing;
- Support from Librarians: corresponds to support from librarians in data management, working with repositories, training researchers and guiding researchers on their data needs;
- Benefits for Society: it is in line with the needs of making data available for the agility to support a certain demand from society, speed in situational analysis and decision making. It includes the scientific community as it contributes to maximize the reuse of data;

- Ethics: indicates the set of values of an area of knowledge and its appropriate behavior or not along with the duties, principles and norms that are compatible with the groups of researchers in their context.

The results show that the framework is suitable for explaining the data sharing behavior of researchers in various fields of knowledge. The framework is going to validated with application to university researchers. It can serve as an instrumental tool to quantitatively measure effects in specific researcher populations or environments.

5   Final Considerations

This article presents theoretical and practical contributions to the study of institutional and personal factors that can influence researchers in sharing their data sets. A framework of factors that could potentially influence the attitude of researchers in sharing or retaining their data sets was developed and presented. The composition of these factors considered elements such as: career, resources, cognitive perceptions, normative and regulatory issues, and social aspects.

The presented framework provides a scenario of basic factors of probable influences that can instigate or inhibit researchers of the various areas of knowledge. Therefore, this study aims to contribute to the direction of future studies on various aspects of data management.

The perception and sharing attitude of researchers may change depending on their region of operation, policies, institutional norms and characteristics of their community. It is likely that studies of the same nature will find different indicators, highlighting the intrinsic particularities of each social context.

References

1. Monteiro, E.C.S.A., Sant'ana, R.C.G., Hernández-Pérez: Direitos autorais de dados científicos no contexto da Ciência Aberta: estudo do repositório de dados do Consórcio Madroño. In: 20th Encontro Nacional de Pesquisa em Ciência da Informação, pp. 1–30, UFSC, Florianópolis (2019). https://conferencias.ufsc.br/index.php/enancib/2019/paper/view/1128/756. Accessed 10 Jan 2020
2. Jorge, V.D.A.: Abertura e compartilhamento de dados para pesquisa nas situações de emergência em saúde pública: o caso do vírus Zika. Universidade Federal do Rio de Janeiro, Rio de Janeiro (2018). https://www.arca.fiocruz.br/bitstream/icict/32604/2/Tese_VanessaJorge.pdf. Accessed 29 Apr 2020
3. Mora, F. (coord.): Compromisos de las universidades ante la Open Science. Madri: CRUE Universidades Espanholas (2019). https://www.crue.org/Documentos%20compart idos/Informes%20y%20Posicionamientos/2019.02.20-Compromisos%20CRUE_OPENSC IENCE%20VF.pdf. Accessed 21 Sept 2020
4. Bezjak, S., et al.: Manual de formação em Ciência Aberta. FOSTER (2018). book.fosterope nscience.eu/pt. Accessed 21 Sept 2020
5. Santos, P.X. (coord.): Livro Verde - Ciência aberta e dados abertos: mapeamento e análise de políticas, infraestruturas e estratégias em perspectiva nacional e internacional. Fiocruz, Rio de Janeiro (2017). https://www.arca.fiocruz.br/handle/icict/24117. Accessed 21 Sept 2020

6. Pontika, N., et al.: Fostering open science to research using a taxonomy and an eLearning portal. In: 15th International Conference on Knowledge Technologies and Data Driven Business, The Open University, Milton Keynes (2015). https://oro.open.ac.uk/44719/2/kmi_fos ter_iknow.pdf. Accessed 21 Sept 2020
7. Ramjoué, C.: Towards Open Science: the vision of the European Commission. Inf. Serv. Use 35, 167–170 (2015). https://pdfs.semanticscholar.org/c47a/f7edc0607c6561390bba2a993c8 9711297c8.pdf. Accessed 21 Sept 2020
8. Bardin, L.: Análise de conteúdo. Rev. e atual. Lisboa: Edições 70 (2010)
9. Fonseca Júnior, W.C.: Análise de conteúdo. In: Duarte, J., Barros, A. (eds.) Métodos e técnicas de pesquisa em comunicação, 2nd edn., pp. 280–304. Atlas, São Paulo (2014)
10. Franco, M.L.P.B.: Análise de conteúdo. 2nd edn. Brasília, DF, Liber Livros (2005)
11. Scott, W.R.: Contemporary institutional theory. In: Scott, W.R. (ed.) Institutions and Organizations, Chap. 3, pp. 32–62. SAGE, London, Thousand Oaks (1995)
12. Ajzen, I.: The theory of planned behavior. Org. Behav. Hum. Decis. Process 52(2), 179–211 (1991)
13. Kim, Y.: Fostering scientists' data sharing behaviors via data repositories. J. Suppl. Pers. Commun. Methods Inf. Process. Manage. 53(4), 871–885 (2017). https://linkinghub.elsevier. com/retrieve/pii/S0306457316305908. Accessed 21 Sept 2020
14. Kim, Y., Zhang, P.: Understanding data sharing behaviors of STEM researchers: the roles of attitudes, norms, and data repositories. Libr. Inf. Sci. Res. 37(3), 189–200 (2015). https://lin kinghub.elsevier.com/retrieve/pii/S0740818815000584. Accessed 21 Mar 2020
15. Kim, Y., Adler, M.: Social scientists' data sharing behaviors: Investigating the roles of individual motivations, institutional pressures, and data repositories. Int. J. Inf. Manage. 35(4), 408–418 (2015). https://www.sciencedirect.com/science/article/pii/S02684012 15000432. Accessed 21 Mar 2020

The Dilemma of Fake News Criminalization on Social Media

Marcio Ponciano da Silva$^{(\boxtimes)}$ (iD) and Angel Freddy Godoy Viera (iD)

Universidade Federal de Santa Catarina, Florianópolis, SC, Brazil
mponcianos@gmail.com, godoy@ufsc.br

Abstract. The present research analyzes the dilemma faced in Brazil about the criminalization of fake news disseminated mainly on social networks. The growth of social networks use around the world promotes more skillful ways of communicating and spreading true or false information in the information society. The objective of this study is to examine the trend of fake news spreading, coined by fake news, and the reaction of Brazil seeking to stop this dissemination through a criminalization process, as a way to confront this phenomenon. This study investigates effects of fake news around the world including Brazil, seeking to examine social changes and the role of Brazilian government in confronting this phenomenon. In Brazil, this battle starts even before a criminal process, as it begins in the criminal investigation carried out by the Brazilian police. However, it is through criminal proceedings that Brazil solves conflicts that arise from criminal conduct defined by law. The dilemma is that no Brazilian law considers fake news crime. In addition, the relationship between the information society and the spread of fake news on social networks is addressed in the research, bringing a reflection about freedom of speech and maintaining the individual responsibility of each individual. Aspects about the challenges of police investigation concerning fake news effects were addressed taking into consideration the issue of related crimes commited on social networks, such as copyright infringement and defamation, benefiting from the reach of information on digital communication networks.

Keywords: Fake news · Police investigation · Criminalization

1 Introduction

The social transformations that have outlined the current information society especially due to the evolution of technological development encourage Brazil to reflect on changes in procedures, aiming to adapt itself to aspects of the new scenario that has been built from these social transformations. Technological development has been an important lever for these changes.

Social networks are an open space for exposition of opinions and manifestations of thought in general. It transports a large volume of information on the communication networks, which is rapidly renewed due to this volume. The freedom of speech is an important right warranted in democratic countries. At the same time, protecting the moral

E. Bisset Álvarez (Ed.): DIONE 2021, LNICST 378, pp. 181–194, 2021.
https://doi.org/10.1007/978-3-030-77417-2_14

integrity and inviolability of constitutional guarantees are duties of the State and in the case of Brazil ensured in its Constitutional Charter.

This study examines these social transformations and the mobilization of the State to outline its role given the effects of these transformations. Faced with the phenomenon of fake news, Brazil faces the dilemma of criminalizing fake news disseminated mainly on social networks. The law proposal of turning fake news into criminal act is ongoing in the Brazilian parliament [1]. About Brazilian police's work activity, the modernization of police investigation procedures and techniques and the criminal process is an urgent need to engage in the fight against fake news.

These transformations in the information society brought changes in social habits especially about advertising in public spaces such as social networks. With this portrait of transformations the effects produced by them came, among them the emphasis on individuality, that its fluid characteristic was labeled as modern and liquid [2].

The awakening of the world by a sort of thirst for knowledge coming from the information absorbed deep in what was conventionally called modernity. Many social transformations resulting from the confrontation with the possibilities of knowing, portray a modern society markedly modified by the reality of a post-war environment, which, as described by Bauman [2], was follow by unprecedented growth in which wealth and security were established. He also states that in this "fluid world" individuals can exercise their freedom of choice.

A democratic country has to deal with possible conflicts between modern liquid behavior and the limit of individual freedom. In this context, there are legal instruments that act to safeguard the maintenance of collective public order, such as the application of a temporary or preventive prison sentence.

Among the challenges examined, the question of copyright, which ends up being violated in various ways and constantly threatened by new means of improper dissemination, characterizes a related crime, that is the crime of piracy. The evolution of how intellectual creations were made available brought more modern means of publishing works, but it also brought problems of copyright.

Among the list of challenges, perhaps the main one is the phenomenon of fake news, and how this phenomenon is affecting both direct freedom of speech and the inviolability of the individual's moral integrity. In addressing the theme manifestations of the country's authorities on the subject are seen. Some public security initiatives are presented, specifically regarding the modernization of criminal investigation, which aims to combat cybercrime.

Throughout the Internet, wide dissemination of documents, images, audios, and others are launched across borders. According to Castells [3], "what characterizes the current technological revolution is not the centrality of knowledge and information, but the application of this knowledge and information, in a feedback loop".

The information technology revolution [3], does not refer only to the volume of information and the speed that its reach has demonstrated, but to how much knowledge has been produced from the expansion of information. This is a discussion about the migration from an information society to a knowledge society. What to do with the information is the object of social debate nowadays.

2 Methodology

This research has exploratory nature. It seeks to examine social changes and the role of Brazilian State struggling with the challenges of the information society arising from these transformations, such as fake news. The work followed the method of bibliographic research, having started with consultations on the Scopus and Web of Science databases, between the years 2018 and 2020, in Portuguese and English, in addition to published literary works, thus seeking to approach the theme and identify important facts for the analysis of the study. The second step was the examination of literary sources that address the proposed theme, and with that constructed the review of the study literature. Another step was to analyze research information that shows the incidence of fake news in the world, and also analyze the draft law in Brazil which is an initiative aiming the criminalization of fake news in the country. On the final, step the final considerations were present.

3 Fake News as a Manipulation Tool

The spread of fake news has no boundaries. Fake news occupies a place of concern around the world. This section discusses the problem of fake news worldwide and Brazil.

3.1 Fake News in the World

The discussion about fake news is present in democratic countries. The global movement called Transparency International published the Global Corruption Barometer - Latin America and the Caribbean [4], where it discusses citizens' opinions and experiences related to corruption.

This survey was carried out between January and March 2019, with over 17 thousand people, and presents issues concerning the fight against corruption in 18 countries. One of the survey's questions asked participating people if they thought corruption had increased in their countries in the past 12 months. Half of the countries had rates above 50%. Table 1 shows which countries were those:

Table 1. Countries with higher than 50%.

Countries	Percent
Venezuela	87%
Dominican Republic	66%
Peru	65%
Trinidad and Tobago	62%
Panama	56%
Brazil	54%
Chile	54%
Honduras	54%
Colombia	52%

The countries that were below this index were: Argentina (49%), Costa Rica (49%), Jamaica (49%), Guatemala (46%), El Salvador (45%), Bahamas (45%), Mexico (44%), Guyana (40%) and Barbados (37%). The theme of this requirement dealt with the rising sensation of corruption. One of Barometer's published findings is that the spread of fake news during elections is one of the causes of political corruption.

The Global Corruption Barometer brings in its first recommendation two points that have been appearing together in discussions around the world: elections and fake news. It proposes the defense by the integrated police in elections and control of fake news, as an anti-corruption and democracy-strengthening measure.

Hendricks and Vestergaard [5] present fake news as an old phenomenon, mentioning the visit of Benjamin Franklin to Paris in 1782, one of the leaders of the American Revolution, to negotiate peace between England and the USA. He says that Benjamin Franklin published a false report to mobilize public opinion in England and Europe. The author presents fake news as a weapon to gain political power and advantages.

The concern with the dissemination of fake news has taken the peace of the world by the fact that this practice interferes in the elections of the countries. Faustino [6], points to the 2016 US elections as a great example of the spread of fake news. The author states that the spread of fake news on social networks was responsible for the final result of that election.

Faustino [6] points out the use of fake news as a political tool, whose objective is the manipulation of public opinion, through social networks. To control the spread of publications that violate rights, the United States of America counts on the Digital Millennium Copyright Act (DMCA). This regulation allows the removal of content from the Internet that violates copyright, fake news and others.

The application of the DMCA takes place under the pressure of strong criticism, as it is compared to a form of censorship. The 1996 Telecommunications Act exempts Internet service providers from blame for content generated by customers.

Another example of fake news was during the period of the great earthquake in eastern Japan, a lot of fake news were spread on Japanese Twitter [7]. During the 2017 German election, several false stories circulated on social media [5]. These stories were polarizing and worked fostering division and distrust.

The United Nations Development Program (UNDP) published a report on its portal that deals with "Accountability in the era of 'disinformation': going beyond transparency in Latin America and the Caribbean" [8]. According to this report, the intentional creation of the dissemination of false information promoted the breaking of the information value chain in Latin America and the Caribbean. Many of these cases are related to facts produced intentionally in the field of politics.

3.2   Fake News in Brazil

Accordingly to a research released by the Brazilian Federal Senate social networks are responsible for influencing the vote of 45% of the population [9]. The research points out that the contribution in influencing opinions comes from technological tools, such as Facebook, WhatsApp, YouTube, Instagram, and Twitter. The result released in this survey indicates that citizens are aware about the fake news problem, but 47% of

Brazilians agree that it is difficult to identify the veracity of information coming from social networks.

Galhardi et al. [10] concluded in their study that WhatsApp is the main channel for sharing fake news, followed by Instagram and Facebook. Also according to these authors, the phenomenon of fake news in Brazil during Covid-19 contributes to discredit science and global public health institutions. This study reports the contribution on the use of the *Eu Fiscalizo* smartphone app, created by researcher Claudia Galhardi, from the National School of Public Health, of the Oswaldo Cruz Foundation (Fiocruz) to face fake news in the contexto of the Covid-19.

Fiocruz is a public foundation linked to the Brazilian Ministry of Health. The launch of *Eu Fiscalizo* had the objective of simplifying communication between society and the government in the face of the proliferation of fake news in the media, especially in social networks.

For Faustino [6], the 2018 elections in Brazil suffered from the same problem as the North American elections in 2016. He states that in the 2018 presidential election there were several instances of fake news involving the two main candidates of that election Jair Bolsonado and Fernando Haddad. The Brazilian Electoral Court in some occasions in these elections determined the withdrawal of content with the dissemination of fake news.

Electoral disputes in Brazil always seem to be on the rise. Internet publications were collected on operations carried out by the Brazilian Federal Police whose theme was elections. In this search, 87 operations were found between 2010 and 2017. Table 2 presents these findings.

Table 2. Brazilian Federal Police operations on elections.

Years	Operations	Prisons	Deponents
2010	2	5	0
2012	3	0	6
2013	1	0	0
2014	3	3	38
2015	3	3	0
2016	52	69	89
2017	23	14	11

The content of Table 2 shows the growth of Federal Police operations over the years. This survey also shows an increasing number of arrests and people carried along to provide further explanation.

In a survey conducted by the Ipsos institute in 2018 [11], 62% of interviewed Brazilians admitted to having falsely believed that a news report was real until they discovered it was false. The world average for this item is 48%, which means that Brazil is above the average.

When asked if people upon hearing the term "fake news" think that these are stories in which the media or politicians only choose facts that support their side of the argument, the answer was only 25%. This index contradicts the result of several surveys that point to the elections as an open field for taking advantage of the use of fake news.

The Reuters Institute produced a report for the year 2019 [12] which gathers information about fake news. According to this report, WhatsApp takes the place of the main social network for exchanging news. In this survey, Brazil appears with 53% of those who responded to using WhatsApp as the main source of news and discussion, alongside Malaysia (50%) and South Africa (49%).

All this information pointed out in research and news show that the concern with the phenomenon of fake news dominated the agenda of countries worldwide. It was no different from Brazil, which has also been reacting to the phenomenon.

4 The State's Duty in the Information Society

Disinformation has a detrimental impact on society, such as diverting the focus of truly important problems. Faced with a negative impact scenario that spreads violence, confusion, and false accusations, it is important to find the role of the State in the information society.

4.1 Context of Application of Police Investigation to Combat Fake News

In the midst of all these transformations in the era of the information society, there is the investigation by police forces that must collect and analyze information from social networks, to curb fake news to preserve collective rights. In Brazil, the product of this investigation is Police Inquiry.

As is known, it is thanks to the technology that information produced in the world has been multiplying. The growth in this volume of information is largely proportional to social networks. For Faustino [6], social networks promote relationships by generating bonds in cyberspace.

The society that the modern world has produced has given each individual its own space of protagonism, where each one can plead their private interests. Along this path, the challenge of achieving satisfaction is in constant flux. Consumerism is no longer about the satisfaction of needs, but of desire [2].

In this context of striking individuality, responsibility is also present. Reflection on freedom, which stands for both good and evil, reveals this coexistence that needs to be conciliated. It is not a question of incompatibility or contradiction, but ambiguity. There is an implicit limit, decisively imposed on individuality, freedom does not overlap society [2].

It is exactly between individual freedom and the collective social order that the invisible limit of the use of information lies, which goes beyond public and private. The State, through police investigation, will act in the incidents that exceed this limit.

On the path to follow and technological evolution, or revolution, resulting from the transformations caused by the information society, the Brazilian State has been seeking to update the form of its procedures. An example of this update is the creation of the electronic process.

For Greco and Martins [13], from the point of view of the observer of the operator in the field of law, it is not possible to predict the advances that may occur in the field of technology in the following years. Although the work of these authors was published in 2001, the same observation is valid, as technology is constantly evolving.

It is in this context that information is collected and produced for police investigation, that is, the information from the police investigation is, at least in part, the product of the information society. The State must focus on social changes. It is up to the jurist to monitor the economic and technological revolution to recognize the changes arising from these transformations and be able to adopt all necessary measures [13].

4.2 Project to Criminalize Fake News in Brazil

Since 1995, there has been a consultative body in Brazil that acts in the integration of Internet-related activities in the country. It is the Internet Steering Committee (Comitê Gestor da Internet - CGI). Resolution CG 003/2009 originated from the CGI's regular meetings, which establishes ten principles related to internet governance in Brazil that came to be known as "The Decalogue of the Internet".

Brazil was already showing signs of concern with the regulation of principles for the use of the internet in the country. Brazilian law 12/965/2014 is known as *Marco Civil da Internet* (Civil Milestone law for the Internet). The previous launch of the Internet Decalogue served as a solid source for the emergence of the Civil Milestone law for Internet, which establishes principles, rights, and duties for the use of the Internet in Brazil.

The phenomenon of fake news results in a great demand that spreads quickly thanks to social networks. As previously seen, the WhatsApp application stands out among those that most provide fake news in Brazil. The theme led the president of the Superior Electoral Court (Tribunal Superior Electoral - TSE), Minister Luiz Fux at the time, to highlight in a lecture at the International Seminar on Fake News how much fake news are harmful to the democratic environment and, above all, to elections [14]. In his speech, he added the intention of being absolutely tireless in the fight against fake news.

The Brazilian Federal Police presented at the Parliamentary Commission of Inquiry (Comissão Parlamentar de Inquérito - CPI) the Tentacles Project, which is an internal initiative to improve the efficiency of cases involving cybercrimes. According to this report, in the case of this police investigation, Project Tentacles avoided up to 582 police investigations [15].

During the 2018 Brazilian presidential elections, the Federal Police signed a partnership with the United States Federal Bureau of Investigation (FBI), with work carried out in Brazil and the United States, to jointly investigate cybercrimes. This measure also included combating the use of fake news [16].

At the end of 2018, the General Law on the Protection of Personal Data (Lei Geral de Proteção de Dados Pessoais - LGPD) came into force, placing Brazil in the group of countries that have their own legislation to defend the privacy of citizens. Faustino [6] states that freedom of speech and privacy are two constitutional principles that are complied with in the LGPD because it deals with personal data.

When talking about the anonymity and identity of the people behind the publications on social networks, Faustino [6] states that the Brazilian Federal Constitution guarantees freedom of expression of thought, but prohibits anonymity. This statement is in line with Bill of Law 2630/20.

Today, Bill 2630/20, which establishes the "Brazilian Law on Internet Freedom, Responsibility and Transparency", is under discussion in Brazil [17]. This project aims to establish rules on transparency in social networks.

The draft of this bill several times mentions "untagged content", referring to the prohibition of anonymity in the dissemination of information. The text proposed in this bill makes it clear that the objective is to make the source of any dissemination of news identifiable.

In addition to providing for the responsibilities and duties of application providers in combating disinformation, the original text of this bill provides for several penalties for application providers. Table 3 presents some of these sanctions included in article 28 of the project.

Table 3. Penalties provided for application providers.

Items	Sanction
I	Warning
II	Fine
III	Temporary suspension
VI	Exercise ban

The gradation of the sanction will depend on some factors, such as how serious the fact is, whether it is a repeat offender, and the economic capacity of the offender. In the final provisions, there is still a provision to insert in the Administrative Improbity Law, Law 8,429/1992, the act of disseminating or competing for the dissemination of disinformation, through inauthentic accounts, artificial disseminators, or artificial disinformation dissemination networks.

Given this, the aforementioned bill deals with the criminalization of the spread of fake news. This bill has already been approved in the Federal Senate. He is currently awaiting agenda for discussion at the Chamber of Deputies.

5   The Problem of Fake News in the Supreme Court of Brazil

In March 2019, after becoming aware of offenses against the Supreme Court (Supremo Tribunal Federal - STF) of Brazil, which is the highest Court in the country, the presiding minister at the time, Dias Toffoli, opened the criminal inquiry (Inquérito Criminal - INQ) 4781 [18], known as the Fake News Survey. The purpose of this investigation is to examine mainly the existence of fake news, but also slanderous denunciations, threats, and others.

The opening of this criminal investigation in the Supreme Court was surrounded by controversy, mainly because there was no indication of act performed that would pose a real threat to the Supreme Court. This was the object of the lawsuit of Arguition of Non-Compliance with Fundamental Precept (Arguição de Descumprimento de Preceito Fundamental - ADPF) [18], which questions the legality of the STF president's act. The ADPF 572 was proposed by the political party named *Rede Sustentabilidade*.

Another question of ADPF 572 was the fact that the investigation is not subject to free distribution, as determined by the Internal Regulation of the STF itself. The president of the Supreme Court directly appointed Minister Alexandre de Morais to conduct the investigation. ADPF 572 has already been judged by the plenary of the Court and the decision was based on the fact that the action was not applicable [18].

The controversy was not only restricted to political parties but extended to the Legislative Branch and even to civil society. The article "Senators criticize STF for validating fake news inquiry" [19] demonstrates the scope of this controversy. According to the article, senators criticized the fact that the Supreme Court opened the case, being a victim and judges of the Court itself.

Also according to the matter, for senator Marcos Rogério (DEM-RO), the initiative of the president of STF is a violation of the accusatory system. The main feature of the accusatory system is the separation between prosecution and trial functions [20]. Although the Supreme Court has repeatedly reaffirmed in its decisions the adoption of the accusatory system, this has not been an obstacle to open an internal investigation that the Court itself will judge.

It is important to remember that in Brazil, legal scholars differ in terms of the criminal system adopted by the country. There are basically three systems of criminal prosecution: inquisitorial, accusatory, and mixed.

In the case of *Instituto Politeia* [21], journalist André Borges Uliano also attacked the initiative of the president of the STF, claiming that it was illegal. In this matter, five reasons are presented that justify the opening of the investigation to be abusive. Table 4 presents these reasons.

Table 4. Reasons that classify the opening of the INQ 4781 in the Supreme Court as absurd..

Reason	Justification
1	The purpose of the investigation is undefined
2	Minister's appointment violates free distribution
3	Lack of Supreme Court attribution for the case
4	Opening in the STF violates the accusatory system
5	The investigation violates freedom of speech

The article was broadcasted with the title "Understanding why the investigation initiated by Dias Toffoli is illegal" [21]. However, since the Fake News Inquiry was opened by the Supreme Court, the many opposing positions were frustrated.

This episode of fake news in Brazil essentially involves politicians who maintain a blog profile to interact with their voters. There are also entrepreneurs involved in the defense of candidates in which they provided support.

The discussion about the legality of the Fake News Inquiry also involved Brazilian President Jair Bolsonaro. Speaking at a live broadcast on 05/29/2020 [22], President Jair Bolsonado stated that the Supreme Court Fake News Inquiry is unconstitutional and has no legal basis. The president argued that political allies were being targeted without justified reasons, as a possible result of retaliation conducted by the Fake News Inquiry.

This episode gave the topic of fake news the ability to bring together the three Branches (Executive, Legislative and Judiciary) of the Federative Republic of Brazil around it with differences of opinion.

6 Initiatives to Combat Fake News

The spread of fake news has driven public institutions, governments, the press, and various sectors of society to adopt measures to face this phenomenon. As a result, initiatives have emerged to combat the spread of false information.

The initiatives involve the public and private sectors, in addition to civil society itself. Given that fake news are at the center of discussion around the world, it may be that the development of measures to combat fake news are evolving.

6.1 International Seminar "Fake News and Elections"

Aiming at the integrity of electoral process, but also at the freedom of speech, the Brazilian Superior Electoral Court created the Consultative Council on Internet and Elections at the end of 2017. This Council was created to develop research and studies aiming at elections of the following year [23].

As a measure to combat fake news, the TSE signed agreements in 2018 with political parties whose commitment was to promote an environment of informational health. TSE also entered into partnerships with political marketing experts, intending to promote transparency in elections.

In 2019, the Brazilian Superior Electoral Court promoted the second International Seminar "Fake News and Elections" (the first Seminar took place in 2018). This 2019 event was supported by the European Union and brought together experts on the topic of fake news. This meeting of experts, which includes political figures, lawyers, journalists and academics, discussed contemporary themes such as electoral law and the limits of advertising, as well as freedom of speech versus crime against honor.

The fourth panel of debates took place under the theme "Tools to fight back fake news". In this theme, some initiatives were mentioned, such as the *Comprova* Project [24], designed to combat disinformation in an election year that works with the voluntary participation of journalists checking the news on social networks.

The result in the 2018 elections, with three months of work, was to identify 147 rumors, where less than 10% were based on real facts. Another initiative is fact-checking agencies (that verifies statements given to the press) and Debunking (checking material that has no origin). These agencies are affiliated and audited by the International Fact-Checking Network (IFCN).

As seen in that Seminar, the use of open-source research techniques allowed the identification of 404 profiles with traces of discrepancies concerning common Brazilian accounts, such as accounts that tweet in another language, but in the electoral period, they publish in Portuguese. In the discussions of this fourth panel, it was clear that the evolution of tools to combat fake news involve the use of Artificial Intelligence (AI).

Among the actions of the Superior Electoral Court in the fight against disinformation in the elections, the monitoring of social networks is one of them. This work was recognized by the Mixed Parliamentary Commission of Inquiry of the Brazilian National Congress [25].

6.2 CNJ Fake News Check Panel

On the initiative of representatives of the National Council of Justice (CNJ), of the magistrates' associations and the superior courts and the press, in April 2019, the Fake News Checking Panel was created to make the population aware about the dangers of fake news [26]. With this initiative, each partner contributes to the project with the tools they have.

This initiative was responsible for the campaign of using the hashtag #FakeNewsNo in the dissemination of posts on social networks, in videos and texts. 2 million tweet impressions were recorded in one month of this campaign. Other hashtags were used, such as #FakeNews and #FakeNewsPerigoReal.

This project counts on the participation of the Brazilian citizen through the e-mail fakenewsnao@cnj.jus.br. The result of this work helps to subsidize the content checking of the Courts involved. The project's interest is to expand the benefit to other Courts.

6.3 Coping with Fake News in Healthcare

In the midst of the coronavirus pandemic called SARS-CoV-2 which caused COVID-19 the world has to deal with the spread of fake news another evil that is also harmful. Content related to fake news has taken up a lot of space on social networks. From false allegations about governments and celebrities, to fake news that Microsoft's co-founder Bill Gates was behind the spread of the virus and that Italians accused the Chinese of bringing the disease to their country. There is no truth in any of this information [27]. The misinformation about COVID-19 ranges from conspiracy theories as a biological weapon in China, to information that coconut oil kills the virus [28].

In Brazil, the Ministry of Health has taken the initiative to make available a WhatsApp contact number to combat Fake News on health [29]. The initiative aims to collect suspicious messages from the population. Thus, viral information will be investigated by technical areas. After analysis, responses classified as false or true are published.

A count of 84 news reports were analyzed during 2020 and had already been released by the Ministry of Health. The news is published with the label "THIS IS FAKE NEWS!", when it comes to fake news, and labeled "THIS IS TRUTH!", when they are true. Among this news, 79 were classified as fake news against 5 news classified as true [30].

7   Final Considerations

This work addressed the aspects of the information society and its relationship with technological evolution that allowed the expansion of the reach of social networks, as well as the effects that resulted from this evolution. One of these effects is the main topic addressed, namely, the spread of false information on social networks, already known worldwide as fake news.

The analyzed information suggests that fake news may be having a detrimental effect on elections worldwide, acting as an instrument of corruption. For this reason, the spread of fake news has posed a threat to elections in democratic countries. This problem requires the defense of political integrity and the democratic system itself.

The analysis also indicates that Brazilians are susceptible to the diffusion of fake news, making Brazil a fertile ground for the spread of fake news. This study presented several aspects of how this theme invaded the Brazilian daily agenda.

The clash over fake news that has been developed points to the criminalization of the practice of spreading fake news. The edition of the Project of the "Brazilian Law of Freedom, Responsibility, and Transparency of the Internet" is currently the main initiative to regulate fake news. However, the outcome of this attempt to characterize fake news as a crime is still uncertain, until the law proposal that is being processed by the Legislative Branch is transformed into law.

If, on the one hand, Brazil does not yet have a law that typifies fake news, the challenges of police inquiry in the information society examined in this study reveal that there is an urgent need to modernize police investigation procedures and techniques. It was seen that there are already initiatives in this direction, such as the development of projects to deal with cybercrimes that makes work more effective and the building of partnerships with other international bodies, as was done with the FBI.

Fake news has left an institutional crisis trail in Brazil among the Branches of the Republic. The expectation is that by regulating a law to deal with fake news these divergences are pacified. This research identified some government initiatives to combat fake news. They complement measures to combat fake news and outline the role of the State in this confrontation.

Measures to confront fake news have been taken around the world. In Brazil, government institutions, the press, volunteer journalists, and even a portion of the population have been engaged in this activity.

References

1. Combate a fake news é tema de 50 propostas na Câmara dos Deputados – Notícias. https://www.camara.leg.br/noticias/666062-combate-a-fake-news-e-tema-de-50-propostas-na-camara-dos-deputados/. Accessed on 30 Oct 2020
2. Bauman, Z.: Modernidade Líquida. Zahar (2001)
3. Castells, M.: A sociedade em rede. Paz & Terra, São Paulo, p. 69 (1999)
4. Pring, C.: Global corruption barometer, Latin America & the Caribbean 2019 - citizens' views and experiences of corruption. (2019)
5. Hendricks, V.F., Vestergaard, M.: Reality Lost: Markets of Attention, Misinformation and Manipulation. Springer International Publishing, Cham (2019). https://doi.org/10.1007/978-3-030-00813-0

6. Faustino, A.: Fake News: A Liberdade de Expressão nas Redes Sociais na Sociedade da Informação. Lura Editorial (2020)
7. Zhao, Z., et al.: Fake news propagates differently from real news even at early stages of spreading. EPJ Data Sci. 9(1), 1–14 (2020). https://doi.org/10.1140/epjds/s13688-020-002 24-z
8. Prestação de contas na era de "desinformação": indo além da transparência na América Latina e no Caribe| PNUD Brasil, https://www.br.undp.org/content/brazil/pt/home/pressc enter/articles/2019/prestacao-de-contas-na-era-de–desinformacao—indo-alem-da-tran.html. Accessed 25 Oct 2020
9. Redes sociais influenciam voto de 45% da população, indica pesquisa do DataSe-nado. https://www12.senado.leg.br/noticias/materias/2019/12/12/redes-sociais-influenciam-voto-de-45-da-populacao-indica-pesquisa-do-datasenado. Accessed 30 Oct 2020
10. Galhardi, C.P., et al.: Fato ou Fake? Uma análise da desinformação frente à pandemia da Covid-19 no Brasil. Ciênc. saúde coletiva. 25(suppl 2), 4201–4210 (2020). https://doi.org/10. 1590/1413-812320202510.2.28922020
11. Global Advisor: Fake News. https://www.ipsos.com/pt-br/global-advisor-fake-news. Accessed 30 Oct 2020
12. Newman, N.: Reuters Institute Digital News Report 2019. P. 156 (2019)
13. Greco, M., Martins, I.: Direito e internet: Relações jurídicas na sociedade informatizada. Editora Revista dos Tribunais, São Paulo (2001)
14. Seminário Internacional sobre Fake News: Luiz Fux afirma que não existe voto livre sem liberdade de opinião. https://www.tse.jus.br/imprensa/noticias-tse/2018/Junho/seminario-internacional-sobre-fake-news-luiz-fux-afirma-que-nao-existe-voto-livre-sem-opiniao-livre. Accessed 30 Oct 2020
15. Comissão Parlamentar de Inquérito: Operação IB2K Projeto Tentáculos. https://www2.cam ara.leg.br/atividade-legislativa/comissoes/comissoes-temporarias/parlamentar-de-inquerito/ 55a-legislatura/cpi-crimes-ciberneticos/documentos/audiencias-publicas/audiencia-publica-dia-20-08.15/dr-stenio-santos-delegado-de-policia-federal-chefe-do-grupo-de-repressao-a-crimes-ciberneticos-sdpf-df. Accessed 20 Oct 2020
16. Aprovada lei que dá à PF atribuição de investigar crimes virtuais contra... – FENAPEF. https://fenapef.org.br/aprovada-lei-que-da-a-pf-atribuicao-de-investigar-crimes-virtuais-contra-mulheres. Accessed 30 Oct 2020
17. PL 2630/2020 - Senado Federal. https://www25.senado.leg.br/web/atividade/materias/-/mat eria/141944. Accessed 25 Oct 2020
18. Plenário conclui julgamento sobre validade do inquérito sobre fake news e ataques ao STF. http://portal.stf.jus.br/noticias/verNoticiaDetalhe.asp?idConteudo=445860. Accessed 30 Oct 2020
19. Senadores criticam STF por validar inquérito das fake news. https://www12.senado.leg.br/ noticias/materias/2020/06/18/senadores-criticam-stf-por-validar-inquerito-das-fake-news. Accessed 26 Oct 2020
20. Avena, N.: Processo Penal. 10. ed, São Paulo (2017)
21. Entenda por que o inquérito instaurado por Dias Toffoli é ilegal. https://www.gazetadopovo. com.br/instituto-politeia/inquerito-toffoli-ilegal/. Accessed on 26 Oct 2020
22. Veja íntegra da live do presidente Jair Bolsonaro de 29/05/20 (2020). https://www.youtube. com/watch?v=r0HkpczjmrE. Accessed 26 Oct 2020
23. Seminário Internacional Fake News e Eleições: anais (2019, Brasília, DF). https://www.tse. jus.br/hotsites/catalogo-publicacoes/pdf/livro-digital-fake-news.pdf. Accessed 30 Oct 2020
24. Projeto Comprova - jornalismo colaborativo contra a desinformação. https://projetocompr ova.com.br. Accessed 17 Jan 2020

25. Relatora da CPMI das Fake News elogia ações do TSE de combate à desinformação – Notícias. https://www.camara.leg.br/noticias/697452-relatora-da-cpmi-das-fake-news-elo gia-acoes-do-tse-de-combate-a-desinformacao/. Accessed 30 Oct 2020
26. Painel de Checagem de Fake News. https://www.cnj.jus.br/programas-e-acoes/painel-de-che cagem-de-fake-news/. Accessed 30 Oct 2020
27. Frenkel, S., et al.: Surge of Virus Misinformation Stumps Facebook and Twitter (2020). https://www.nytimes.com/2020/03/08/technology/coronavirus-misinformation-social-media.html
28. Pennycook, G., et al.: Fighting COVID-19 misinformation on social media: experimental evidence for a scalable accuracy-nudge intervention. Psychol. Sci. 31(7), 770–780 (2020). https://doi.org/10.1177/0956797620939054
29. Fake News. https://antigo.saude.gov.br/fakenews/?start=20. Accessed 30 Oct 2020
30. Novo Coronavírus Fake News. https://antigo.saude.gov.br/component/tags/tag/novo-corona virus-fake-news. Accessed 30 Oct 2020

Online Data Processing Technologies

Evaluating the Effect of Corpus Normalisation in Topics Coherence

Luana da Silva Sousa$^{(\boxtimes)}$, Vinicius Melquiades de Sousa ,
Rogerio de Aquino Silva , and Gustavo Medeiros de Araújo

Engineering and Data Science Lab, Federal University of Santa Catarina, Florianópolis, Brazil
gustavo.araujo@ufsc.br

Abstract. Probabilistic topic models are extensively used to better understand the content of documents. Due to the fact that topic models are totally unsupervised, statistical and data driven, they may produce topics not always meaningful. This work is based on the hypothesis that, since LDA takes into account the number of occurrences of words, we could affect the quality of topics by semantically normalising the text, where each concept would be represented by the same word. We can find a formal description of lexemes found in text using a knowledgebase and extract the several forms of mentioning a lexeme to normalize a corpus. We use topic coherence metric, as it represents the semantic interpretability of the terms used to describe a particular topic, to quantify the influence of semantic corpus normalisation in topics. The first tests on the semantic normalisation framework of texts showed prominent results, and shall be investigated in depth in future.

Keywords: Corpus normalisation · LDA · Topic coherence · Ontology · Natural language processing

1 Introduction

Extracting useful information from large collections of text documents has become more challenging in recent years.

Understanding and modeling the content of documents can be very useful in many applications, such as information retrieval, natural language processing (NLP), document classification, text summarization, etc. [2].

The foundation in statistics and its capability to be extended and combined with other models make probabilistic topic model one of the most used algorithms to deal with these problems [2]. Topic modeling is a form of finding latent semantic structure within a collection of documents, and probabilistic models, such as Latent Dirichlet Allocation (LDA), have become the standard method employed [6, 20]. The intuition is that pairs of descriptor terms that co-occur frequently or are close to each other within a semantic space are likely to contribute to higher levels of coherence for a specific topic [20]. LDA model has been criticized for favoring highly frequent, general words in topic descriptors [20]. Due to the fact that topic models are totally unsupervised, statistical and data driven, they may produce topics not always meaningful [2].

© ICST Institute for Computer Sciences, Social Informatics and Telecommunications Engineering 2021
Published by Springer Nature Switzerland AG 2021. All Rights Reserved
E. Bisset Álvarez (Ed.): DIONE 2021, LNICST 378, pp. 197–208, 2021.
https://doi.org/10.1007/978-3-030-77417-2_15

Higher interconnectivity between information sources has the potential of increasing the utility of information. By connecting unstructured information in text documents with structured semantic data available on the internet, facts from this huge Web of Data can be used to enhance several tasks such as information extraction [25], information retrieval [27, 28], text classification [10], feature extraction [13], etc.

The goal of this work is to present the first results on a text semantic normalisation framework. Our work was based on the hypothesis that, since LDA takes into account the number of word occurrences, we could affect the quality of topics by semantically normalising the text, where each concept would be represented by the same word. If the same concept is represented by two different words in different texts, the algorithm would probably struggle more to find coherent topics. We can find a formal description of lexemes (unit of meaning, composed of one or more words) found in text using a knowledge base (KB) and extract the several forms of mentioning a lexeme to normalise our corpus. The topic coherence measure is used to address the semantic interpretability of the terms used to describe a particular topic [20], and it is the measure we used to quantify the influence of semantic corpus normalisation in topics.

1.1 Contributions

- This paper proposes a framework to semantically normalise texts, and show experimental results on topic modeling task using two widely used datasets;
- Topic coherence improvement compared to traditional LDA.

1.2 Organization of the Work

The work is organized as follows: Sect. 2 presents a background content of the methods approached in this paper. Section 3 brings related work and compares our approach to others in literature. Section 4 exposes the proposed method to semantically normalise text and how it was applied to topic modeling. Section 5 presents the results and a discussion. And finally, Sect. 6 concludes our work.

2   Methods

2.1 Topic Modeling and Topic Coherence

A topic is a probability distribution over words and documents are mixtures of topics. Hence, a topic model can be considered a generative model for documents [24]. A more formal description of the Topic Modeling problem using LDA model is described as follows.

In LDA, it is assumed that there are K underlying topics from which the documents are generated and that each topic is represented as a multinomial distribution over the V words in the vocabulary. Therefore, a document is generated by sampling a mixture of these topics and then sampling words from that mixture [6].

A document with N words $d = w_1, \ldots, w_N >$ is generated by the following process:

1. The mix of topics θ is sampled from a Dirichlet distribution $(\alpha_1, \ldots, \alpha_k)$;
2. For each of the N words, a topic $z_n \in \{1, \ldots K\}$ is sampled from a $Mult(\theta)$ distribution, where $p(z_n = 1 \vee \theta) = \theta_i$;

3. Each word w_n is sampled, conditioned to the z_n-th topic, from the multinomial distribution $p(k \vee z_n)$.

It is possible to think of θ_i as the degree that a topic refers to a document. So, the probability of a document is the following mix:

$$p(d) = \int_{\theta} \left(\prod_{n=1}^{N} \sum_{z_n=1}^{K} p(w_n|z_n; \beta)p(z_n|\theta) \right) p(\theta|; \alpha)d\theta \qquad (1)$$

Where $p(\theta; \alpha)$ is Dirichlet, $p(z_n \vee \theta)$ is a multinomial distribution parametrized by θ, and $p(w_n|z_n; \beta$ is a multinomial distribution over words. This model is parametrized by the parameters $\alpha = \langle \alpha_1, \ldots, \alpha_k \rangle$ and a matrix β with dimensions $K \times |V|$. The per-word topic assignment, per-document topic distribution and topics are all latent variables and are not observed. The only observed variable is words within the documents, to infer the hidden structure (latent variables) with statistical inference [26].

As a way of making it clearer, Fig. 1 depicts the word distribution over the topics and a topic distribution over documents. As it was said before, a document is a distribution of topics and a topic is a distribution of words.

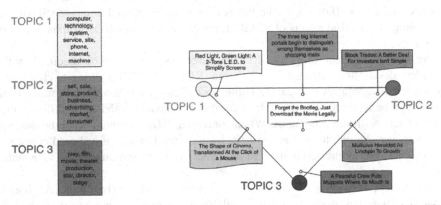

Fig. 1. Word-Topic distribution on the left and Document-Topic distribution on the right. These three topics represent the first three topics from a fifty topic LDA model trained on articles from the New York Times [8].

The topic quality (quality means interpretable and meaningful), measured as topic coherence, is based on the hypothesis that words with similar meaning tend to co-occur within a similar context. Each topic distribution contains every word but assigns a different probability to each of the words. The words with the highest probabilities within a topic are those that tend to co-occur more frequently. So, the top 10 or 15 high-probability words are usually used to interpret and semantically label the topics [26].

Researchers use several metrics of model fit, such as perplexity or held-out likelihood. However, such measures are only useful for evaluating the predictive model and do not address the explanatory goals of topic modeling [8]. The task of quantifying the coherence of a set of topics have been studied to remedy the problem that topic models give no guaranty on the interpretability of their output [23].

Many measures of coherence have been proposed recently, based on approaches that include co-occurrence frequencies of terms within a reference corpus [16, 19, 23]. A recent study [23] systematically and empirically explored the multitude of topic coherence measures and their correlation with available human topic ranking data. Their approach revealed a new coherence measure, called C_V, which achieved the highest correlation compared with all human ranking data. Hence, this study adopts the C_V coherence measure for topic coherence calculations.

2.2 Semantic Web

The idea of Semantic Web was described in 2001 by Tim Berners-Lee et al. as "A new form of web content that is meaningful to computers". In this new form of web content, introduced as an extension of the current web, information is given a well-defined meaning, where computers and people can work in cooperation [4].

The Semantic Web is based on the Resource Description Framework (RDF), a formal language for describing structured information [15]. An RDF document describes a formal specification of an arbitrary domain. This specification is modeled by a directed, labeled graph where each edge represent a link between two resources, represented by the graph nodes [17]. The link is expressed as RDF triples (*subject, relation, object*). Uniform Resource Identifiers (URI) are used to identify RDF resources and relations. To access and query RDF graphs the Protocol And RDF Query Language (SPARQL) was developed [21]. The results of SPARQL queries can be new RDF graphs or sets of resources.

The relationships and properties RDF resources may have can be specified by the vocabulary description language RDF Schema (RDFS) [7]. RDFS allows to create custom defined vocabularies to organize knowledge. Since URIs enable to identify RDF resources globally, it seems reasonable to combine vocabularies shared by different creators and across different domains. When shared, an RDF vocabulary can be denoted as an *ontology*. An ontology is an explicit, formal specification of a shared conceptualization and defines the terms used to describe and represent an area of knowledge [14].

The concept of ontology brings us to the Linking Open Data (LOD) project. It aims to identify datasets in the web that are available under open licenses, re-publish these datasets in RDF and interlink them with each other [5]. The term Linked Data refers to a set of principles to publish and interlink structured data on the web. One of the ontologies available on the web is YAGO (Yet Another Great Ontology - https://yago-knowledge.org/). YAGO is a large semantic knowledge base, derived from Wikipedia, WordNet, WikiData, GeoNames, and other data sources [22]. Currently, YAGO knows more than 17 million entities (like persons, organizations, cities, etc.) and contains more than 150 million facts about these entities. SPARQL queries are used in this work to query Yago Knowledge base in order to fetch alternative words for the same lexeme.

2.3 Named Entity Linking

Named Entity Linking can be described as the task of identifying lexemes in a text and linking them to the entity they name in a knowledge base, such as DBPedia. Before going too deep, an introduction of terminology and concepts is established. The term *entity* refers to something which is cognitively representable. An entity *mention* refers to the part of the text where a reference to an entity is made. It is also called *lexeme*, which is the basic unit of meaning. The *surface form* is a specific syntactic representation of the lexeme (the exact character string). A *knowledge base entity* refers to a representation of the entity, usually identified by an *URI* [28].

Now, let K be a formal knowledge base, $d \in D$ a document of the corpus D, $W \subseteq d$ the words of document d, $M \subseteq 2^W$ the set of entity mentions, and $m = (s, l, d, c) \in M$ denote an entity mention in a document d with start position s, length l and confidence score $c \in [0, 1]$. The *named entity linking problem* can be described as this [28]:

Definition 1 (Name Entity linking Problem)

- An extraction function $f_{ex} : W \rightarrow M$ to extract the entity mentions M from a document set D.
- A mapping function $f_{map} : M \rightarrow 2^W \cup NIL$ to compile a list $C \in 2^K$ of potential knowledge base entity candidates for every lexeme.
- A scoring function $f_{score} : C \rightarrow R$ to calculate a score, which indicates the degree of certainty that the candidate URI is to be selected as the correct one.
- A selection function $f_{sel} : C \rightarrow K$ to select the right candidate according to the calculated scores.

The degree of ambiguity is indicated by the size of the candidate list C. Hence, the *disambiguation task* is described by putting the mapping, scoring and selection functions together. The entire *context* is observed when processing the analysis items in the implementation of these functions. Just like in communication theory and linguistics the context is essential when interpreting pieces of information, in NEL it is as well. Examining context is crucial for NEL, because some context items can be very decisive when interpreting the context information [28].

There are some options of automated entity linking, and one of them is DBpedia Spolight (https://www.dbpedia-spotlight.org/) [18]. It is and open source project developing a system for automatic annotation of DBpedia entities in natural language text. It provides an interface for phrase spotting (recognition of phrases to be annotated) and disambiguation (entity linking) as well as various output formats (XML, JSON, RDF, etc.) in a REST-based web service [9]. DBpedia Spotlight is used in this work as the tool to find resources (URIs corresponding to formal descriptions of a concept) in text. These resources have several information and metadata about the concept, as well as links to other knowledge bases.

3   Related Work

There are related works of a type of topic modeling called of knowledge-based topic modeling. The main difference with the known knowledge-based topic modeling [3, 12] is that in this work the knowledge-based content is nor on the sampling neither on the inference steps [11, 29], it is a preprocessing step, applied to the input text.

Furthermore, there have been lots of works trying to solve different NLP tasks using semantics [18]. Short text classification was dealt by [12]. They exposed the use of DBpedia ontology to better represent short texts, so that semantically similar texts with no words in common can have similar context [12]. Their approach consisted in three steps: (i) identify concepts in text using DBpedia Spotlight and annotate them as resources; (ii) select the concepts with higher similarity; and (iii) extract additional knowledge, like categories, types or topics of identified concepts. The main dissimilarity between [12] and this word is that they added the additional knowledge (additional words) to the text and did not normalise the text. Moreover, they tested their hypothesis in a classification task, and not in a topic modeling context.

[29] proposed a knowledge-based topic modeling based on multi-relational knowledge graphs. They proposed a method that models document-level word co-occurrence with knowledge encoded by entity vectors automatically learned from external knowledge graphs. In other words, they do not consider only lexemes recognized in text, but from triples in external knowledge graphs. Our work is different from [29] because they add semantic knowledge into the generative process and not in preprocessing. Yet to the best of our knowledge, this is the first work that semantically normalise documents using a semantic replacement methodology.

4 The Proposed Method

There are two big steps that compose this method: (i) semantic corpus normalisation and (ii) topic modeling. The first one is the main contribution of this work, where we semantically normalised a corpus in order to benefit from the explicit semantics of Linked Data to evaluate the effect on the coherence of topics; while the second is the method used to show how semantically normalised texts affect semantic coherence of NLP tasks such as topic modeling using small texts.

4.1 Corpus Normalisation

Figure 2 depicts the normalisation method architecture, where each box is explained in the following paragraphs.

The normalisation is composed of two steps: (i) Resource extraction and (ii) Transformer. This first step is to find all lexemes in texts and associate them with resources from DBpedia. Lexemes are annotated by the process of NEL, using the DBPedia Spotlight annotation tool. Once we have the possible resources mentioned in the text and its respective URIs, we can find additional knowledge related to this resource. We decided to search in another knowledge base in order to find potential alternative surface forms, such as other labels used to describe that resource. The Yago KB is used in this step as the authors found more options in *alternate Name* and *label* fields in this KB. The second step is to create a replacement data structure, where all possible labels of a resource would be replaced by only one. Finally, the last step is to replace them all.

Since the resource extraction is achieved by making HTTP requests and SPARQL queries for each document separately, it is modularized and convenient to parallelize. The documents are saved in a database, each one with an associated unique identifier. We

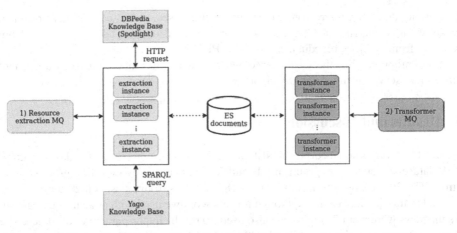

Fig. 2. Normalisation Architecture. The yellow blocks refer to the resource extraction step, as the blue block refers to the transformation step.

used Elasticsearch (https://www.elastic.co/) [1] as database. This identifier is used to keep track of which documents had already been processed. The RabbitMQ (https://www.rab bitmq.com/) tool is used to coordinate the extraction, creating a queue of documents to be processed. The resource extraction module runs in several instances (left side of Fig. 2), in order to accelerate the extraction. Each instance consumes from the queue, represented in the Fig. 2 as *Resource extraction MQ* in order to know which text it should process next.

The extraction works in this way for each text: first, it annotates all resources found in text using DBpedia Spotlight; second, for each resource, it makes a SPARQL query to Yago Knowledge base searching for alternative labels and the labels registered for that resource; lastly, it aggregates all possible labels for a resource and save them in the database.

The transformer step, which is done once all resource extraction is over, collects all resources and labels and organize them in a big mapping list. The mapping list maps all possible labels of a concept to a main label, which is going to replace all possible mentions of that concept. Once the mapping list is built, a regex substitution task is performed in order to make all substitutions. A queue is used to manage all texts that are being processed, similar to the resource extraction phase.

4.2 Topic Modeling

The first step to extract topics is the preprocessing one. The following preprocessing is done: (i) remove invalid characters and punctuation; (ii) lowercase; (iii) tokenize (transform text into a word vector); (iv) remove stopwords (too common words that do not aggregate meaning); (v) form bigrams (composed words, e.g. "United States") and (vi) lemmatize (remove word inflections, returning it to its root form, e.g. "said" to "say"). Besides usual stopwords, a list of too frequent words is removed too. From 20-Newsgroup: from, subject, re, edu, use, not, would, say, could, _, be, know, good, go, get, do, done, try, many, some, nice, thank, think, see, rather, easy, easily, lot, lack, make, want, seem, run, need, even, right, line, even, also, may, take, come; and from Reuters: from, subject, re, edu, use, say, inc, -PRON-.

After preprocessing, the vocabulary of words is ready to compose the word-document matrix that serves as input to LDA algorithm.

5   Results and Discussion

We used two very known corpus of NLP tasks: 20-Newsgroups (https://scikit-learn.org/0.19/datasets/twenty_newsgroups.html) and Reuters (https://www.nltk.org/book/ch02.html). The 20-Newsgroups has more than 18.000 newsgroups posts on 20 topics. Its is divided in training and testing, although for this work we used both as an unique dataset. As the news were from 20 topics, we also used 20 for the hyper-parameter of topics. The Reuters Corpus contains more than 10.000 news documents totaling 1.3 million words. The documents have been classified into 90 topics, and grouped into two sets, called "training" and "test". However, for this work we use both training and test set to extract the topics. Also, as the original corpus was annotated in 90 topics, we used 90 for the hyper-parameter of topics.

In Fig. 3 we can see the distribution of words per document in each corpus used. There is a bigger variety of sizes in 20-Newsgroup corpus, as well as the document length mean is higher before preprocessing. After preprocessing the remaining useful words were similar between both corpora.

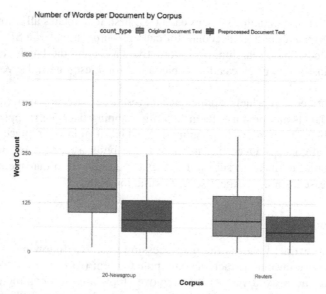

Fig. 3. Number of words per document by corpus. The red boxplot shows the counter of all words, separated only by spaces. The blue one shows the preprocessed documents, where stopwords and bigrams were built. This preprocessing is the same the documents are exposed before topic modeling algorithm.

In Table 1 there are examples of both corpora. The 20-Newsgroup has a form of e-mails, short texts and Reuters has a form of article documents. It can be seen on Fig. 3 that after preprocessing, 20-Newsgroup has lost more words than Reuters, because a big amount of characters were not letters or digits, which are removed on the preprocessing step.

Table 1. Document examples of each corpus.

20-Newsgroup	Reuters
"From: Edwin Gans Subject: Atheism Nntp-Posting-Host: 47.107.76.97 Organization: Bell-Northern Research Lines: 1"	"AMERICAN CENTURY \< ACT > RESTATES EARNINGS\n American Century Corp said it\n has restated its earnings for the fiscal year ended June 30,\n 1986 to provide an additional five mln dlrs to its loan loss\n allowance, causing a restated year-end net loss of 14,937,000\n dlrs, instead of 9,937,000 dlrs.\n The company said the change came after talks with the\n Securities and Exchange Commission on the company's judgement\n in considering the five mln dlrs collectible.\n In the note to its 1986 financial statement, American\n Century said it considered the five mln dlrs collectible,\n making its loan loss provision less than required.\n The company said in spite of the SEC decision, it still\n feels its allowance for possible loan losses at June 30, 1986\n was adequate and that it has considered all relevant\n information to determine the collectability of the five mln dlr\n receivable.\n But, it said continued disagreement with the SEC staff\n would not be in its best interest.

After a minimal analysis of the corpora used, the resources and possible labels were extracted from text and saved into the database. With all possible labels saved, the mapping list was built and used to transform the texts. The results for this experiment are shown on Table 2. The topic coherence for 20-Newsgroup corpus decreased with the corpus normalisation, as for Reuters corpus the coherence increased from 0.456 to 0.475.

Table 2. Topic coherence of the top 10 words in topic using C_V measure.

Dataset	Original corpus	Normalised corpus
Reuters	0.456	0.475
20-Newsgroup	0.672	0.667

As it can be seen by the results in Table 1, there is a positive effect on topic coherence on Reuters corpus, while on 20-Newsgroup it seems to have decreased the metric. From this results, we can leverage a number of hypothesis for these differences: (i) the size of the documents matters, because in small texts it is more difficult to get resources due to the fact that there is little context for the algorithm to disambiguate resources; (ii) the nature of text, as 20-Newsgroup has an e-mail like writing and Reuters is more article like; or (iii) the completeness of the knowledge base in specific topics. On the first hypothesis on the size of documents we can say that when a document it too small, the algorithm cannot be confident enough that a lexeme corresponds to a resource, so it does not capture it. Although 20-Newsgroup has a higher length of documents, both corpora are small, with a mean of less than 200 words per document. Also, by the Fig. 3 we can see that the number of valid words decrease much more on 20-Newsgroup than on Reuters corpus. Hence, we can infer that, although the total number of words is bigger on 20-Newsgroup, the number of valid lexemes to the algorithm to extract resources is very close to Reuters corpus. Besides that, as the context matters, and the context is the set of words around a lexeme, it is very difficult to the NEL algorithm to link a useful resource to the lexemes in text if only just a few words are valid.

This leads to the second hypothesis on the nature of text. It can be seen by Table 1 that the texts have very different natures. An e-mail like text is much more prone to have symbols and initials or acronyms, as seen in the first text of 20-Newsgroup of Table 1. On Reuters, it can be seen that the text is more fluent and without many symbols.

Related to the third hypothesis, we can explore in the future implementations a more complete log to track the resources that exist in the KB or not. The authors noticed during the execution of tests that many resources from Yago linked in the DBpedia page were not available anymore.

6 Conclusion

In this work we presented a framework to semantically normalise texts using resources from the Semantic Web. Our framework was tested in a topic modeling problem using two known corpora in order to have the first results and take insights for improvements.

The framework for normalisation is capable of improving the topic coherence of one of the corpora being tested.

So, the first tests on the semantic normalisation framework of texts showed prominent results and shall be investigated in depth in future. The authors plan to test this normalisation framework on a larger corpus from scientific articles or Wikipedia pages, in order to improve the analysis on the first and second hypothesis.

References

1. ELASTICSEARCH (2019). https://www.elastic.co/pt/
2. Allahyari, M.: Semantic Web Topic Models: Integrating Ontological Knowledge and Probabilistic Topic Models. Ph.D. thesis, University of Georgia (2016)
3. Allahyari, M., Kochut, K.: Semantic tagging using topic models exploiting wikipedia category network. In: 2016 IEEE Tenth International Conference on Semantic Computing (ICSC), pp. 63–70. IEEE (2016)

4. Berners-Lee, T., Hendler, J., Lassila, O., et al.: The semantic web. Sci. Am. **284**(5), 28–37 (2001)
5. Bizer, C., Heath, T., Idehen, K., Berners-Lee, T.: Linked data on the web(ldow2008). In: Proceedings of the 17th International Conference on World WideWeb, pp. 1265–1266 (2008)
6. Blei, D.M., Ng, A.Y., Jordan, M.I.: Latent dirichlet allocation. J. Mach. Learn. Res. **3**(Jan), 993–1022 (2003)
7. Brickley, D., Guha, R.V., McBride, B.: RDF schema 1.1. W3C Recomm. **25**, 2004–2014 (2014)
8. Chang, J., Gerrish, S., Wang, C., Boyd-Graber, J.L., Blei, D.M.: Reading tea leaves: How humans interpret topic models. In: Advances in Neural Information Processing Systems, pp. 288–296 (2009)
9. Daiber, J., Jakob, M., Hokamp, C., Mendes, P.N.: Improving efficiency and accuracy in multilingual entity extraction. In: Proceedings of the 9th International Conference on Semantic Systems, pp. 121–124 (2013)
10. De Melo, G., Siersdorfer, S.: Multilingual text classification using ontologies. In: European Conference on Information Retrieval, pp. 541–548. Springer (2007)
11. Doshi-Velez, F., Wallace, B., Adams, R.: Graph-sparse IDA: a topic model with structured sparsity. arXiv:1410.4510 (2014)
12. Flisar, J., Podgorelec, V.: Document enrichment using dbpedia ontology for short text classification. In: Proceedings of the 8th International Conference on Web Intelligence, Mining and Semantics, pp. 1–9 (2018)
13. Garla, V.N., Brandt, C.: Ontology-guided feature engineering for clinical text classification. J. Biomed. Inform. **45**(5), 992–998 (2012)
14. Gruber, T.R.: Toward principles for the design of ontologies used for knowledge sharing? Int. J. Hum.-Comput. Stud. **43**(5–6), 907–928 (1995)
15. Hitzler, P., Krotzsch, M., Rudolph, S.: Foundations of Semantic Web Technologies. Chapman and Hall/CRC (2009)
16. Lau, J.H., Newman, D., Baldwin, T.: Machine reading tea leaves: automatically evaluating topic coherence and topic model quality. In: Proceedings of the 14th Conference of the European Chapter of the Association for Computational Linguistics, pp. 530–539 (2014)
17. Manola, F., Miller, E., McBride, B., et al.: RDF primer. W3C Recomm. **10**(1–107), 6 (2004)
18. Mendes, P.N., Jakob, M., Garcia-Silva, A., Bizer, C.: Dbpedia spotlight: shedding light on the web of documents. In: Proceedings of the 7th International Conference on Semantic Systems (I-Semantics) (2011)
19. Newman, D., Lau, J.H., Grieser, K., Baldwin, T.: Automatic evaluation of topic coherence. In: Human Language Technologies: The 2010 Annual Conference of the North American Chapter of the Association for Computational Linguistics, pp. 100–108 (2010)
20. O'callaghan, D., Greene, D., Carthy, J., Cunningham, P.: An analysis of the coherence of descriptors in topic modeling. Exp. Syst. Appl. **42**(13),5645–5657 (2015)
21. Prud'hommeaux, E., Seaborne, A.: SPARQL query language for RDF. W3C Recommendation, W3C. Retrieved on 16 Nov 2009 (2008)
22. Rebele, T., Suchanek, F.M., Hoffart, J., Biega, J., Kuzey, E., Weikum, G.: YAGO: a multilingual knowledge base from wikipedia, wordnet, and geonames. In: The Semantic Web - ISWC 2016 - 15th International Semantic Web Conference, Kobe, Japan, 17-2 Oct 2016, Proceedings, Part II, pp. 177–185 (2016). https://doi.org/10.1007/978-3-319-46547-019
23. Röder, M., Both, A., Hinneburg, A.: Exploring the space of topic coherence measures. In: Proceedings of the Eighth ACM International Conference on Web Search and Data Mining, pp. 399–408 (2015)
24. Steyvers, M., Griffiths, T.: Probabilistic topic models. Handb. Latent Seman. **427**(7), 424–440 (2007)

25. Suganya, G., Porkodi, R.: Ontology based information extraction-a review. In: 2018 International Conference on Current Trends towards Converging Technologies (ICCTCT), pp. 1–7. IEEE (2018)
26. Syed, S., Spruit, M.: Full-text or abstract? Examining topic coherence scores using latent dirichlet allocation. In: 2017 IEEE International conference on data science and advanced analytics (DSAA), pp. 165–174. IEEE (2017)
27. Vallet, D., Fernández, M., Castells, P.: An ontology-based information retrieval model. In: European Semantic Web Conference, pp. 455–470. Springer (2005)
28. Waitelonis, J.: Linked Data Supported Information Retrieval. Ph.D. thesis, Karlsruher Institut für Technologie (2018)
29. Yao, L., et al.: Incorporating knowledge graph embeddings into topic modeling. In: Thirty-first AAAI Conference on Artificial Intelligence (2017)

Fostering Open Data Using Blockchain Technology

Simon Tschirner[1](✉)(iD), Mathias Röper[1](iD), Katharina Zeuch[1](iD),
Markus M. Becker[2](iD), Laura Vilardell Scholten[2](iD), and Volker Skwarek[1](iD)

[1] University of Applied Sciences Hamburg, Ulmenliet 20, 21033 Hamburg, Germany
{simon.tschirner,mathias.roper,katharina.zeuch,
volker.skwarek}@haw-hamburg.de
[2] Leibniz Institute for Plasma Science and Technology, Felix-Hausdorff-Str. 2,
17489 Greifswald, Germany
markus.becker@inp-greifswald.de
http://www.haw-hamburg.de, https://www.leibniz-inp.de

Abstract. While open science is growing in popularity and especially publishing as open access is common and has proven its success in today's research, sharing of research data in terms of open data is still lacking behind. INPTDAT provides a platform to share research data in the field of plasma technology. In the course of this, the project QPTDat aims to increase the incentives to publish, share, and reuse research data, following FAIR principles and fostering the idea of open data. QPTDat identified the following main success factors: Authors need a secure proof of authorship to guarantee that they are credited for their work; a proof of data integrity, to ensure that reused data has not been modified or fabricated; a convincing system for quality curation, to ensure high quality of published data and metadata; and comprehensive reputation management, to give an additional incentive to share research data. This paper discusses these requirements in detail, presents use cases and concepts for their implementation using blockchain technology and finally draws a conclusion regarding utilisation of blockchain technology in the context of open data, summarising the findings in form of a research agenda.

Keywords: Open data · Open science · Research data reuse · Blockchain technology

1 Introduction

In research fields such as plasma technology, data-driven science relies on research data that is findable, accessible, interoperable and reusable [6]. *FAIR* research data, as defined by [29], increase research data reusability [6] and is a first step towards avoiding a phenomenon that is called a reproducibility crisis [1].

The work was funded by the Federal Ministry of Education and Research (BMBF) under the grant marks 16QK03A and 16QK03C. The responsibility for the content of this publication lies with the authors.

The interdisciplinary data platform for plasma technology INPTDAT[1] aims to bring the *FAIR* principles to research in plasma technology. It provides data publications with a unique digital object identifier (DOI), a plasma source catalogue, faceted search options and API based access to (meta)data [2].

This paper proposes an architecture extending INPTDAT to a blockchain-supported open science platform for plasma technology, adding aspects as proof of authorship, data integrity, quality curation, and reputation management. The presented solution initially focuses on plasma technology. However, the principles are described in a sufficiently abstract way to be easily adaptable to further scientific disciplines.

Open science is a term comprising several aspects – mainly open access, open data, licensing, uniqueness and citation tracking (cf. [25]) – with the aim of opening up scientific structures to make research available to a broader audience [20, p. 9]. Open access usually means publication of research articles to be read free of charge, while open data refers to the availability of data produced during the research process. Open access has now been around for about two decades, becoming quite successful (cf. [21]). Publishing open access has several advantages for researchers, e.g. higher visibility and increased number of citations (cf. [17]).

Open data specifically aims to foster the research process by increasing the number of published research data sets in general and an earlier publication of data during the research process. Additionally, the transparent publication of original data intends to reduce publications with tampered data. However, despite many programs and efforts supporting open science, the success of motivating researchers to share their raw data is still limited, especially when addressing early publication of data in the research process.

As stated by [3], blockchain technology offers new possibilities to open science. Providing de-centrality and information distribution, all participants are aware of data and actions on the blockchain [11, p. 546]. This way, it equips open science systems with transparency and independence from a single ruling authority (cf. [23, p. 8]). Furthermore, blockchains use cryptographic functions to store and chain data. These functions ensure that the data have not been manipulated, changed or dismissed, but maintains integrity. It is even possible to use a blockchain without making data instantly public (cf. [3, p. 14]), but it still allows the author to proof data integrity when the data is published later on.

This article provides a further research agenda towards a comprehensive blockchain-based system for open science. On the path, it evaluates the feasibility of the sketched approach by providing a system architecture in Sect. 2, use cases as starting points for implementation in Sect. 3, and challenges in Sect. 4.

---

[1] https://www.inptdat.de – latest access: February 6, 2021.

2   Architecture

The proposed system aims to implement the ideas of open access, open data, identification and citation tracking. The main focus is to design a system that motivates researchers to engage in open data and research data sharing.

In this section, the main goal will be broken down into a few main requirements, giving further motivation to explore the blockchain-based approach. Besides, it examines the current applications of blockchain in science. The suggested architecture combines conventional systems for research data management and web databases for accessing data sets with the proposed blockchain-based system. It provides the basis for an open science system implementing the requirements.

2.1   Requirements

The following requirements result from a literature review, focusing on open science, open data and their combination with blockchain technology. Discussions and workshops with researchers experienced in open science or from the domain of plasma technology have validated and further refined these requirements.

The first requirement, aimed at encouraging the sharing of research data early in the research process, is (1) ensuring authorship. Researchers are more likely to share their data if they can prove authorship and when they are assured to be associated with their data. Related to this point is the requirement that (2) different access rights have to be available. E.g. data from a collaboration with a commercial partner might need to remain private for a certain period or permanently.

Several requirements are needed to guarantee a quality standard: On one hand, (3) data integrity has to be ensured, meaning that data did not alter in between its creation until its reuse. This covers the intentional manipulation of data as well as unintentional changes. Important is that researchers can be sure that data, when reused or cited in a future research item, has been in the same form as at the time of recording. If a platform takes measures to ensure data integrity, it is more trustable – (4) this includes critical metadata, e.g. those needed for reproducibility. On the other hand, (5) published research items themselves should meet defined quality standards. In this way, users of the platform can trust the published content, as they can in a peer-reviewed journal with a good reputation.

To function as a base for researchers to explore and reuse existing data, (7) the system needs to follow the FAIR principles. One requirement on open data is to (8) allow for proper citation of research data. Therefore, the unique identification of research data is needed, which allows citation tracking (cf. [25]).

A system needs to give its users the right incentives to add their high-quality content. Following the tradition of financial applications based on blockchain technology, some researchers suggest establishing an alternative way to fund and merchandise research (e.g. [15]). The approach presented here sees reputation as the main incentive for researchers. Work is submitted to conferences and journals to share knowledge, make an impact and eventually to increase reputation. Therefore, (6) the system should add another mean to generate impact and gain reputation.

Finally, there is an implicit requirement, leading back to the motivation to use blockchain technology. Its mechanics provide transparency and security. The system should provide proof of authorship, guarantee data integrity and quality and monitor reputation. All these points need a secure, manipulation-proof implementation.

The requirements in summary:

1. Ensure authorship
2. Facilitate different access rights
3. Enable the validation of data integrity
4. Guarantee the integrity of significant information (metadata)
5. Ensure basic quality of data and metadata
6. Implement a robust reputation system
7. Published items must be searchable, accessible, interoperable and reusable (FAIR)
8. Research data must be citable

2.2   Related Work

Leible et al. [16] give a systematic review of the general suitability of blockchain technology for the field of open science. They conclude that open science can benefit from various aspects blockchain technology offers, especially from its tamper-proof recording of transactions and its decentralised nature, introducing a new level of trust. Nevertheless, the review states to consider a blockchain as one component of many. All of these have to be combined in a meaningful manner to create successful solutions fostering open science.

Tennant et al. [26] conducted an extensive multi-disciplinary study about innovations in the peer review domain. They see blockchain technology suitable as an enabler for new approaches, such as new incentive systems and authentication/certification methods for research data to prevent fraud and protect authorship.

Specifically dedicated to the area of science is the Bloxberg-blockchain[2] that has been developed by Max Planck Digital Library in 2019. It offers smart contracts for certification and verification of research data, governance and voting and consensus mechanisms.

Current approaches to blockchain-based open science systems mostly target subsets of these requirements or have a commercial background.

CryptSubmit [9] approaches the issue in research processes of not having persistent evidence of data existence. Therefore, it creates this evidence using trusted timestamps on the Bitcoin blockchain.

Focusing on the scientific publishing process, a blockchain-based system including document submission, review and publication, including a reputation

---

[2] www.bloxberg.org – latest access: November 6, 2020.

management has been proposed by [27]. Pluto offers a decentralised research network for publishing research data in general. It includes its verification, peer review process and a token-based reputation system [19].

Frankl [7] presents a commercial open science platform specialised in cognitive assessments. It offers blockchain-based data management and an app-based data sharing marketplace, using a token-based incentivisation mechanism.

ARTiFACTS[3] is a commercial blockchain-based platform to create a proof-of-existence of research data. Also included are the possibilities to get citations on work in progress and linkage of corresponding data via metadata. Research-Hub proposed the ResearchCoin[4] to represent the scientific reputation of an individual. It resembles a currency, earned by community votes.

2.3 Related Concepts

For those new to the topics, this section briefly introduces two key concepts used throughout this paper: blockchain technology and hash values of data.

Blockchain Technology combines cryptography, data management, peer-to-peer (P2P) networking and consensus mechanisms. It creates a trusted environment for the verification, execution and recording of transactions between parties. In the context of "cryptocurrencies", transactions mainly are of financial nature, while in other contexts, they can be used to add information to the blockchain. One way to look at a blockchain is to regard it as a ledger with its content organised in blocks, following an append-only logic. It is distributed and synchronised among the network peers (called nodes). Cryptography – especially cryptographic hashes – ensures the ledger's integrity. Some blockchains allow deploying code, which's execution can be triggered by transactions. Executable code deployed to a blockchain is called smart contract. Smart contracts can implement functions and data structures on the blockchain and enable the creation of decentralised applications [31, p. 3 ff.].

To summarise, the main properties of blockchain technology in the context of this paper are: (1) they work append-only, meaning that information once stored on the blockchain, can usually not be altered or removed later on, (2) they are handled by a P2P network, meaning that no centralised authority is needed or even desired, and (3) smart contracts add additional logic, leading to decentralised applications.

Hash Functions are mathematical functions taking input data of any length and produce an output of fixed size. This output is called a hash or hash value, comparable to a fingerprint of the input data. Hash functions are one-way functions and always generate the same hash out of the same input data.

---

[3] www.artifacts.ai – latest access: November 6, 2020.
[4] www.researchhub.com/paper/819400/the-researchcoin-whitepaper – latest access: November 6, 2020.

Hash values are of a certain length, while the input data is usually not limited. Thus, in theory, different input data could create the same hash value. But since hash functions aim for an even distribution of all possible inputs among the possible hash values, a slight change to the input results in an entirely different hash value. Therefore, the chance that two inputs lead to the same hash (this is called a collision) is very low [12, p. 246 f.]. Also, it is practically impossible to find a corresponding input which leads to a specific hash value.

2.4   Proposed Architecture

The proposed architecture is an open science system with a blockchain backbone. It lies in the nature of blockchain-systems, that storage is expensive, due to its redundancy. Therefore, blockchain-based systems strive to minimise data stored on-chain, but to store most of the data off-chain. Usually, integrity of off-chain data is secured by storing a hash of the data on-chain. This principle is also used by the proposed system architecture (cf. Fig. 1) that handles data in three different categories and stores them accordingly in different places: (1) research data, stored on the researcher's personal computer or affiliations research data management system (RDM), (2) metadata, stored on a traditional web platform (INPTDAT), and (3) secured data, meaning hash values of research data or metadata, stored on the blockchain (BC Network).

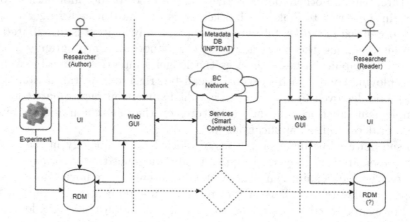

Fig. 1. The proposed architecture, combining a common metadata database and a blockchain-based solution

Six use cases have been identified to meet the main requirements. They will be deployed as services implemented by smart contracts on the architecture. The desired functionality will be shortly presented in the following, mapping the use cases to the architecture. In the subsequent section, the use-cases are discussed in-depth.

In the RDM, researchers can store their data, e.g. retrieved from an experiment. Although RDMs usually provide a user interface (UI), possibly a web interface, all three data storage should allow interaction via the same user interface (Web GUI). That way, users do not have to cope with several different systems. The BC network offers the service to certify research data stored in the RDM. Certification of data is then the basis for data verification by readers later on (cf. Sect. 3.1). From this moment, authorship and integrity of the research data can be proven; note that the data does not have to be made public immediately. This can happen later, by publishing the record on the web platform (INPTDAT).

Metadata databases allow fellow researchers and other interested persons to find interesting data sets. Therefore, researchers should add relevant metadata to the metadata database on research data publication (cf. Sect. 3.2). Next, an automated curation process (cf. Sect. 3.4) should assure the data quality. If necessary, this is complemented by a manual review process that can be supported by the blockchain-based system (cf. Sect. 3.5). Besides, other reused research data may receive additional reputation (cf. Sect. 3.6).

When readers identify interesting research data via the web interface, they can request a copy. Depending on the configuration or author's settings, a record can be copied directly from the author's RDM to the reader's RDM (or personal computer). Alternatively, the system prompts the author for the allowance of data sharing (cf. Sect. 3.3). After a successful transfer, data can be again checked for integrity, using a BC service and the data's reputation can get increased by a read.

3   Use Cases

Based on the general architecture, this section presents use cases meeting the requirements. The detailed description of the use cases helps to evaluate if the architecture supports them. The goal is to determine the degree, to which blockchain technology is useful for their implementation. Directions for further research are identified based on a discussion of the state-of-the-art and open challenges regarding each use case.

3.1   Certification and Verification

The central use case for applying a blockchain-based solution is the certification of research data. The aim is to certify that a specific researcher or a group of researchers has possessed specific research results at that time. If researchers certify their data directly after collection, this establishes authorship. The suggestion is to store authorship information on a blockchain. Thus, after certification, it will be almost impossible to manipulate authorship.

Additionally, the certified research item has to be stored. However, placing the research item itself on-chain is not feasible. First, the amount of data stored on the blockchain needs to be as small as possible. Second, information stored on a blockchain typically is publicly available. Whilst, somewhat contradicting the main idea of open data, sharing research data might not be desired under certain conditions, e.g. very early on in the research process, before a proper analysis. Therefore, usually, only the hash of the data is stored on-chain. Storing a hash gives a sufficient compromise of a minimal amount of stored data and security. Figure 2 shows this use case.

Researchers can proof their authorship by providing the research item and the blockchain address where its hash-value was stored during certification. The verification function compares the given information. It also confirms the exact date of certification, because every stored block includes a verified timestamp. The verification function further ensures data integrity: If research data got modified after certification, verification would fail, since the hash-value of the presented research data would not match the stored one. Ensuring data integrity fosters research data sharing and reuse, as this avoids future investigations being carried out based on altered or incorrect reference data.

At its core, this use case has been implemented many times before. Several services exist, where users can prove authorship on certain data, e.g. by adding a hash to a Bitcoin transaction (see e.g. OriginStamp [10]). This use case is also the first use case implemented by Bloxberg (cf. [13]).

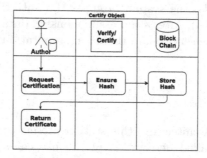

Fig. 2. When a user requests certification of a research item, the verify/certify service is responsible for creating a trustworthy hash value of the research item. The hash value will be stored tamper-protected on the blockchain.

Seeming relatively simple, a couple of design decisions imply further challenges. One challenge is related to authorship; a research item can have one or several authors, whose identities should be validatable. At the moment, in Bloxberg, authorship is just a text string that is hashed and stored together with the hash-value of the research item. This simple solution has some disadvantages. First, it is not possible to evaluate the provided identity. Thus it is not possible to prevent someone from adding information under a false identity. Second, it is not easily possible to include a unique identity. Third, if authorship would be stored trackable on the blockchain, this can lead to issues related to the General Data Protection Regulation (GDPR; for GDPR-issues regarding blockchain technology, see e.g. [14]). These issues – identity and GDPR-conformity – are research objectives on their own and placed on the research agenda developed throughout this paper. For now, the recommendation is to use an external identity-provider that delivers a solution using a GDPR-compliant structure, see [30] for such.

Another challenge is tamper-resistance of the hashing mechanism. It must be infeasible to generate data to match a certain hash-value. Otherwise, dishonest researchers could add any hash-value to the blockchain and later fabricate data around a core data-set. The created data would then result in the same hash value stored on the blockchain and would, in consequence, be verified by the system, even if it is not the data that has originally been certified. Until today, SHA-256 is, correctly implemented (see e.g. [8]), seen as sufficiently tamper-proof. Still, technical progress might imply that hash-algorithms that are seen as secure today might not be secure in a few years from now (cf. [18]). A sustainable blockchain solution might require the implementation of mechanisms that allow for a later exchange of the used hash-algorithm.

Additionally, research items processed by the hash-algorithm have to be kept private but still hashed with the correct algorithm. Privacy is hard to guarantee if data has to be sent to the entire blockchain to get hashed. In turn, it is hard to secure the intended execution of the hash-algorithm, if it is only executed locally on the researcher's system. An interesting but payment-oriented approach is the usage of off-chain state channels as implemented by Perun [4].

3.2 Publication Process

The second use case covers the step of publication (see Fig. 3). It could be publishing research data that has already been certified but otherwise kept private or the publication of research articles. When a (data) publication is accepted, it will be published via the conventional metadata database (INPTDAT) and becomes publicly available for search on the web.

When authors request publication, the proposed platform will forward the request to the quality curator described in Sect. 3.4 (UC4). When the quality control passed successfully, the research item is ready for publication. Metadata usually supports the possibility to reuse data. In plasma science, e.g., metadata should document the experimental setup and environmental conditions under which the data has been collected. This information is critical for reproducibility and to relate different pieces of work to each other. This importance implies that essential metadata should be immutable afterwards. Therefore, on publication, even metadata will be secured by a hash-value written to the blockchain.

The last step projects the classical citation of research articles to research items. I.e. other research items that have been influential for the item to be published are gaining reputation. Reputation is handled by the use case (UC5) in Sect. 3.6.

This use case is straight forward and does not add significant tasks to the proposed research agenda. A conventional web platform will implement the main functionality. The contribution of blockchain technology in this use case lies in the additional protection of critical metadata.

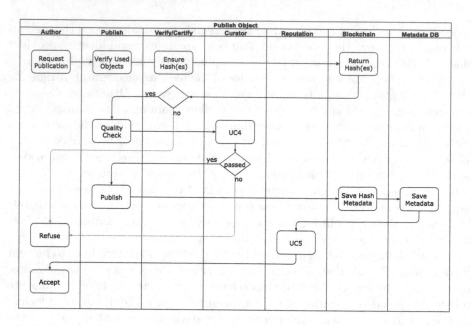

Fig. 3. For publication, (1) validation of all relevant certificates (of cited or reused work), (2) ensuring the quality of the research item, and (3) incrementation of the related reputation counters (e.g. of cited work).

3.3 Access

A published research item (see Fig. 4) will be findable and (indirectly) accessible via the conventional web database. The author can give access restrictions to an item. It can either be publicly available, i.e. accessible without any constraints, or available on request, meaning that a reader needs to request the item from its author.

The system can directly resolve a link to the location of publicly available items (e.g. on the author's RDM); it is immediately possible to create a copy. Otherwise, the author receives a request, e.g. in the web UI. If he or she grants access, the requesting researcher gets a unique, one-time link for download. In both cases, after completion, the verification smart contract validates the copy. It will only be possible to access the copy after successful validation. Via the reputation-service, the author receives a virtual incentive for his or her read reputation. If verification fails, the system removes the copy and informs the author and reader. Possible issues are that the item got corrupted during transfer, that data integrity got violated, or simply that the host is off-line. The author will have the opportunity to resolve the situation, e.g. by providing the item again. If access fails continuously, this can have an impact on the item's reputation. It has to be validated manually. Ultimately, it might be de-listed.

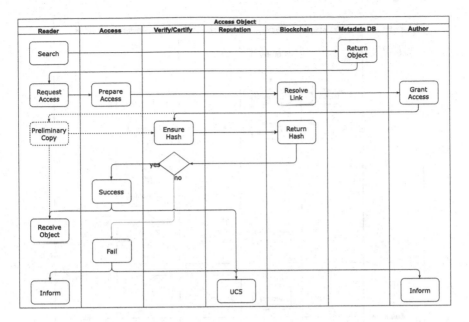

Fig. 4. The steps to resolve access to a research item.

As for the previous use case, access is a relatively straight forward implementation, that mostly depends on other use cases (namely verification and reputation) or the conventional metadata database INPTDAT.

3.4   Curation

The quality of available research items is a crucial success factor for a platform providing research data. If data is inconsistent or the process of its creation not well-documented, it is not reproducible and its reuse value low. Including such items on a platform will be frustrating for its users. They would have to skip through a couple of data sets to identify the content of reasonable quality. As the aim is to establish a platform with a strong reputation, positioning it as a real alternative to commercial websites, to guarantee a certain level of data quality is a primary requirement. Traditionally, a review process involving human reviewers is the curation process of choice. While this process has its clear advantages, it is usually lengthy and tedious. The presented approach aims to support this process with automation, if possible.

This paper suggests an algorithmic analysis based on certain key-information as metadata. Realistically, this can only result in a pre-check of research publications, e.g. monitoring the bibliography. When the object to review is data, already available in a machine-readable form, the premises for (partly) automation are positive. The next section covers classical peer-review.

The automatic curation (cf. Fig. 5) is initiated by the publication use case shown in Sect. 3.2. At this point, the integrity of the current data set is already

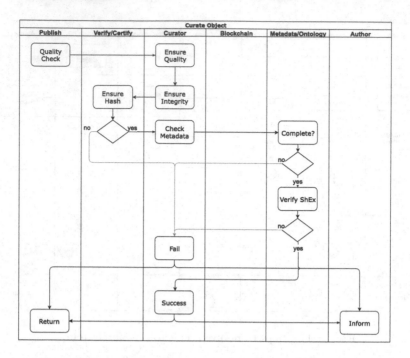

Fig. 5. The suggested process to execute automatic data curation.

confirmed (cf. Fig. 3). Now, the system cycles through (cited) data sets that have influenced the current research, to confirm their integrity, too. To do so, the reused data sets and related blockchain transactions (those data sets' certificates), have to be submitted. If this check succeeds, the curator continues to check the metadata of the current data set.

In general, metadata gives further context to a data set and thus, e.g., making it easier to find and interpret. Available criteria for metadata differ a lot in different disciplines. A distinctive example exists within the COMBINE community, dealing with standardisation and modelling in clinical and biomedical domains [28]. Here, comprehensive standards exist, e.g. to represent models used for data generation and processing. Given such a basis, it would eventually be possible to analyse models and data automatically.

Such achievements are usually results of decades of collaboration. It is one aim of the research project "Quality assurance and linking of research data in plasma technology – QPTDat", to foster the process of generating a joint, comprehensive metadata schema for plasma technology. Given such a schema would open up two possibilities of automated checks: (1) metadata can be checked for completeness, (2) metadata can be tested for soundness of the given metadata. The current approach suggested within QPTDat is to use shape expressions (ShEx) [24] for such a quality check.

Usage of blockchain technology to support automated quality curation of research data seems promising. Based on the certification and verification, it is possible to prove the integrity of reused data sets. It is furthermore possible to protect critical metadata from undesired alteration. Finally, the system allows logging the status of a research item's quality check.

A challenge is to develop a system for automated quality curation in detail and its meaningful connection to the blockchain – due to computational complexity, it would not be possible to run the check on the chain, meaning that each node performs the calculation. A solution offering transparency and tamper-resistance for the automated quality check is still desirable. However, even a comprehensive metadata schema will not fully replace a review by a human expert. Therefore, consideration of mechanisms for human interaction is necessary.

3.5   Peer-Review

The automated quality curation described in the previous section is especially interesting for the sharing of research data. However, substantial peer-reviews of research articles are usually a strong advantage of traditional paper processing offered by conferences and journals, guaranteeing the quality of published research items. To publish articles, a peer-review is still necessary. Even though peer-review is not a primary concern in the presented research project, it is still briefly outlined here.

It is possible to trigger peer-review during the publication use case. The latter will be stalled, until a result of the peer-review is available. First, suitable reviewers, e.g. having a positive reputation and experience in the field, are selected. Advantageous is a track record of good quality reviews submitted within the desired time frame. Reviewers should also not have conflicts of interest with the item to review. The system then notifies the reviewers and waits for their acceptance. The reviewers then have time to react and finish their review. After an agreed time or if a reviewer resigns, it is possible to reassign. A sufficient number of reviews leads to a verdict like accept, reject or request for changes.

Designing the peer-review-process with blockchain technology needs a clear distinction between the parts of the process that require a blockchain architecture and those better supported by traditional technology. This distinction depends, e.g., on the philosophy, one follows regarding open peer review [22]. One could choose to design a system, where only critical data, such as review scores, are collected and stored on the blockchain, or a system where the reviews themselves are stored, or even one where the complete review process is modelled and managed via smart contracts. The Decentralized Science project is currently working on a conversion of the classical peer review process onto a blockchain. Preliminary results support the claim, that transparency and decentralisation provided by blockchain technology is an enabler for the shift towards open access [27].

3.6   Reputation Management

The last main requirement is a robust reputation management system. Citations of scientific articles and publication in journals of high reputation are still among – if not the – most important reputation factors for scientists. This leads to three implications: (1) Principle incentive for the researchers to use a blockchain-based system for open science will be to gain additional reputation. (2) Designing reputation management should probably centre around citations. And (3) a newly designed approach allows to include further factors to represent researchers' reputation, factors that often might be less visible, as in the case of research data sharing, the amount of data shared, quality and reuse of that data.

Finding a suitable, motivating, and fair new reputation index is a research question on its own and out of scope for this article. Instead, the following four factors are recommended as elementary values for a probable reputation index:

- Citations or number of reuses: How often research items of a researcher get reused and cited by others.
- Reads: How many times their research items get downloaded and read.
- Number of publications: The number of research items published.
- Number of reviews: The number of reviews given for research items of other scientists

An optional fifth factor could be the quality of the submitted/published research items. Possibly, the number of citations/reuses already indicates this factor.

The use case diagram depicted in Fig. 6 contains three independent tasks. First and upmost, the retrieval of reputation scores, which itself could be an array containing four values representing the factors mentioned above. The scores are stored verifiable on the blockchain and can be accessed, e.g. for inclusion in result lists on INPTDAT. The middle and bottom tasks are increments of the reputation, based on reads or publications. During publication, even the reputation score of the cited/reused research objects is increased.

An advantage is that a blockchain can store these values with high integrity and that their accumulation is stored transparently. However, this system will put a lot of stress on the tamper-resistance of the smart contracts that manage the reputation counters. Here the scepticism expressed by Leible et al. [16] has to be heard and accepted as a challenge.

The next challenge is to find the right representation of the reputation counters on the blockchain. Here, it will be particularly problematic to manage reputation related to the researchers' identity. If reputation scores are stored related to research items, these have to be connected to the correct author, to allow proper aggregation of reputation. If it is stored related to authors, its calculation and the items adding to it need to be transparent. Even ways to alter the score calculation at a later date have to be considered. Be it that errors or fraud have been revealed or for reasons connected to GDPR. It even has to be considered that attackers might generate arbitrary content to hide tampering with reputation indicators.

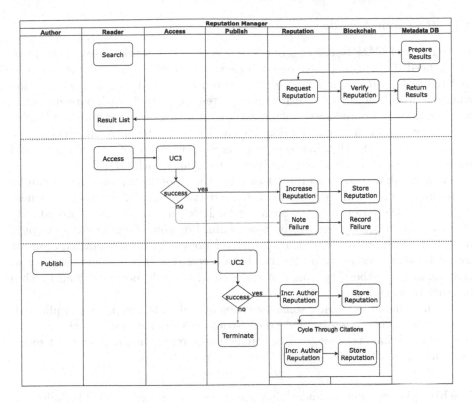

Fig. 6. The suggested process to execute automatic data curation.

4   Results

Compared to other work, describing possibilities to support science and publication of research results with blockchain technology in general, this work takes a step closer to its implementation. The purpose is to investigate the potential of blockchain technology to foster open data. Special attention was paid to the opportunity of increasing the attraction of sharing research data, especially prepublication. For this purpose, the main requirements, which resulted from the related QPTDat research project, have been listed. The main contribution of this paper lies in the proposed architecture and use cases. These show a starting point for the implementation of a blockchain system supporting open data. The implication is that the application of blockchain technology in research data sharing seems promising. The almost tamper-proof data structure and the transparent and decentralised nature of blockchain networks have the potential to replace common, centralised, and often commercial structures in the publication of research results in favour of an open science approach. However, a complete implementation is still pending, and additionally, quite some challenges still lie ahead. An overview of the latter is given in the subsequent section.

4.1   Research Agenda

Identity and GDPR are closely related. While knowing the researchers' identity is needed, regarding their research items and, to some extent, even reviews and reads, GDPR requires that identities and all related personal information can be removed on request. Authorship of a research item is such personal information. In this case, the nature of blockchain technology making it almost impossible to remove or alter information once added to the blockchain, which is one of the main reasons why this technology is feasible for proof of authorship and integrity of research data, makes it harder to comply with GDPR.

A thorough discussion of the relation between blockchain and GDPR can be found in [5]. One mentioned solution is to store the personal data off-chain, where it could be deleted on request. Such external identity management would only need to store a pseudonym on the blockchain. Editable off-chain storage would handle the relation between identity and pseudonym. A consequent requirement would be that each entry on the blockchain comes with a unique identifier so that removing authorship from one research item would not revoke authorship of other research items.

Even if there are already some solutions available, this topic is complicated. Solving the issue of identity and GDPR-conformity might impact other advantages of blockchain technology. The connection to reputation management seems particularly problematic.

Hashing is the central underlying mechanism to prove the authorship and integrity of research items via blockchain. There are two main issues. (1) During implementation, it has to be made sure that the hash algorithm is exchangeable in the future, in case the used algorithm becomes insecure. This would eventually require a complete re-hashing of all certified research data, which is a challenge by itself. Due to its computational complexity, (2) the hash function cannot be executed on the (complete) blockchain network (by using a smart contract). However, this is not even desired, as that would mean that the research data cannot be kept private (which is a requirement, e.g. for sensitive data pre-publications). However, performing the hashing off-chain could increase the risk of manipulated hash values entering the system, which eventually could cause problems.

Another open question is how to perform complete integrity checks for research items, including the integrity checks of reused data. First of all, a basic technical solution would require an author to provide all reused data sets. Additional details, e.g. certificates, would be needed to verify integrity via the blockchain automatically. Specific challenges lie in the confirmation that data has been reused soundly. E.g. that the considered data sample is representative of the whole data set. This might currently be a limitation of automatic quality curation, requiring human experts as reviewers.

Details of Reputation Management. Regarding the reputation system, basic general questions are: which data structure should be used to store reputation, how should it be managed on the blockchain and (how) should a reputation index be calculated.

Security is a central topic that is part of almost all items on the research agenda. Data structures and smart contracts will have to be developed and tested very carefully to prevent tampering with the system. The proof of authorship using hash values is a relatively well-explored field and thus security risks are easier to avoid.

However, the situation regarding reputation management is quite the opposite. Reputation has been identified as one of the key incentives to use the proposed open science system. It must be impossible to manipulate reputation scores. Therefore, very cautious development and testing of the related data structures and smart contracts is mandatory.

Possible scenarios might be: attackers taking over reputation scores which do not belong to them (a question related to the identity management), reputation not being appropriately registered, faked research items or identities could be used to gain additional reputation, to name a few.

Others. Two additional topics should be part of the research agenda: The detailed design of the used blockchain itself and challenges in the area of peer-review (as far as not included in the previous topics). However, this paper does not discuss these topics. Instead, the usage of Bloxberg as blockchain is recommended. Bloxberg provides basic functionality as well as a strong community and network. With its scientific background, it is for the time being an excellent choice for the proposed system. The Bloxberg consortium handles challenges related to the blockchain infrastructure. The Decentralized Science project, mentioned earlier, covers the field of peer-review well.

4.2   Success Factors

Even a perfect technical solution cannot guarantee adoption by its potential users. Without adoption, the main incentive, gaining reputation from the publication of research data, will be non-existent. Resembling the chicken or the egg dilemma, this leads to the last two main requirements for the system to be developed. (1) It has to be easy to use. In the best case, the system integrates seamlessly with the researcher's workflow. QPTDat aims to add the functionality directly in a solution for research data management. (2) The researchers need to see clear benefits of the solution. The aim has to be to design the final platform useful right away. Combining the most important of the before mentioned requirements: The system is easy to use, so researchers add their data. Added data is of high quality and as such easy to find and reuse. Ideally, the quality curator would help researchers to improve data quality. Finally, scientists will reuse the published research data and increase their reputation from their data publication – and, all of this is done on a transparent, decentralised platform, open to everyone.

5   Summary and Conclusion

This paper has presented requirements towards an open science platform aiming to foster open data in plasma technology. The proposed architecture shows the integration of such a platform with a blockchain structure. Several presented use cases depict possible solutions to the blockchain integration and outline further challenges. A summary of these challenges suggests the direction of future efforts in research and development. These are the challenges that have to be faced to finally fulfil the requirements and lead to an open science system that, with the help of blockchain technology, fosters sharing of research data even early on in the research process. The platform aims to increase willingness to share research data by giving researchers the security of authorship, lowering the threshold to reuse data, ensuring data quality and integrity, and eventually giving deserved reputation for researchers sharing their research data. A brief analysis of security issues implies that security of reputation management is a concern to be further considered.

References

1. Baker, M.: Is there a reproducibility crisis? Nature **533**(7604), 452–454 (2016). https://doi.org/10.1038/533452a. http://www.nature.com/articles/533452a
2. Becker, M.M., Paulet, L., Franke, S., O'Connell, D.: INPTDAT - a new data platform for plasma technology, October 2019. https://doi.org/10.5281/zenodo.3500283. https://doi.org/10.5281/zenodo.3500283
3. Bell, J., LaToza, T.D., Baldmitsi, F., Stavrou, A.: Advancing open science with version control and blockchains. In: 2017 IEEE/ACM 12th International Workshop on Software Engineering for Science (SE4Science), pp. 13–14. IEEE (2017). https://doi.org/10.1109/SE4Science.2017.11. http://ieeexplore.ieee.org/document/7964307/
4. Dziembowski, S., Eckey, L., Faust, S., Malinowski, D.: Perun: virtual payment hubs over cryptocurrencies. In: 2019 IEEE Symposium on Security and Privacy (SP), pp. 106–123, May 2019. https://doi.org/10.1109/SP.2019.00020. ISSN 2375-1207
5. Finck, M.: Blockchain and the General Data Protection Regulation: Can Distributed Ledgers be Squared with European Data Protection Law?: Study. European Parliament (2019)
6. Franke, S., Paulet, L., Schäfer, J., O'Connell, D., Becker, M.M.: Plasma-MDS, a metadata schema for plasma science with examples from plasmatechnology. Sci. Data **7**(1), 439 (2020). https://doi.org/10.1038/s41597-020-00771-0
7. Frankl: Frankl - An open science platform (2018). https://docsend.com/view/gn8t7k9. Library Catalog: docsend.com
8. Gilbert, H., Handschuh, H.: Security analysis of SHA-256 and sisters. In: Matsui, M., Zuccherato, R.J. (eds.) SAC 2003. LNCS, vol. 3006, pp. 175–193. Springer, Heidelberg (2004). https://doi.org/10.1007/978-3-540-24654-1_13
9. Gipp, B., Breitinger, C., Meuschke, N., Beel, J.: CryptSubmit: introducing securely timestamped manuscript submission and peer review feedback using the blockchain. In: 2017 ACM/IEEE Joint Conference on Digital Libraries (JCDL), pp. 1–4. IEEE (2017). https://doi.org/10.1109/JCDL.2017.7991588. http://ieeexplore.ieee.org/document/7991588/

10. Hepp, T., Schoenhals, A., Gondek, C., Gipp, B.: OriginStamp: a blockchain-backed system for decentralized trusted timestamping. IT Inf. Technol. **60**(5–6), 273–281 (2018)
11. Janowicz, K., et al.: On the prospects of blockchain and distributed ledger technologies for open science and academic publishing. Semant. Web**9**(5), 545–555 (2018). https://doi.org/10.3233/SW-180322. https://www.medra.org/servlet/aliasResolver?alias=iospress&doi=10.3233/SW-180322
12. Kizza, J.M.: Guide to Computer Network Security. CCN. Springer, Cham (2017). https://doi.org/10.1007/978-3-319-55606-2
13. Kleinfercher, F., Vengadasalam, S., Lawton, J.: Bloxberg - the trusted research infrastructure [whitepaper]. Technical report, Max Planck Digital Library, February 2020. https://bloxberg.org/wp-content/uploads/2020/02/bloxberg_whitepaper_1.1.pdf
14. Kunde, E., et al.: Faktenpapier Blockchain und Datenschutz. Technical report, Bitkom c.V. (2017)
15. Lehner, E., Hunzeker, D., Ziegler, J.R.: Funding science with science: cryptocurrency and independent academic research funding. Ledger **2**, 65–76 (2017)
16. Leible, S., Schlager, S., Schubotz, M., Gipp, B.: A review on blockchain technology and blockchain projects fostering open science. Front. Blockchain **2**, 16 (2019). https://doi.org/10.3389/fbloc.2019.00016. https://www.frontiersin.org/article/10.3389/fbloc.2019.00016
17. McKiernan, E.C., et al..: How open science helps researchers succeed. eLife **5**, e16800 (2016). https://doi.org/10.7554/eLife.16800
18. Mosca, M.: Cybersecurity in an era with quantum computers: will we be ready? IEEE Secur. Priv. **16**(5), 38–41 (2018).https://doi.org/10.1109/MSP.2018.3761723. Conference Name: IEEE Security Privacy
19. Network, P.: Pluto - breaking down the barriers in academia [whitepaper]. Technical report, Pluto Network (2018). https://assets.pluto.network/Pluto_white_paper_v04_180719_1355_BSII.pdf
20. OECD/OCDE: Making Open Science a Reality (2015). https://doi.org/10.1787/5jrs2f963zs1-en. https://www.oecd-ilibrary.org/science-and-technology/making-open-science-a-reality_5jrs2f963zs1-en. Series: OECD Science, Technology and Industry Policy Papers
21. Piwowar, H., et al.: The state of OA: a large-scale analysis of the prevalence and impact of Open Access articles. PeerJ **6**, e4375 (2018). https://doi.org/10.7717/peerj.4375. https://peerj.com/articles/4375
22. Ross-Hellauer, T.: What is open peer review? A systematic review. F1000Research **6** (2017)
23. Rossum, J.V.: Blockchain for research. Technical report, Digital Science (2017). https://doi.org/10.6084/M9.FIGSHARE.5607778.V1. https://digitalscience.figshare.com/articles/Blockchain_for_Research/5607778/1. Artwork Size: 2269031 Bytes
24. Staworko, S., Boneva, I., Gayo, J.E.L., Hym, S., Prud'Hommeaux, E.G., Solbrig, H.: Complexity and expressiveness of ShEx for RDF. In: 18th International Conference on Database Theory (ICDT 2015) (2015)
25. Taylor, S.J.E., et al.: Open science: approaches and benefits for modeling & simulation. In: Proceedings of the 2017 Winter Simulation Conference. WSC 2017, IEEE Press (2017)
26. Tennant, J., et al..: A multi-disciplinary perspective on emergent and future innovations in peer review [version 3; peer review: 2 approved]. F1000Research **6**(1151) (2017). https://doi.org/10.12688/f1000research.12037.3

27. Tenorio-Fornés, A., Jacynycz, V., Llop-Vila, D., Sánchez-Ruiz, A., Hassan, S.: Towards a decentralized process for scientific publication and peer review using blockchain and IPFS. In: Proceedings of the 52nd Hawaii International Conference on System Sciences (2019)

28. Waltemath, D., et al.: The first 10 years of the international coordination network for standards in systems and synthetic biology (combine). J. Integr. Bioinform. **17**(2–3) (2020)

29. Wilkinson, M.D., et al.: The FAIR guiding principles for scientific data management and stewardship. Sci. Data **3**(1), 160018 (2016). https://doi.org/10.1038/sdata.2016.18. http://www.nature.com/articles/sdata201618

30. Wirth, C., Kolain, M.: Privacy by blockchain design: a blockchain-enabled GDPR-compliant approach for handling personal data. In: Proceedings of 1st ERCIM Blockchain Workshop 2018. European Society for Socially Embedded Technologies (EUSSET) (2018). https://doi.org/10.18420/blockchain2018_03. https://dl.eusset.eu/handle/20.500.12015/3159

31. Xu, X., Weber, I., Staples, M.: Architecture for Blockchain Applications. Springer (2019). https://doi.org/10.1007/978-3-030-03035-3

A Roadmap for Composing Automatic Literature Reviews: A Text Mining Approach

Eugênio Monteiro da Silva Júnior(✉) and Moisés Lima Dutra

PGCIN, Federal University of Santa Catarina, Florianópolis, SC, Brazil
eugenio.monteiro@posgrad.ufsc.br, moises.dutra@ufsc.br

Abstract. Due to accelerated growth in the number of scientific papers, writing literature reviews has become an increasingly costly activity. Therefore, the search for computational tools to assist in this process has been gaining ground in recent years. This work presents an overview of the current scenario of development of artificial intelligence tools aimed to assist in the production of systematic literature reviews. The process of creating a literature review is both creative and technical. The technical part of this process is liable to automation. For the purpose of organization, we divide this technical part into four steps: searching, screening, extraction, and synthesis. For each of these steps, we present artificial intelligence techniques that can be useful to its realization. In addition, we also present the obstacles encountered for the application of each technique. Finally, we propose a pipeline for the automatic creation of systematic literature reviews, by combining and placing existing techniques in stages where they possess the greatest potential to be useful.

Keywords: Systematic review · Text mining · Automation

1 Introduction

It is remarkable that the scientific production keeps growing at an accelerated rate. According to [9], at August 2018 there were $33,100$ active English-language peer-reviewed journals, which published together 3 million papers per year, resulting in an annual growth of approximately 5%. The large number of publications on certain topics means that writing literature reviews consumes many hours of human work, since it requires the analysis of several texts. Although information technology tools have facilitated the access to a myriad of journals around the world and have made the search process more streamlined, the human effort to find potentially useful information when a large number of documents is retrieved is still too high. According to [22], an experienced reviewer can evaluate on average two abstracts per minute and, in the case of more complex topics, each abstract may require several minutes to be evaluated. This time multiplied by hundreds or even thousands papers results in a total of many hours

E. Bisset Álvarez (Ed.): DIONE 2021, LNICST 378, pp. 229–239, 2021.
https://doi.org/10.1007/978-3-030-77417-2_17

of work, when considering only the initial stage of selecting the relevant papers. The evolution of Artificial Intelligence (AI) techniques observed in recent years, especially in the subarea known as Natural Language Processing (NLP), allows us to envisage scenarios in which these modern techniques and their associated tools can be used to enhance the process of creating literature reviews, from an automatic composition approach.

This paper aims to present an overview of the current scenario of the application of AI techniques for the automatic creation of literature reviews. Furthermore, we propose a general pipeline resulting from the combination of these techniques, in order to highlight the challenges and possibilities currently existing in this area of research. The main contribution of this work is to present the current possibilities for automating systematic reviews of literature and how they can be put to work together to facilitate the reduction of the operational workload of researchers during the conduct of a literature review.

2   Literature Review

Before thinking about automating a literature review process, it is necessary to know how it is traditionally conducted. Therefore, this section aims to conceptualize and briefly describe how to manually create a literature review.

There are several types of literature review, each one with its own objectives [7]. Among these types, the state-of-the-art and the systematic reviews (SR) stand out, as they are better known. A state-of-the-art review considers mainly the most current research in a given area or on a given topic. It often summarizes current and emerging trends, research priorities and standards in a particular field of interest. This review aims to provide a critical survey of the extensive literature produced in recent years, along with a synthesis of current thinking in the area. It may offer new perspectives on an issue or point out an area that needs more research [5]. Systematic reviews are a widely used method to gather the results of multiple studies in a reliable manner. According to [6], as a research method, systematic reviews are undertaken according to explicit procedures. The term "systematic" distinguishes them from reviews undertaken without clear and accountable procedures. According to [7], a SR seek to gather all available knowledge on a given topic with the guarantee of being transparent in reporting their methods to facilitate other researchers to replicate that process. Another function of a systematic review is to identify research gaps, in order to develop new ideas [11].

There is no consensus regarding how many steps the production of a systematic review can be divided into. Several different proposals in this regard can be found in the literature. While some authors propose only 3 steps, others like [21] suggest 15 steps. In this work, we consider the 4 steps shown below. According to [1], those steps are usually part of a review process:

- **Searching:** extensive searches are carried out to locate as much relevant research as possible according to a query. These searches include scrutinizing electronic databases, scanning reference lists, and searching for published literature.

- **Screening:** it narrows the scope of search by reducing the collection to only the documents that are relevant to a specific review. The aim is to highlight key evidence and results that may impact on the policy.
- **Mapping:** the Evidence for Policy and Practice Information and Co-ordinating Centre (EPPI-Centre)[1] has pioneered the use of "maps" of research as a method to both understand research activity in a given area and as a way of engaging stakeholders and to identify priorities for the focus of the review.
- **Synthesizing:** it correlates evidence from a plethora of resources and summarizes the results.

The process of preparing a systematic review is both creative and technical. It is worth mentioning that there is a natural dichotomy of tasks: creative tasks are performed during the development of the core question to be answered and the protocol to be applied, while technical activities can be performed automatically following exactly the applied protocol [21].

There are some standards for the development of systematic reviews in a traditional way that can serve as guides for the automation process, such as the PRISMA (Preferred Reporting Items for Systematic Reviews and Meta-Analyses) statement and the PICO (Patient, Intervention, Comparison, and Outcome) framework. PRISMA consists of a checklist with 27 items and a four-phase flowchart to help authors improve the reporting of systematic reviews and undertake meta-analysis. It is focused on randomized trials, but it can also be used as a basis for reporting SR from other types of research, particularly evaluations of interventions. PRISMA may also be useful for critical appraisal of published SR [16]. Regarding the PICO framework, according to [4], it can be used to develop a well-formulated research question with a clear statement of objectives. Some other standard models that are worth mentioning are PEO (Patient, Exposure, and Outcome) and PIO (Patient, Intervention, and Outcome). They are used to formulate the inclusion and exclusion criteria defined to select relevant studies, in order to answer the research question.

3   Text Mining

Text data mining or text mining is a derivation of data mining that, instead of working with numerical and structured data, works with textual data. The main difference between regular data mining and text mining is that in text mining the patterns are extracted from natural language text rather than from structured databases of facts. Databases are designed for software applications to process them automatically; text is written for people to read. There are no programs that can "read" text as humans do and there is no evidence that they will exist in the near future. Many researchers think it will require a full simulation of how the mind works before we can write programs that read the way people do [8].

---

[1] A specialist center for: (i) developing methods for systematic review and synthesis of research evidence; and (ii) developing methods for the study of the use research. https://eppi.ioe.ac.uk.

However, there is a research field called computational linguistics (also known as Natural Language Processing - NLP) that is making great progress in carrying out small sub-tasks in text analysis. For example, it is relatively easy to write a program to extract sentences from a paper or book that, when shown to a human reader, appear to summarize its content [8]. The main methods of NLP used in systematic reviews are text classification and data extraction. The classification methods look for models that can automatically associate documents (abstracts, full texts or parts of these texts) with previously defined categories. The data extraction methods try to identify parts of the text or individual words/numbers that correspond to a variable of interest [15]. Since scientific production is mostly presented in textual form, AI techniques specifically aimed at processing textual data have a wide field of application to aid in the production of literature reviews.

4 Automating the Creation of Systematic Reviews

Since the production of a SR are both creative and technical, it is expected that all stages considered technical are subject to automation. Indeed, the idea of automating the steps of a systematic review is not exactly new. According to [10], the first paper to propose the use of Machine Learning (ML) to this purpose was published in 2005. From that year on, several works were published regarding the application of computational techniques in each of the SR stages. One good way to observe the evolution of this idea is by reading systematic reviews published on this subject. In 2015, while [11] published a review that exclusively covers works of data extraction in SR, [18] dedicated to review papers related to automatic identification of relevant studies.

There are several methods for implementing text mining and related tasks. The methods currently considered the most relevant to support systematic reviews are: automatic term recognition (ATR), text clustering, text classification, and text summarization [20]. There is also a large amount of software applications specifically developed to assist in the production of systematic reviews. The SR Toolbox[2] website provides a list of various available tools for supporting systematic reviews of literature.

It is important to highlight that all the technical steps of the review process can be automated through some computational technique with the main objective of reducing the human workload.

4.1 Challenges and Opportunities

After defining the theme of the research and the inclusion criteria, the first technical stage of a SR is the search for correlated studies. Ideally, 100% of the existing studies on the topic should be retrieved. Text mining can help by suggesting possible query terms. Even if the researcher already has found some documents that meet his/her inclusion criteria, he/she can always use a term

---

[2] http://www.systematicreviewtools.com/.

recognition service that suggest new terms and concepts to be used in a new query [20]. According to [1], term extraction improves the search strategy by creating additional metadata that can increase accuracy by automatically identifying key phrases, concepts or technical terms, within the documents. The improvement in the set of search terms has the potential to expand the coverage of the results, which may be sufficient in cases where the object of study is very specific. Thus, the number of papers retrieved is relatively small, but the entire literature is covered. In some other cases, the number of papers retrieved is very high, which makes it even more challenging to find works that are really relevant to the subject to be searched. Consequently, it is possible to think about methods that can be applied to the set of retrieved papers in order to find among them those that are really relevant.

In a systematic review, the term *screening* refers to the manual process of sifting through, at times, thousands of titles and abstracts that are retrieved from database searches. In order to improve reliability, the titles and abstracts are often screened by two people. This is a very labour-intensive task and adds considerably to the review's cost and time [20]. This is a stage where machine learning techniques can be very useful, by means of filtering not only titles and abstracts, but also full texts. According to [12], the first study to consider this possibility was [3]. According to [20], there are two ways to use text mining to automate this step: the first aims to prioritize the list of items for manual screening so that the studies at the top of the list are those most likely to be relevant; the second one uses the studies manually labeled (included/excluded) as a training dataset, so that the system can "learn" to automatically classify the other works. Apparently, conducting this step in a semi-automatic manner can bring many benefits to the researcher. However, [15] highlights that the main limitation of the automatic screening of abstracts is the fact that it is not clear at what point it is "safe" for the reviewer to the interrupt the manual screening. Even systems that, instead of providing a definitive and dichotomous classification, provide classifications based on probabilities are not free from the risk of loss. For example, a paper that has received a low probability may be relevant, and if a researcher chooses to stop screening in a certain threshold of probability, this paper may not be included in the results.

Another way to find relevant studies to a literature review is by citation mining. As an example, the study of [2] proposes a method to systematically mine the various types of citation relations between papers to retrieve documents that may be related to the topic searched by a specific systematic review. The author's proposal is conceptual and was conducted manually. This method, according to this author, despite having potential for automation, had some limitations related to the available databases API's that made it impossible to create a computational algorithm at that time. Currently, the existing databases API's provide more and more information about the indexed papers, which makes it possible to write algorithms that can automatically retrieve related papers through the citation mining technique. Moreover, [19] stresses it is possible to think about the integration of the aforementioned content-based methods with the citation-based methods, in order to create a more efficient model for retrieving relevant papers.

Once the set of relevant studies are identified and retrieved, the next step is to extract the useful information present in each one of them. According to [21], extracting data from texts is one of the most time consuming tasks in a systematic review. Therefore, there are already several works whose objective is to automatically extract data from texts. According to [15], when considered specifically reviews of randomized controlled trial (RCT), there are only few prototypes of platforms that make these technologies available, such as ExaCT[3] [13] and RobotReviewer[4]. For basic science reviews, the NaCTeM (the United Kingdom National Center for Text Mining) has developed several systems that use structured models to extract concepts such as genes and proteins from texts. Since the desired information can be present in several sections of the paper, extracting it can become a complex cognitive task. Consequently, even partial automation can reduce the time required to complete this task, as well as reduce errors and save time [21].

One obstacle for achieving better data mining models is the lack of training data. ML systems need a dataset with manually assigned labels in order to adjust model parameters. Associating labels with individual terms in documents to enable the training of data-extraction models is an expensive task. EXaCT, for example, was trained on a small set (132 in total) of full-text papers. RobotReviewer was trained by using a much larger dataset, but the 'labels' were semi-automatically induced, using a strategy known as 'distant supervision'. This means the annotations used for training were imperfect, thus introducing noise to the model [15]. Recently, [17] released the EBM-NLP dataset, which comprises about 5000 abstracts of RCT reports manually annotated in detail. This may provide training data helpful for the development of data extraction models [15].

The last step of a SR that can be assisted by text mining techniques is the synthesis of information. According to [15], although the software tools to support the synthesis of revision data have been around for a long time (especially for performing meta-analyzes), the methods for automating it are beyond the capabilities of ML and NLP.

Furthermore, it is also possible to think about ways to automatically summarize the texts that were selected for review, by extracting information from the full texts of the papers and not just from their abstracts. For this, there is the technique that creates automatic text summaries. According to [14], this technique either generates a summary for a single document at once or for multiple documents together (MDS - multi-document summarization), by extracting the most relevant information found within the texts. Automatic summarization is quite important in systematic review processes, as it condenses the information that was discovered and classified and thus provides a solution to the information overload problem [20].

---

[3] https://exact.cluster.gctools.nrc.ca/ExactDemo/.
[4] https://www.robotreviewer.net/.

The use of MDS methods offers the benefits of reducing the overwork on the reviewer, as well as enabling an overview of a body of research. However, the proper place and use of such summarization must be established for it to offer the greatest benefit, regarding the current state of the art. This is partly an issue for a system designer but also partly an issue of training and experience for the reviewer. Thus, running a MDS on a large collection of texts from many domains, on many subjects, would probably not be a useful exercise and would indicate a lack of understanding about getting the most from summarization. However, if the reviewer have previously produced a cluster or a classification of documents, then it makes sense to apply the MDS, since documents in a cluster or class can reasonably be expected to have something in common, consequently, the results would be meaningful [20].

Finally, there are the Natural Language Generation (NLG) technologies, which can be used to automatically write specific paragraphs of the review, such as a description of the types of documents retrieved, results of the evaluation, and summary of the conclusions [21]. Currently existing techniques are not able to produce perfect texts like those written by humans. However, the automatically generated text can serve as basis for the text to be written manually by reviewers, e.g. avoiding errors in data transfers from multiple sources. Importantly, this kind of technology still has a lot to be improved. Thereby, in search for more integrated tools to automate systematic reviews, researchers should be aware of the new text generation methods that are emerging.

4.2   A Pipeline for Creating Systematic Literature Reviews

The intention of this paper is not to present a strict rule of how an SR should be automated, nor to indicate specific tools or technologies for that purpose. The objective here is just to highlight, based on what has already been presented in the scientific literature on this topic, the stages of the literature review pipeline with the greatest potential for automation. We try to indicate what should be the focus of researchers on AI, when they go to work within this theme. Disregarding the steps that naturally involve a creative process and, consequently, must be performed by humans, the next paragraphs focus on the operational tasks that are part of the reviewing process. In Fig. 1, we propose a pipeline that combines several techniques used by different projects to automatically generate a literature review. This pipeline shows not only a sequence of technical steps required for the creation of an automatic literature review, but also the respective AI techniques that can be useful in each phase.

In the searching phase, computational techniques can help in suggesting terms to maximize the amount of documents retrieved, however, the human operator remains essential to carry out the process. Thus, this phase is considered to have a medium automation potential. In spite of that, the works found in the literature propose techniques with great potential for increasing the degree of automation in this stage. Machines usually perform this task better than humans do. Besides, scientific databases are increasingly providing structured data on references, which facilitates the automation process. In addition, it is

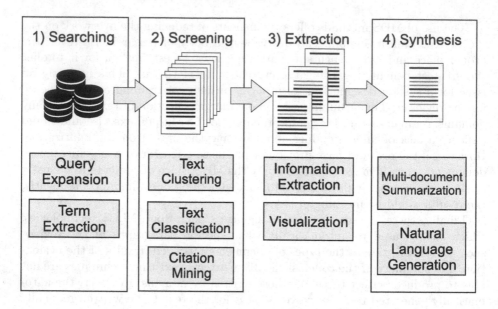

Fig. 1. A pipeline for automatically creating systematic literature reviews

possible to work on the direct extraction of references from papers (i.e. PDF files). Therefore, this is a step that deserves more attention and investment from researchers.

As for the screening stage, we consider that it still has a low potential for automation, currently. That statement comes from the fact that the conversion of texts into vectors while preserving semantic relations is still incipient, among other causes, due to the small number of positive training examples available. Text classification methods are extremely dependent on a good conversion of texts into vectors, since their accuracy in classification is highly impacted by the extraction of text characteristics. Usually, ML-based classification methods depend on training data to 'learn' patterns. In systematic reviews, the number of papers labeled as 'included' is less than the size of the set of 'excluded' papers, which makes it difficult to properly adjust ML-based models due to this imbalance of sets. As previously mentioned, given the risk of losing potentially relevant papers during the screening stage, researchers may not feel secure in delegating the exclusion of much of the retrieved papers to an automatic classification process. Thus, we believe it is still necessary to develop new methods for extracting more precise text characteristics, so that it is possible to consider that automatic sorting as a secure time saver for researchers.

The extraction and synthesis steps present a great potential to be automated. In these stages, computational techniques operate by extracting and organizing important information from texts. Various techniques for extracting certain data from texts are being developed and can be applied at this stage of automatic creation of literature reviews. Especially for medical reviews, which are already more standardized, there is great potential for applying these techniques. Nevertheless, as the natural language processing keeps evolving, it is possible to imagine for a near future the extraction of texts for automatically creating literature reviews to be applied to some other areas of knowledge, such as the social sciences.

5 Conclusions and Next Steps

Automating literature reviews is a promising research field because the number of published papers grows every year. The large amount of available texts makes human work difficult in writing scientific literature reviews. For this reason, the development of computational tools to assist researchers in this purpose continues to arouse interest in the scientific community. It is important to highlight that, due to the many limitations of the existing computational techniques, there are still no definitive/standardized tools to help in the automatic creation of systematic reviews.

In this way, the papers found in the literature only present specific/partial solutions for certain stages of the construction of a systematic review. The supervised methods, despite being very useful in some of these phases, face the problem of lack of data for training. Consequently, these techniques present less potential for development in SR. As for the unsupervised methods, there are greater possibilities. Summarization, visualization and document clustering are examples of tasks that can help researchers deal with the large number of publications available, without relying on previously-labeled databases for training. For this reason, the development of computational models that will contribute to the reduction of human workload, especially during the operational stages of SR, can provide more agility to the process of generating scientific knowledge. This article brings together some existing initiatives aimed at this purpose. In the stages of search, screening, extraction and synthesis, some computational techniques have already been used in order to facilitate the reviewer's work.

As for future work, the computational implementation of the proposed pipeline will be carried out. Ideally, this implementation will use mainly unsupervised methods to avoid relying on training data, which is still very scarce. We intend to use existing algorithms for grouping, extracting and synthesizing information, available in the literature, that best adapt to the scenario that is being worked on. Our ultimate goal is to achieve a complete solution to automate the operational steps of a systematic literature review. As a subsequent step to the development of the prototype, we intend to test it in application scenarios from different areas of knowledge, and make it available for specialized researchers in these areas to qualitatively evaluate the results obtained.

References

1. Ananiadou, S., et al.: Supporting systematic reviews using text mining. Soc. Sci. Comput. Rev. **27**(4), (2009). https://doi.org/10.1177/0894439309332293

2. Belter, C.W.: Citation analysis as a literature search method for systematic reviews. J. Assoc. Inf. Sci. Technol. (2015). https://doi.org/10.1002/asi.23605
3. Cohen, A.M., et al.: Reducing workload in systematic review preparation using automated citation classification. J. Am. Med. Inform. Assoc. **13**(2), (2006). https://doi.org/10.1197/jamia.M1929
4. Davis, D.: A practical overview of how to conduct a systematic review. Nurs. Stand. **31**(12), (2016). https://doi.org/10.7748/ns.2016.e10316
5. Dochy, F.: A guide for writing scholarly articles or reviews for the Educational Research Review (2006)
6. Gough, D., Thomas, J., Oliver, S.: Clarifying differences between review designs and methods. Syst. Rev.**1**, 28 (2012). https://doi.org/10.1186/2046-4053-1-28
7. Grant, M.J., Booth, A.: A topology of reviews: an analysis of 14 review types an associated methodologies. Health Inf. Libr. J. (2009). https://doi.org/10.1111/j.1471-1842.2009.00848.x
8. Hearst, M.A.: What is text mining? https://people.ischool.berkeley.edu/~hearst/text-mining.html. Accessed 12 Oct 2020
9. Johnson, R., Watkinson, A., Mabe, M.: The STM Report: An Overview of Scientific and Scholarly Publishing. 5ª edição. STM: International Association of Scientific, Technical and Medical Publishers, The Hague (2018)
10. Jonnalagadda, S., Petitti, D.: A new iterative method to reduce workload in systematic review process. Int. J. Comput. Biol. Drug Des. **6**(1–2), 5–17 (2013). https://doi.org/10.1504/IJCBDD.2013.052198
11. Jonnalagadda, S.R., Goyal, P., Huffman, M.D.: Automating data extraction in systematic reviews: a systematic review. Syst. Rev. (2015). https://doi.org/10.1186/s13643-015-0066-7
12. Khabsa, M., Elmagarmid, A., Ilyas, I., Hammady, H., Ouzzani, M.: Learning to identify relevant studies for systematic reviews using random forest and external information. Mach. Learn. **102**(3), 465–482 (2015). https://doi.org/10.1007/s10994-015-5535-7
13. Kiritchenko, S., et al.: ExaCT: automatic extraction of clinical trial characteristics from journal publications. BMC Med. Inform. Decis. Mak. **10**, 56 (2010). https://doi.org/10.1186/1472-6947-10-56
14. Mani, I.: Automatic Summarization, John Benjamins, Amsterdam (2001)
15. Marshall, I.J., Wallace, B.C.: Toward systematic review automation: a practical guide to using machine learning tools in research synthesis. Syst. Rev. (2019). https://doi.org/10.1186/s13643-019-1074-9
16. Moher, D., et al.: Preferred reporting items for systematic reviews and meta-analyses: the PRISMA statement. PLoS Med. (2009). https://doi.org/10.1371/journal.pmed.1000097
17. Nye, B. et al.: A corpus with multi-level annotations of patients, interventions and outcomes to support language processing for medical literature. In: Proceedings of the 56th Annual Meeting of the Association for Computational Linguistic (2018)
18. O'Mara-Eves, A., et al.: Using text mining for study identification in systematic reviews: a systematic review of current approaches. Syst. Rev. (2015). https://doi.org/10.1186/2046-4053-4-5
19. Sarol, M.J., Liu, L., Schneider, J.: Testing a citation and text-based framework for retrieving publications for literature reviews. In: BIR 2018 Workshop on Bibliometric-Enhanced Information Retrieval (2018)
20. Thomas, J., McNaught, J., Ananiadou, S.: Applications of text mining within systematic reviews. Res. Synth. Methods (2011). https://doi.org/10.1002/jrsm.27

21. Tsafnat, G., et al.: Syst. Rev. Automat. Technol. Syst. Rev. (2014). https://doi. org/10.1186/2046-4053-3-74
22. Wallace, B.C., et al.: Semi-automated screening of biomedical citations for systematic reviews. BMC Bioinform. (2010). https://doi.org/10.1186/1471-2105-11-55

Interactive Domain-Specific Knowledge Graphs from Text: A Covid-19 Implementation

Vinícius Melquíades de Sousa[(⊠)] and Vinícius Medina Kern

Universidade Federal de Santa Catarina, Florianópolis, Brazil
`v.m.kern@ufsc.br`

Abstract. Information creation runs at a higher rate than information assimilation, creating an information gap for domain specialists that usual information frameworks such as search engines are unable to bridge. Knowledge graphs have been used to summarize large amounts of textual data, therefore facilitating information retrieval, but they require programming and machine learning skills not usually available to domains specialists. To bridge this gap, this work proposes a framework, KG4All (Knowledge Graphs for All), to allow for domain specialists to build and interact with a knowledge graph created from their own chosen corpus. In order to build the knowledge graph, a transition-based system model is used to extract and link medical entities, with tokens represented as embeddings from the prefix, suffix, shape and lemmatized features of individual words. We used abstracts from the COVID-19 Open Research Dataset Challenge (CORD-19) as corpus to test the framework. The results include an online prototype and correspondent source code. Preliminary results show that it is possible to automate the extraction of entity relations from medical text and to build an interactive user knowledge graph without programming background.

Keywords: Knowledge graphs · COVID-19 · Information retrieval software · Natural language processing · Personalized analytics

1 Introduction

Shannon's Mathematical Theory of Communication [19] is understood as the Information Science debut [4]. Ever since Shannon's work the field has evolved into a number of sub-fields, following the advances in society. One of such fields is Information Retrieval, which was considered to be the Information Science main core [17]. It started in the 1970's and its focus was on the creation of retrieval indexes and the physical allocation of information. As the technological development took place the focus shifted towards information processing and

Funded by CNPq, Research Productivity Grant 314140/2018-2.

efficient information retrieval, digitally speaking [5]. The use of knowledge graphs to represent human knowledge, and therefore as a way into information retrieval, has been receiving attention both from academia and industry. A knowledge graph can be defined as a structured representations of facts, in the form of entities and relations and its semantic description [10]. A knowledge graph is composed by triplets in the form (head entity, relation, tail entity), Fig. 1 depicts an example of a knowledge graph where on the left-side it is presented the triplets and on the right-side the representation in a graph form.

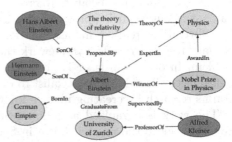

(Albert Einstein, **BornIn**, German Empire)
(Albert Einstein, **SonOf**, Hermann Einstein)
(Albert Einstein, **GraduateFrom**, University of Zurich)
(Albert Einstein, **WinnerOf**, Nobel Prize in Physics)
(Albert Einstein, **ExpertIn**, Physics)
(Nobel Prize in Physics, **AwardIn**, Physics)
(The theory of relativity, **TheoryOf**, Physics)
(Albert Einstein, **SupervisedBy**, Alfred Kleiner)
(Alfred Kleiner, **ProfessorOf**, University of Zurich)
(The theory of relativity, **ProposedBy**, Albert Einstein)
(Hans Albert Einstein, **SonOf**, Albert Einstein)

Fig. 1. Example of knowledge graph. Extracted from [10].

The construction of knowledge graphs can be classified into two main groups: (i) manually/curated or (ii) automatic/semi-automatic. The first group consists of allocating domain specialists to annotate, in accordance with a set of rules, the entities, relations and descriptions [22]. Manually constructed knowledge graphs are time consuming and tends to advance at a slower pace than information development. On the other hand, automatic/semi-automatic knowledge graphs are built upon a workflow, usually starting from a text corpus, from which entities and relations are inferred. Automatic/semi-automatic constructed knowledge graphs are able to keep up with the information creation, at the cost of (i) quality, that is, the entities and relations are not as accurate as when the knowledge graph is manually annotated [9] and (ii) having to deal with engineering challenges, such as data acquisition and storage, text parsing, information extraction, etc. While companies such as Google and Microsoft have the necessary resources to solve these challenges, smaller organizations and independent researchers are required to have programming skills in order to be able to use the advances of research in the information retrieval through knowledge graphs [18]. In other terms, the use of machine learning in information retrieval through knowledge graphs results in an increase on the complexity demanded to make use of such advances. The higher the complexity, the more limited is the number of people capable of making use of the gains allowed by those advances [8,14].

The information accessibility and availability for possible users is one of the tasks that Information Science is responsible for [15], as the general view of the information process, from creation to utilization, is a core activity of the area [2]. Domain specialists is a particular group of users, with real needs, that could benefit from using knowledge graphs. They are not usually proficient in programming/machine learning optimization skills and, at the same time, their information needs are not fulfilled by regular knowledge frameworks, such as google [11]. Therefore, if: (i) domain specialists cannot assimilate, through human cognition, the information in the same pace that the information is created [9]; (ii) regular knowledge frameworks are not sufficient to fulfill the domain specialists information needs; and (iii) domain specialists do not have the technical skills in order to make use of algorithms that would allow them to process and interact with a large amount of information. Then, it can be stated that a framework that allowed domain specialists to create and interact with their own knowledge graphs without requiring programming skills would be a step towards narrowing the information creation and assimilation gap. The present work depicts the preliminary results towards a framework that aims to assert the previously stated problem. In other words, it is presented the preliminary work of a framework that aims to allow domain specialist to make use of the advantages of Knowledge Graphs research by creating its own knowledge graph.

The work is organized as follows. Section 2 presents the used methods in order to achieve the results shown in Sect. 3. A discussion about the results is found in Sect. 4 and, finally, Sect. 5 concludes the present work.

2   Methods

This section presents the methods that were used in order to create the presented results. The Subsect. 2.1 depicts the search result for similar works, followed by the Subsect. 2.2 that presents the general overview of the proposed framework. Subsect. 2.3 explains the NLP technique that was used to build the knowledge graph. And finally, Subsect. 2.4 is responsible for justifying the use of network visualization.

2.1   Similar Works

In order to execute a search for similar works at least three search parameters have to be defined: (i) Scientific Bases; (ii) Keywords; and (iii) inclusion and exclusion criteria. Such definitions are as follows. (i) Searched Scientific Bases are: Web of Science, Scopus, IEEE Xplore and Association for Computing Machinery Digital Library (ACM). (ii) Chosen keywords: Knowledge Graph, text OR corpus and Graphical Interface OR Web Application. (iii) The inclusion criteria is listed:

1. Present a framework to build a knowledge graph from text corpus;
2. Present a form of interacting with the knowledge graph;
3. Make the source code or the framework available for use.

And, finally, the exclusion criteria:

1. Not Present a framework to build a knowledge graph from text corpus;

2. Not Present a form of interacting with the knowledge graph;
3. Not Make the source code or the framework available for use;
4. Not being an scientific paper;
5. Not being in English or Portuguese

The search resulted in seventy-two (72) retrieved papers, after removing duplicate papers a total of sixty-nine (69) paper abstracts were read by the authors. For each abstract it was attributed the inclusion and exclusion criteria. Figure 2 depicts the distribution count of paper for each criteria combination. The papers placed within the black rectangle refers to the papers to which were attributed at least one inclusion criteria and none exclusion criteria. That is, these are the works considered to be similar to the present one. The work myDIG: Personalized illicit domain-specific knowledge discovery with no programming [11] was the only one that was classified as a similar work by the criteria defined.

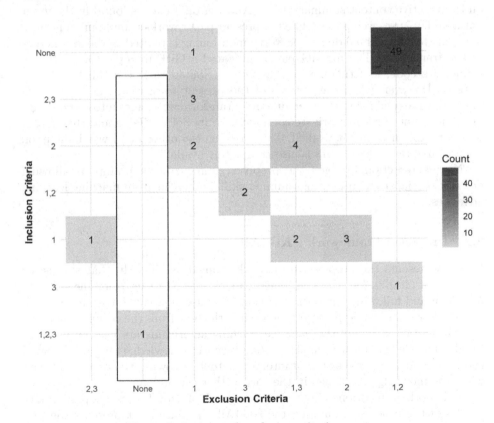

Fig. 2. Inclusion and exclusion criteria count

The work was developed at the Information Sciences Institute of University of Southern California, and presents a framework that allows investigative domain specialists to build and interact with their own knowledge graphs from the web pages. As one would expect, there are similarities and differences between myDIG and the present work.

The main similarity is found in the problem to be solved. Both works acknowledges that domain specialists struggle to keep up with the information creation. At the same time, the advances of data processing with Machine Learning, that would allow a way to narrow the gap between information creation and assimilation, requires programming and machine learning skills, that is not commonly found in domain specialists, restricting the number of domain specialists that can make use of such advances.

On the other hand, the main difference is found in the user profile. Both works have in mind domain specialists. However, while myDIG is focused in a case where the user has a well defined idea of what she is looking for, the present work focuses on the step where the domain specialist needs to have an overview of the knowledge relation in his corpus, that is, an easy to assimilate and interactive content summarization. Another difference is found in the input data, myDIG uses web pages while the present paper works is build upon natural language text. One final difference worth mentioning is related to the availability of the framework. The myDIG paper indicated a GitHub repository with the framework code, and therefore is was not attributed to it the third exclusion criteria. However, when the authors of the present work read the full myDIG paper it was explained that the engine that transform web pages into a knowledge graph is maintained by a private company and its not available and therefore it was not explained how it worked. This work on the other hand was built upon open source technologies and is also completely available[1].

Next sub-section presents the proposed framework that aims at allowing domain specialists with no programming skills to benefit from machine learning advances.

2.2   Proposed Framework - KG4All

Figure 3 presents the proposed framework, named as KG4All, that stands for Knowledge Graphs for All. The image can be read starting from the left side user icon and following the lines direction. The user uploads a corpus to a web application. This web application then sends the text from the corpus to a backend. This stage is where the machine learning algorithms are used in order to build a knowledge graph from the texts. Once the Knowledge Graph is created the web application makes use of interactive tools, allowing the user to interact with the knowledge contained in the corpus that was uploaded.

This work, as mentioned in the last paragraph of Sect. 1, presents preliminary results of the process of building the KG4All. Specifically, it presents the first

---

[1] Web interface code: https://github.com/viniciusmsousa/kg4all. Data Processing workflow: https://github.com/viniciusmsousa/KG4All-data-processing-explained. At the current stage these components are not connected in the application, as explained in Sect. 3.

results of the elements inside the black rectangle. That is, the *corpus* element, highlighted with red in Fig. 3, has not been implemented yet.

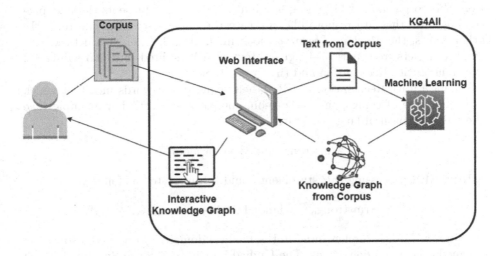

Fig. 3. Proposed framework (Color figure online)

A few practical considerations should be made. The choice to build a web application was made as the result of the following reasoning: The utilization of digital a tool is necessary mainly due to the fact that large amount of data processing is only possible through computers. Therefore, the real decision to be made is whether to build an web application or a smartphones app. The authors have chosen to build the web application for the following reasons. First, given the authors background build an web application presented less technical challenges. And secondly, people tend to be more productive on personal computers when compared to smartphones [1]. In order to build the presented framework the authors used the Shiny R package [7], which is a framework to build web application using the statistical programming language R [16]. Examples of others apps built with the framework can be found in the maintainer official gallery web page[2]. The main advantage of the framework is that it allows the creation of fully functional web application with in a relatively simple structure. The *en_core_sci_sm* model from the SciSpacy [13] python package was the choice to build the NLP tasks, that are explained in the Subsect. 2.3. Finally, as explained in Subsect. 3.1, the implementation was made using the metadata file from the COVID-19 Open Research Dataset Challenge (CORD-19) [3] and the raw data used for the results presented can be found in the link[3].

With that in mind the rest of this section presents the steps taken in order to achieve the preliminary result, shown in Sect. 3.

---

[2] https://shiny.rstudio.com/gallery/.

[3] https://drive.google.com/drive/folders/1YAHpv4-93rqMy94CyP830fRzN81Cwk9_?usp=sharing.

2.3   NLP

The objective is to allow the user to upload it's own corpus into KG4All, then from this corpus a knowledge graph is built. This section presents the text processing tasks that are responsible to create the triplets set from the texts. In other words, the *Machine Learning* block in the Fig. 3. The general task, *i.e.*, extract triplets from natural language text, can be splited into two sub-tasks: (i) Name Entity Recognition and (ii) Entity Linking.

Name Entity Recognition (NER) labels sequences of words in a text which are the names of things, such as person, company, etc. [21]. For example take the following natural language statement:

Armstrong landed on the moon.

After a NER processing this statement could be annotated as follows:

Armstrong$_{person}$ landed on the moon$_{location}$.

Since KG4All is implemented in the medical domain it is needed a source to get medical entities definitions. The Unified Language Medical System (UMLS) [6] provides just that. A few examples are shown in Table 1 and the full database with the definitions and relations from UMLS used in this work can be found in this link[4].

Table 1. Examples of medical entities from the UMLS.

Entity type	Entity name
Intellectual Product	Clinical Trial Objective
Virus	Avipoxvirus
Cell Component	Azurophilic granules
Temporal Concept	Priority
Bird	Aves
Intellectual Product	Report (document)
Population Group	Donor person

Therefore, an example of a NER annotated medical text could look like:

The report$_{Intellectual\ Product}$ on the Avipoxvirus$_{Virus}$ is the current priority$_{Temporal\ Concept}$.

The second sub-task is called Entity Linking which aims at finding a relation between two entities [21]. For example, by reading the statement **Armstrong landed on the moon** the human cognition interprets the semantic meaning

---

[4] https://drive.google.com/drive/folders/1kEw1_rJA7pI5VycmaXBVwbN0XMWUM sST?usp=sharing.

and concludes that there is a link between the entities **Armstrong** and **moon**. And this link is **landed_on**. Finally, this knowledge can be represented in a triplet form as:

(Armstrong, landed on, moon)

Entity linking aims at using algorithms to detect these relations. The algorithms usually integrate three steps to link entities [21]:

1. Entity mention spotting: Detects mentions in the text of multiple entities;
2. Entity mention mapping: Lists the possibles entities from a formal knowledge base;
3. Candidate Selection: Selects, based on a criteria, which candidates are indeed linked with the mentioned entity.

Therefore, by completing this two sub-task it is possible to build a knowledge graph from text. An example of the data processing workflow developed by authors to create a KG from Medical text can be found on this github page (prepared by the authors)[5]. It presents the use of the SciSpacy [13] which is a open source python [20] framework dedicated to dealing with scientific texts from medical domain. The framework allows a large range of tasks, but the purpose of this research it was focused on the Named Entity Recognition and Entity Linking.

Once the data processing workflow is completed it is possible to create tools to allow the user to interact with the knowledge graph without having to program anything, this is the topic of the Subsect. 2.4.

2.4   Knowledge Graph Visualization

Having the extracted triplets from the corpus the next step towards the proposed framework is to allow a user to interact with the knowledge graph. As shown in Sect. 1, a knowledge graph has a network structure, *i.e.*, nodes (the entities) connected by edges (the relations).

Producing and examining a network plot is often one of the first step in a network analysis, since its overall purpose is to allow a better understanding of the underling structure in the data [12]. Figure 1 is an example of how a network can be visualized in order to reveal the underlying structure of the data. The use of aesthetics can enhance certain feature from the data in a better visual form. Color the nodes to indicate different types of nodes and change the edge size to depict the relation strength or count are two ways to do so. Therefore, the first interaction element implemented in the KG4All is the tool that allows the user to view a network graph from the knowledge graph extracted form the corpus, by selecting a document of interest.

In summary, this section started demonstrating, in Subsect. 2.1, the research gap in allowing domain specialists to benefit from the advances in the machine

---

[5] https://github.com/viniciusmsousa/KG4All-data-processing-explained/blob/main/01DataProcessingExplained.ipynb.

learning and natural language processing research in order to interact with a large number of documents. Second, Subsect. 2.2, presented and explained in a high level a possible way towards fulfilling the gap previously mentioned with the KG4All framework. Thirdly, Subsect. 2.3 explained the tasks of entity recognition and entity linking, which are the tasks that the presented work relies upon. And finally, the current sub section justified the choice of using network graphs to create an interactive knowledge graph as kick start to the web application. The next session presents the results obtained as this research evolves.

3   Results

This section presents the functional prototype of the KG4All framework. As stated before the KG4All source code is open source, currently it is present in two github repositories[6] due to the fact that the Web Application is not integrated with the Machine Learning back end yet. And the application can be accessed throught the link viniciusmsousa.shinyapps.io/KG4All[7]. The prototype current main features are: (i) detects the relations within the abstracts from the COVID-19 Open Research Dataset Challenge (CORD-19) [3] and (ii) connect these relations to the UMLS relations mapping. The following of this presents the domain implementation and test corpus in Subsect. 3.1. Next, the web interface components in Subsect. 3.2. The triplets display component in Subsect. 3.3 and finally the interactive graph in Subsect. 3.4.

3.1   Domain Implementation

The Covid-19 pandemic breakout in early 2020 and changed most people's life. The subject became an important topic in the international organizations agendas. Due to the fact that this research was taking place during the pandemic peak in Brazil the authors decided to implement the proposed framework in the Medical Domain. Specifically, it was chosen to develop a data processing workflow that works with medical texts, based in the Unified Medical Language System [6] and tested it to generate the results using the abstracts of papers related to the coronavirus extracted from the COVID-19 Open Research Dataset Challenge (CORD-19) [3], which is the result of a response coordinated by the White House to make available the scientific publications related to the coronavirus.

3.2   Web Interface Components

Figure 4 presents the KG4All Web Interface, *i.e.*, the interface that the domain specialist interacts with. It is composed mainly by three components marked in the figure. Component 1 is the search bar that allows the user to search for a

---

[6] Web application: https://github.com/viniciusmsousa/kg4all. Data Processing: https://github.com/viniciusmsousa/KG4All-data-processing-explained.

[7] https://viniciusmsousa.shinyapps.io/KG4All.

desired document using words. Component 2 presents the triplets extracted from the abstract of the document selected in component 1. And finally, component 3 presents the interactive graph visualization of the relations.

The Subsect. 3.3 explains the KG4All's triplets component.

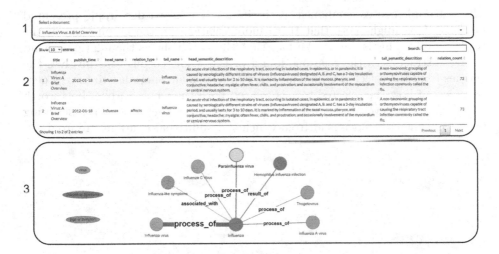

Fig. 4. KG4All Web Interface ('Influenza Virus: A Brief Overview')

3.3 Triplets Component

The triplets are shown in component 2 of the KG4All. Figure 5 is a zoom in component 2. Each row each represents on relation found in the text. The relation in itself is in the columns *head_name*, *relation_type* and *tail_name*. The other columns of the table presents additional information about the relation. We highlight that the column entitled *relation_count* depicts the number of times that the relation occurred in the whole corpus.

	title	publish_time	head_name	relation_type	tail_name	head_semantic_description	tail_semantic_description	relation_count
1	Influenza Virus: A Brief Overview	2012-01-18	Influenza	process_of	Influenza virus	An acute viral infection of the respiratory tract, occurring in isolated cases, in epidemics, or in pandemics; it is caused by serologically different strains of viruses (influenzaviruses) designated A, B, and C, has a 3-day incubation period, and usually lasts for 3 to 10 days. It is marked by inflammation of the nasal mucosa, pharynx, and conjunctiva; headache; myalgia; often fever, chills, and prostration; and occasionally involvement of the myocardium or central nervous system.	A non-taxonomic grouping of orthomyxoviruses capable of causing the respiratory tract infection commonly called the flu.	73
2	Influenza Virus: A Brief Overview	2012-01-18	Influenza	affects	Influenza virus	An acute viral infection of the respiratory tract, occurring in isolated cases, in epidemics, or in pandemics; it is caused by serologically different strains of viruses (influenzaviruses) designated A, B, and C, has a 3-day incubation period, and usually lasts for 3 to 10 days. It is marked by inflammation of the nasal mucosa, pharynx, and conjunctiva; headache; myalgia; often fever, chills, and prostration; and occasionally involvement of the myocardium or central nervous system.	A non-taxonomic grouping of orthomyxoviruses capable of causing the respiratory tract infection commonly called the flu.	73

Showing 1 to 2 of 2 entries   Previous 1 Next

Fig. 5. Triplets component ('Influenza Virus: A Brief Overview')

For example, from the first line of the figure it can be seen that the entity **Influenza** (head entity) is a **process of** (relation) the **Influenza Virus** (tail entity). Next step in to connect this entities with other entities found in the corpus, through the UMLS. This is presented in the Subsect. 3.4.

3.4   Interactive Graph

The last implemented component is the interactive knowledge graph, presented in Fig. 6. One might note that there more relations in the graph than in the triplets table. This is due to the fact that the graph shows the relations found in the abstract with the relations found in the whole corpus that involves the entities from the selected text. This allows the domain specialist to have a general view of how the selected text is related with the whole corpus. It is worth mentioning that even though it is not implemented, the authors are studying ways to explicit the document from which the additional relations are from.

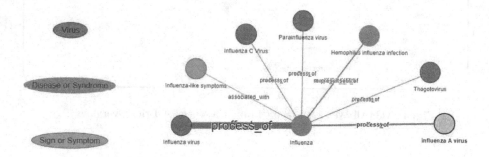

Fig. 6. Interactive knowledge graph ('Influenza Virus: A Brief Overview')

As is expected in network visualizations KG4All use some aesthetics to add more information to the relations. The nodes color demonstrates to which group each node belong to. For example, **Influenza** is classified by the UMLS as a **Disease or Syndrome**. On the other hand **influenza A virus** is classified as a **virus**. Another aesthetics used in the interactive graph is the edge (or link) size. It proportional to the number of times that the relation was present in the corpus.

Section 4 presents considerations about the results, improving directions that are in the authors workflow.

4   Discussion

A few considerations about KG4All itself. First, the current implementation uses a selected dataset to create the interactive knowledge graph, however this is a temporary implementation. Once the upload interface and the integration between the web interface and the back end model is done, KG4All will have an upload interface where the users will be allowed to upload their own medical corpus. Second, there are two factors that impacts KG4All computing cost: (i) The model used to detect the medical entities and (ii) the size of the corpus that is submitted to the data processing pipeline. The model that is current being used is the *en_core_sci_sm* [13] and once it is loaded it uses 132 MiB of memory. And the dataset used to create the prototype, with 81.354 medical abstracts, used around 10 GB while running on win10 with intel i7. It is worth noting that in practical use the authors expect smaller corpus, for instance, the result OS a scarch in scientific articles database. Third, the use of machine learning algorithms to extracted the triplets cannot guarantee that all the entities relations present in the text will be extracted. How ever, as shown in the SciSpacy paper [13] the amount of relations detected are not insignificant, providing a reasonable summarizing of the knowledge present in the corpus. And, finally, there are both some implementations as well as corrections to be made on the current state. For example, a way to explicit from each document the entity was extracted, when it is a relations that is not in the selected document is a implementation to be made. In some cases there are overlapping of the edges name, which is a correction in the back log.

Besides the practical differences from the myDIG [11] work, explained in Subsect. 2.1, the authors believes that KG4All complements the myDIG, in the sense that the same issue, gap between domain specialists information assimilation and creation, is being addressed. And contributing for a different group of information users by focusing in an open source tool for knowledge graph creation and interaction.

5   Conclusion

The present work has argued that there is a gap between information creation and assimilation. This gap impacts the domain specialists, a group that the traditional information tools does not satisfies their information necessities. It has also been argued that the research advances in the field of information retrieval through knowledge graph using machine learning algorithms is evolving and provides ways to narrow the information gap. However, such advances are relied on a high level of computational and mathematical complexity. This high complexity results in the need of programming and machine learning skills in order to make use of the advances. Such skills are not commonly found in domain specialists. It was presented the KG4All prototype, which is a framework that will allow users to upload their own corpus and interact with a knowledge graph created from the corpus without the need of programming skills. The domain that the KG4All is implemented is the medical one, due to the fact that the Covid-19 pandemic took place while this research was taking place. There are other works proposing solutions to the same problem however with a differences in the target domain specialists user profile, and therefore, the present work contributes to the research on how to make the advances in knowledge graph through machine learning usage.

References

1. Adepu, S., Adler, R.F.: A comparison of performance and preference on mobile devices vs. desktop computers. In: 2016 IEEE 7th Annual Ubiquitous Computing, Electronics Mobile Communication Conference (UEMCON), pp. 1–7 (2016). https://ieeexplore.ieee.org/document/7777808
2. Agarwal, R., Dhar, V.: Editorial-big data, data science, and analytics: the opportunity and challenge for IS research. Inf. Syst. Res. **25**(3), 443–448 (2014). https://doi.org/10.1287/isre.2014.0546
3. AI, A.I.F. https://www.kaggle.com/allen-institute-for-ai/CORD-19-research-challenge
4. Araújo, C.A.A.: Correntes teóricas da ciência da informação. Ciência da informação **38**, 192–204 (2009). http://www.scielo.br/scielo.php?script=sci_arttext&pid=S0100-19652009000300013&nrm=iso
5. Araújo, C.A.A.: Fundamentos da Ciência da Informação: correntes teóricas e o conceito de informação. Perspectivas em Gestão Conhecimento **4**(1), 57–79 (2014). https://periodicos.ufpb.br/ojs2/index.php/pgc/article/view/19120
6. Bodenreider, O.: The unified medical language system (UMLS): integrating biomedical terminology. Nucleic Acids Res. **32**(1), D267–D270 (2004)
7. Chang, W., Cheng, J., Allaire, J., Xie, Y., McPherson, J.: shiny: Web Application Framework for R (2020). https://CRAN.R-project.org/package=shiny, R package version 1.5.0
8. Elbashir, M., Collier, P., Davern, M.: Measuring the effects of business intelligence systems: the relationship between business process and organizational performance. Int. J. Accounting Inf. Syst. **9**(3), 135–153 (2008). https://www.scopus.com/inward/record.uri?eid=2-s2.0-51249116446&doi=10.1016%2fj.accinf.2008.03.001&partnerID=40&md5=f6748444fd6918d43aa33b5de2c118d3, cited By 254
9. Hoyt, C.T., et al.: Re-curation and rational enrichment of knowledge graphs in Biological Expression Language. Database 2019 (2019). https://academic.oup.com/database/article/doi/10.1093/database/baz068/5521414
10. Ji, S., Pan, S., Cambria, E., Marttinen, P., Yu, P.S.: A Survey on Knowledge Graphs: Representation, Acquisition and Applications (2020). https://arxiv.org/abs/2002.00388
11. Kejriwal, M., Szekely, P.: myDIG: personalized illicit domain-specific knowledge discovery with no programming. Future Internet **11**(3) (2019). https://www.mdpi.com/1999-5903/11/3/59
12. Luke, D.A.: A User's Guide to Network Analysis in R. UR, Springer, Cham (2015). https://doi.org/10.1007/978-3-319-23883-8
13. Neumann, M., King, D., Beltagy, I., Ammar, W.: ScispaCy: fast and robust models for biomedical natural language processing. In: Proceedings of the 18th BioNLP Workshop and Shared Task, Florence, pp. 319–327. Association for Computational Linguistics (2019). https://www.aclweb.org/anthology/W19-5034
14. Olszak, C., Ziemba, E.: Approach to building and implementing Business Intelligence systems. Interdisciplinary J. Inf. Knowl. Manag. **2**, 135–148 (2007). https://www.scopus.com/inward/record.uri?eid=2-s2.0-77749242597&partnerID=40&md5=fd70fbb98a2ddee0b6daf68f28050db5, cited By 81
15. Pinto, A.L., Silva, A.M., Sena, P.M.B.: Ontologias baseadas na visualização da informação das redes sociais. Prisma.com (Portugual) **24**(13), 5–24 (2010). https://www.brapci.inf.br/index.php/res/v/68060

16. R Core Team: R: A Language and Environment for Statistical Computing. R Foundation for Statistical Computing, Vienna (2020). https://www.R-project.org/
17. Saracevic, T.: Ciência da informação: origem, evolução e relações. Perspectivas em Ciência da Informação 1(1) (1996). http://portaldeperiodicos.eci.ufmg.br/index.php/pci/article/view/235
18. Sen, S., Li, T.J., Team, W., Hecht, B.: WikiBrain: Democratizing Computation on Wikipedia (2014). https://doi.org/10.1145/2641580.2641615
19. Shannon, C.E.: A mathematical theory of communication. Bell Syst. Tech. J. 27(3), 379–423 (1948). https://onlinelibrary.wiley.com/doi/abs/10.1002/j.1538-7305.1948.tb01338.x
20. Van Rossum, G., Drake, F.L.: Python 3 Reference Manual. CreateSpace, Scotts Valley, CA (2009)
21. Waitelonis, J.: Linked Data Supported Information Retrieval. Ph.D. thesis, Karlsruher Institut für Technologie (2018)
22. Yuan, J., et al.: Constructing biomedical domain specific knowledge graph with minimum supervision. Knowl. Inf. Syst. 62(1), 317–336 (2020). https://link.springer.com/article/10.1007/s10115-019-01351-4

A MapReduce-Based Method for Achieving Active Technological Surveillance in Big Data Environments

Daniel San Martin Pascal Filho, Douglas Dyllon Jeronimo de Macedo[iD],
and Moisés Lima Dutra[✉][iD]

PGCIN, Federal University of Santa Catarina, Florianopolis, Brazil
{douglas.macedo,moises.dutra}@ufsc.br

Abstract. Technological Surveillance systems stand out as a structured way to assist organizations in monitoring their internal and external technological environments, in order to anticipate changes. However, since the volume of digital data available keeps growing, it becomes increasingly complex to keep this type of system running without proper automation. This paper proposes an automated MapReduce-based method for technological Surveillance in Big Data scenarios. A prototype was developed to monitor key technologies in specialized portals in the Furniture and Wood sector, in order to illustrate the proposed method. The proposal was evaluated by industry experts, and the preliminary results obtained are very promising.

Keywords: Technological surveillance · MapReduce · Big Data · Ontologies

1 Introduction

Throughout history, technological changes have created, transformed and destroyed markets. The monitoring of the main technological trends plays a key role in increasing the competitiveness of companies. However, the accelerated pace of technological development and the steady reduction of time between different innovation cycles increase complexity and the efforts required to maintain an up-to-date picture of the whole technological scenario [23].

In search of solutions, a wide variety of approaches to technological monitoring have been proposed. They range from interviews, through simple keyword-based Internet searches, to sophisticated text mining systems in search of hidden patterns [7]. Among those, Technological Surveillance (TS) has gained prominence. TS seeks to establish a monitoring process that ranges from collecting data to communicating insights about it to decision makers. That allows organizations to have an overview of the technological changes from different sources of information. Nevertheless, maintaining a functional and active monitoring system to evaluate real Big Data scenarios composed of patents, scientific papers,

E. Bisset Álvarez (Ed.): DIONE 2021, LNICST 378, pp. 254–271, 2021.
https://doi.org/10.1007/978-3-030-77417-2_19

books, and spreadsheets, to name a few, is still one of the biggest challenges faced by organizations.

In this sense, the objective of this work is to propose a method to identify the use of existing technologies in Big Data scenarios, especially in those environments densely populated by large volumes of data. In a nutshell, we intend to answer the following research question: how to automatically identify a set of technologies of interest whose use is increasing in popularity on Web portals?

The remainder of this work is organized as follows. In Sect. 2, a theoretical framework of reference is given, which depicts the concepts of technological surveillance, and also talks about information, domain, Big Data scenarios, and the MapReduce model. Section 3 presents the methodological procedures used to develop this work. In Sect. 4, the proposed method is described, along with the prototype developed to implement it and the experimental results achieved. Finally, Sect. 5 presents the final considerations and suggestions for future works.

2 Theoretical Framework

2.1 Technological Surveillance

Technological Surveillance (also known as Technological Vigilance or Technological Watch) arose from the need to monitor the internal and external environments of organizations in a structured and systematic way, in order to map possible changes in technological scenarios.

Table 1 presents the most relevant TS definitions in the context of this work. They converge in terms of the proposed/existing stages of the surveillance process, raging across collecting information, analyzing it, and communicating the results obtained to interested parties. Palop and Vicente [21] point out the key reasons that lead organizations to adopt a TS system: (i) the need to anticipate changes in order to avoid competitive disadvantage, especially in scenarios where technology can be a differential; (ii) cost reduction; (iii) the progress of the organization in relation to the market; (iv) the need to innovate; and (v) new possibilities to establish cooperation with other organizations. In summary, it can be stated that Technological Surveillance is a method used to monitor and evaluate different sources of information, mostly of which to be used as an input for patent registration processes. That is, it searches for evidences that could signal changes in the technological scenario in which an organization is inserted.

In the literature, proposals for automated technological monitoring are presented by using patent bases [17, 29], such as the database of the USTPO (US Patent and Trademark Office) [9, 18], scientific papers [1, 11], and publications available on Web portals [26]. The techniques used by the authors are mainly focused on the extraction of terms from text documents, by means of calculating the TF-IDF (term frequency–inverse document frequency) and applying the LDA (Latent Dirichlet Allocation) algorithm for topic modeling [17, 18, 29]. Besides, several techniques for document clustering are applied, notably the k-means algorithm [9, 18], in order to detect and recognize outstanding technologies.

Table 1. Definitions of technological surveillance.

Definition	Authors
Method of collecting, analyzing, disseminating, communicating and using information for decision making, including competitive intelligence or similar terms	Palop and Vicente (1999) [21]
A structured system to coordinate the activities of information retrieval, analysis and dissemination, both inside and outside the organization, according to an organizational plan and strategy	Salgado Batista et al. (2003) [4]
Technological surveillance goes through the stages of diagnosis, research and capture of information, analysis of information, valuation of relevant information, information dissemination and communication, by offering guidance to decision making	OVTT [20]
A technology surveillance model consists of a set of processes: identification of needs, a definition of sources and means of access to information; search, processing and validation of information; and valuing of information, results, measurement and improvements	AENOR [3]
A systematic process that aims to identify, organize and correlate the results of technological prospecting in order to make them useful to the organization's strategies	ABNT [2]

However, these proposed methods limit their field of monitoring, since most organizations are immersed in Big Data scenarios, in which they need to store and monitor a large volume of data in a variety of formats, such as text documents, database content, audio, video, spreadsheets, patents, clicks, physical-device data, internal systems data, Web portal content, event records, news-website content, social-media content, scientific publications, among others.

2.2   The Information and Its Sources

From the point of view of Information Science (IS), according to Buckland's view [5], information in the process of Technology Surveillance can be understood as knowledge and the analyzed documents as "things" that can reduce uncertainties. Therefore, the ability to evaluate the quality of the documents and correctly delimit their sources is fundamental to this process, being one of the factors that most impact surveillance results, according to León et al. (2006) [16].

According to quality criteria presented by Muñoz et al. (2006) [14], sources of information can be divided into formal, such as articles or books, and informal, such as talks or visits to fairs. A TS system should work with formal sources of information. As far as processing is concerned, sources of information can be classified as "electronically available" and "not electronically available".

The quality of information sources can be analyzed by means of applying specific criteria and indicators. In her book "Information Sources on the Internet", Tomaél (2008) [28] lists as possible criteria to be used: information architecture and its intrinsic aspects (content *versus* user need), the credibility of the information source (contextual handling aspects), and information representation (conciseness and consistency) and its sharing aspects.

2.3 Domain

The concept of domain was introduced in the IS context by Birger Hjørland (1995) [12]. Before that, it was already widely used in Computer Science, in which was disseminated in the mid-1980s [19]. A domain is a group that shares an ontology, undertakes common research or work, and also engages in discourse or communication, formally or informally [27].

The analytical domain paradigm of Information Science states that the best way to understand the concept of information in IS is studying the domains of knowledge as communities of thought or discourse, which are part of the division of labor of society [12]. According to the domain analysis, if we wanted, for example, to develop a software application for Brazilian Geography analysis, we should not focus on certain users, but instead call a geographer specialized in Brazilian Geography to help us. Going against this premise, this research used information produced by experts from a particular industry sector to model ontologies containing the key technologies and critical resources to their area.

2.4 Big Data

The information of interest in Technological Surveillance systems are usually collected from different sources and possess different formats. TS systems need to deal with thousands of unstructured digital documents, as well as huge volumes of data that, due to the high level of information digitization, keep growing. In this work, we consider these configurations as part of Big Data scenarios.

Among the several existing Big Data definitions found in the literature, two stand out: the definitions 3 "Vs" and 5 "Vs". The 3 "Vs" definition was adopted by authors such as [13] and [15], which characterize it as being: variety, and velocity. The definition of the 5 "Vs" adds two new features to them: veracity and value [30]. There are still other definitions composed by more "Vs", and even other letters. Actually, there is no consensus at all about the concept of Big Data. In Technological Surveillance scenarios, data is only useful if it has value and can be checked for accuracy. Consequently, in this work we adopted the definition of 5 Vs to define Big Data scenarios.

Unlike traditional database management systems (DBMSs), such as MySQL or Oracle, which were built to work with structured data in the form of relational tables, Big Data data often varies in structure. Therefore, operations as text mining are often common and required to process or retrieve information. Partitioning data across multiple computers or nodes may also be required to manage the large volume of data or allow it to be processed and analyzed in an acceptable time frame.

2.5 MapReduce Processing in Big Data Scenarios

MapReduce was conceived in Google as a solution to optimize its Web search engine in 2003, based on the paradigm of functional computing and programming languages such as LISP. Between 1999 and 2004, Google engineers developed dozens of algorithms to process data in order to generate more data. In 2008, the model was already scaled, processing 20 petabytes daily [6]. According to Dean and Ghemawat [6], the programming model MapReduce is an associated implementation useful for a wide variety of real world tasks and was designed to handle large datasets. Its structure enables parallel processing in large collections of data through computer clusters. Currently, it is the main programming method used to work in Big Data scenarios.

Through it, it is possible to elaborate simple algorithms capable of being executed both in a personal computer and in distributed environments composed of thousands of computers, allowing the construction of scalable solutions. Otherwise, the processing would be sequential, possibly presenting performance problems in environments with large volumes of data. Conceptually, the Map function receives a collection of input data and applies a function on them, generating a new collection. The Reduce function receives a collection of input data and applies a function on them, providing a reduction in the size of the collection.

Figure 1 presents the application of the MapReduce model for counting words. The input sentences are splitted into words, for which the value 1 is assigned during the Map step. In the Reduce step, similar words have their values summed and assigned. As a result, a list is generated with the words and the total number of their occurrences in the input text.

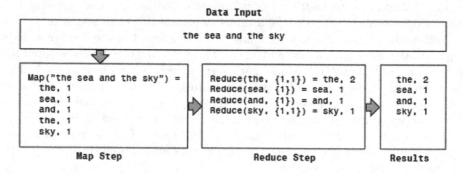

Fig. 1. MapReduce example.

3   Methodological Procedures

This work is an applied research, as it aims to generate knowledge for the application of methods aimed at solving a specific problem involving local interests. Regarding the objectives, among the classifications proposed by Gil [10], this work conducts an explanatory research, which seeks to understand the causes and effects of the phenomena under study, in a structured, quantitative and lateral way in experimental methods. To overcome the limits of the quantitative research [8], this work also makes use of a qualitative approach. When validating the proposal with specialists, a qualitative evaluation of the result is necessary. Quantitative approaches are also used in the data analysis process, during the experiments.

This work uses experimental and bibliographic procedures, too. The bibliographical research seeks to explain a problem from theoretical references published in documents, such as books, periodicals, scientific papers, without the elaboration of hypotheses. This approach is important mainly in the theoretical foundation, by bringing knowledge to the authors and structuring the basis for the construction of the remaining knowledge of this work. According to Gil [10], the experimental research consists in determining a study object, selecting the variables that would be able to influence it, and defining the ways of controlling and monitoring the effects produced by the variables on the object.

The elaboration of this research takes place through the definition of a research problem and a research question, in order to establish the paper's goals. In addition, in order to support the early research process, topics like technology surveillance, information, domain, Big Data, and MapReduce are studied. Besides, an analysis of related works is carried out to compose the necessary theoretical basis of the work, which is used to objectively interpret and analyze the subject, as well as produce conclusions based on proven theories.

During the analysis of the related works, opportunities for improvement are identified. Subsequently, they are included in this work's main proposal. To validate it, a functional prototype is built, by using MongoDB technology as the data repository, Protegé as the software tool for ontology modeling, Python as the programming language for capturing, parsing and analyzing data, and Javascript as the programming language for implementing the MapReduce model. To validate this prototype, we present the results obtained from specialists who pointed out the correct alignment of the results in relation to the market.

4   The Proposed Method

Table 2 summarizes the proposed method for Technological Surveillance in Big Data scenarios. It aims to automatically identify key technologies and assess their popularity in large volumes of documents collected from formal sources of digital information.

It is based on the TS cycle proposed by Sánchez Torres and Palop Marro (2002) [22]. The main steps of the proposed method are: (1) **Planning**. Identification of the needs and objectives of the organization; (2) (3) **Collection** and **Organization**. Gathering and filtering of the data, in order to select it and eliminate what is not necessary; (4) **Intelligence**. Qualification of information and its alignment with the organization's strategies; and (5) **Communication**. Dissemination of the results within the organization for the purpose of decision making.

The proposed method (Table 2) delimits a series of activities required to enable the automation process. The **Planning** step requires human intervention, which involves experts, decision makers, and anyone else interested in the outcome of the process, as it must be aligned with the strategic objectives of the organization. At this stage, a domain analysis is applied, by means of delineating the domains of interest and the technologies that should be monitored. The sources of digital information, such as Web portals, are also evaluated by using criteria indicated by Tomaél (2008) [28]. Finally, the technological knowledge of the domain is modeled in the form of ontologies, used to represent the terms to be located in the collected texts. For example, if there is interest in monitoring technologies related to energy production, the Energy domain could be modeled, by creating classes such as Renewable Energy and Non-renewable Energy. The Renewable Energy class, in turn, could aggregate subclasses like Solar Energy, Wind Energy, or Biomass Energy, which would represent the searched terms. This approach is not only useful because it maintains the hierarchy of concepts, but also because it facilitates the grouping of terms by levels of abstraction, i.e., the superclasses.

Table 2. The proposed method for technological surveillance.

Step	Main Features
1. Planning	Domain analysis. Ontology modeling. Assessment of digital sources of information
2. Collection	Web Crawlers. Web Scrapers. Query APIs. Gathering of digital data
3. Organization	Data wrangling. Storage in non-relational databases
4. Intelligence	MapReduce application. Identification of key technologies. Technology counting. Construction of time series. Creation of charts
5. Communication	Report generation. E-mail sending. Business Intelligence tools. Website feed

The **Collection** step is responsible for automatically collecting publications from predefined sources and make them available for the next step. It demands the construction of web robots, known as Web Crawlers [25]. Web crawlers gather data through a process known as Web Scraping. Moreover, publication data can also be collected either through APIs (Application Programming Interfaces) provided by the content portals, or by the directly collecting of digital documents from specialized repositories. All the collected data are stored in its raw form.

During step 3, **Organization**, a data wrangling process is performed. This process consists of extracting content from the collected publications, such as titles, dates, entities, among others. Each piece of extracted content is converted to a structured data type, like text, date, time, or numeric, in order to allow its further manipulation. Subsequently, they are organized and indexed in appropriate databases. In addition, the duplicated documents are removed in this step.

The texts extracted from the collected documents still represent a challenge for the analysis process, mainly due to its high dimensionality. Much of the information contained in the publications are useless. Some words are not relevant to the TS process. Therefore, it is essential to identify and extract useful data from the texts. Figure 2 shows an example of text reduction during the content extraction process. On the left is the original document collected. On the right side are the data stored in the database, after the data wrangling process. In the end, the reduced data refer to technologies identified with the analyzed scenario.

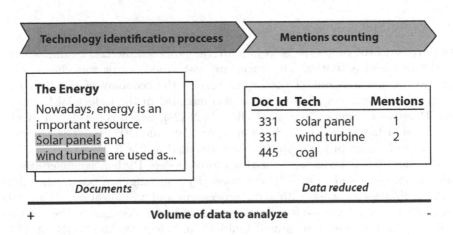

Fig. 2. Dimensionality reduction.

In Big Data contexts, moving large amounts of data for processing is quite computationally expensive. MapReduce functions possess the advantage that they can be executed in parallel on each cluster node where the data is. In this sense, the **Intelligence** step uses a MapReduce algorithm to identify key technologies, in order to enable the TS system to cope with Big Data. The algorithm generates a data subset with less dimensionality, which contains the technologies found in the source publications.

Once the volume of technologies has been quantified, it is possible to build time series that correlate the mentions of the technologies identified with their publication dates. Next, graphs that summarize the calculations performed can be generated to facilitate the analysis of the specialists.

Finally, the communication stage aims to satisfy users' information needs through the generation or automatic update of by-products of technological surveillance. In the previous steps, data were collected, analyzed and transformed, making them more understandable. Now, the interpretation of the final users must be facilitated by the enrichment and formatting of this new information, through the generation of reports containing the produced charts. Other examples of TS by-products are: (i) sending alerts by e-mail, (ii) updates to a specific BI system, (iii) and updates to a certain section of a website dedicated to the mining subject.

The proposed method defines a clear sequence of steps to create an automated system, which is designed to monitor a set of technologies of interest in electronically available information sources. This monitoring also includes checking how popular these technologies are on the Web. In the next section, we present a prototype that implements the proposed method and runs it on a given set of information sources.

4.1   Prototype

The purpose of the prototype is to validate the proposed method by implementing the proposed activities. The Furniture and Wood sector was chosen as a scenario because it is one of the key sectors in the economy of the Brazilian state of Santa Catarina. This sector is also included in the Industrial Development Program of Santa Catarina (PDIC 2022) [24], which is developed by the Federation of Industries of the State of Santa Catarina (FIESC). The architecture and main steps of the prototype are presented in Fig. 3.

The main technologies used for this prototype are Python language (along with its libraries) and MongoDB database. Python language ecosystem offers a set of libraries to work with data processing and visualization. MongoDB is a multi-platform open source database designed for documents classified as NoSQL. In it, documents are stored in JSON-style format. MongoDB was chosen because of its ability to scale horizontally, by adding new instances in other computers. Moreover, it possesses orientation to documents, since it allows the modeling of the various data collected during the TS process, such as scientific publications, patents, documents come from trade show and events, etc. MongoDB makes the representation of those documents closer to the analysts' reality, i.e. flexible for analysis. Besides, it provides the possibility of data processing through Map and Reduce functions.

Planning. In the Planning stage, a domain analysis was carried out on the Furniture and Wood sector, present in the PDIC 2022 notebook. Moreover, a discussion with FIESC specialists was held. Subsequently, it was modeled a domain ontology containing the key technologies listed for the this economy sector. Each technology was connected to a superclass that is used to group similar concepts. The ontology modeling was done in the Protegé tool, and the output was an OWL (Web Ontology Language) file. The selection of information sources was

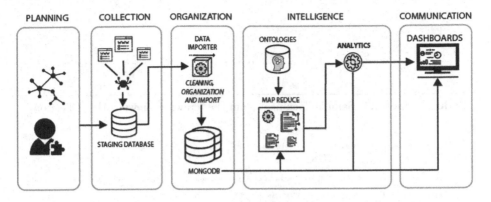

Fig. 3. Prototype architecture.

based on interviews with FIESC specialists, and also by reusing those already monitored by the FIESC Observatory, such as the Web portals: For Mobile[1], Furniture News[2], Mega Movers[3], Wood Business[4], and Woodworking Network[5].

Collection. A set of web crawlers was developed in the Python language to collect publications from the selected web portals. The main Python library used was BeautifulSoup[6], which offers specialized functions to access web pages and capture the desired content. Web crawlers work in two steps. First, they browse the websites where the publications are listed, collect their links, and store them. After, they visit each URL to scan the documents. Their titles, texts, and publication dates are collected and stored in a specific collection in the MongoDB database, as shown below.

Organization. The content extracted from the URLs visited are stored as documents in a collection called "crawled_news_urls". These documents contain metadata, e.g. an indicator of whether the content of a given document was extracted or not and its URL source. The content captured in the second stage of the Collection stage constitute another type of document, and are stored in a collection called "publications". Code snippets 1.1 and 1.2 below show examples of each type of document.

---

[1] http://www.formobile.com.br.
[2] http://www.furniturenews.net.
[3] http://www.megamoveleiros.com.br.
[4] http://www.woodbusiness.ca.
[5] http://www.woodworkingnetwork.com.
[6] Python package used to analyze documents in HTML and XML formats.

```
1  {
2      "_id"  :  ObjectId("5b4ff6202e77ca12b95c98c8"),
3      portal_name"  :  "Wood Business",
4      "domain"  :  "FurnitureAndWood",
5      "portal_url"  :  "https://www.woodbusiness.ca/industry-news/news/Page-8",
6      "news_url"  :  "http://www.woodbusiness.ca/industry-news/news/the-new-bioeconomy-adding-value-to-
            biomass-4655",
7      "extracted"  :  1
8  }
```

Code Snippet 1.1. Sample document from the "crawled_news_urls" collection.

```
1  {
2      "_id"  :  ObjectId("5b4ff9272e77ca132655b833"),
3      "portal_name"  :  "Wood Business",
4      "domain"  :  "FurnitureAndWood",
5      "portal_url"  :  "https://www.woodbusiness.ca/industry-news/news/Page-2",
6      "news_url"  :  "http://www.woodbusiness.ca/industry-news/news/canfor-appoints-dianne-watts-to-board-
            of-directors-4980",
7      "title"  :  "Canfor appoints Dianne Watts to board of directors",
8      "pub_date"  :  ISODate("2018-06-08T00:00:00.000Z"),
9      "text"  :  "'We are excited to have Dianne join our board and look forward to the experience,
            knowledge and fresh perspective that she will bring,' said Don Kayne, (...) company."
10 }
```

Code Snippet 1.2. Sample document of "publications" collection.

Intelligence. At this stage, we applied the MapReduce algorithm to identify the key technologies present in the documents collected. The algorithm is written in JavaScript and considers the ontologies terms and their relationships. A Python function was used to convert the OWL file content into a JSON-based dictionary within the JavaScript code.

The Map function converts the publication texts to lowercase and use a regular expression to look for technology mentions in them. As an output, this function returns the publication date, the founded technology and its superclass, and the number "1", which indicates that at least one term was found in the analyzed document. Then, the Reduce function adds the same terms found and returns the total value associated with the label "term_count".

The execution of this algorithm was done in the MongoDB environment, and generated a new collection of documents called "TechnologyMapping". Each document of this collection has a format as shown in the code snippet 1.3. In it,"pub_date" is the month in which the document was published, "technology" means the key technology found, "superclass" means the ontology class to which the technology belongs, and "value" is the document number in which the technology was mentioned that month.

```
1  {
2      {
3          "_id"  :  {
4          "pub_date"  :  "2017-03-01",
5          "technology"  :  "Composite Wood",
6          "superclass"  :  "Biotechnology"
7          },
8      "value"  :  1.0
9      }
10 }
```

Code Snippet 1.3. Output from the MapReduce Processing.

With the publication date and the amount of technology mentions obtained, we built a monthly time series by highlighting the number of documents in which a key technology appeared over the months. It is known that a process is stationary when the characteristics do not undergo changes over time, i.e. a process that develops randomly in time and oscillates around the mean. The stationary time series is the inverse of a non-stationary time series, in which growth or fall trends occur and, then, the mean and variance are related to time.

In order to know the unit root thesis for each time series constructed (formed by the number of documents that mention a technology monthly), the Augmented Dickey-Fuller Test was performed, which helps in detection of non-stationary time series, and verifies if the series used follow a steady stochastic process. This type of process would indicate that there would be no significant variations in the number of key technologies mentioned, and there could be a mature technology. On the other hand, a non-stationary series may indicate a tendency for growth or a drop in interest in a particular key technology.

Finally, graphs of the series were constructed and a polynomial of order 2 was adjusted, in order to facilitate the visualization and analysis of the specialists. The difference between the values $y0$ and $y1$ was calculated, representing the number of mentions, in percentage, in order to give an idea of the size of the variation and whether it was positive or negative.

Communication. In order to communicate the results to specialists, the system creates a report containing a summary of the main data, such as the number of documents captured, the ten most mentioned technologies, and the associated charts.

4.2 Experimental Results

During the initial stage of Planning, an OWL file containing the ontologies modeled for the Furniture and Wood sector was produced. This file comprises key technologies organized in the form of sheets. These technologies are (in alphabetic order): 3D printing, Additive Fabrication, Additive Manufacturing, Augmented Reality, Automation, Automation and Robotics, Biomass, Biotech, Biotechnology, Certifications, Composite Wood, Construction Wood, Distinctive Design, Engineered Wood, Hardboard, Health and Safety, High Density Fiberboard, Information and Communication Technology, Information Technology, Liquid Wood, Medium Density Fiberboard, Medium Density Particleboard, Multifunctional Furniture, Nanotechnology, Optimized Furniture, Rapid Prototyping, Robotics, Sensory Design, Single Households, Smart Furniture, Strategic Design, Virtual Reality, Waste Management, and Wood Frame.

In the Collection stage, web crawlers were constructed to scrape publications dated between June 1st, 2017 and June 1st, 2018. Table 3 shows the quantity of publications collected from the portals For Mobile, Wood Business, Furniture News, and Woodworking Network. After collected, these publications were organized and saved as MongoDB documents, as detailed above.

Table 3. Number of publications collected in each portal.

Portal	Publications collected
For Mobile	30
Wood Business	190
Furniture News	198
Woodworking Network	2500
Total	**2.918**

During the Intelligence stage, we synthesized indicators such as the number of documents in which the monitored technologies were mentioned (Table 4) and the volume of documents that quoted each class of the key technologies of the ontology (Table 5). In this way, it is possible to see which key technologies and classes were most cited in the period.

Table 4. Number of documents in which the technologies were cited in the period.

Key technology	Number of documents
Automation and Robotics	132
Wood for building	72
Biomass	31
Composite wood	31
Information and Communication Technologies	28
Virtual Reality	18
Health Safety	12
MDF	8
3D Printing	7
HDF	6
Nanotechnology	6
Augmented Reality	6
Waste Management	3
Strategic Design	1

Further details are presented in the Figs. 4 and 5, which show the total of documents that mentioned each key technology and the ontology-defined classes over the monitored period of time.

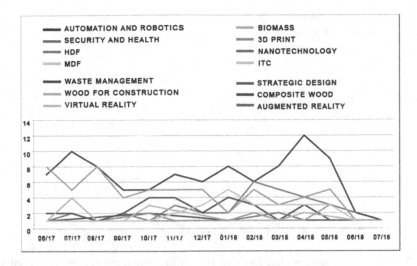

Fig. 4. Total of documents that mention the key technologies, per month.

When analyzing Fig. 4, one can observe that "Automation and Robotics" had a significant increase of citations between February 2018 and April 2018, experiencing a dropping ever since. In both Fig. 4 and 5, it is possible to notice that the volume of citations of most of the terms that oscillate do not show a visible tendency.

There are obvious difficulties related to the evaluation of trends by means of using exclusively the visual graphic presented in Fig. 4. To overcome that, the behavior of each monitored key technology was separately evaluated, in order to verify if its time series presented only oscillations around the same axis, tending to be stationary, or if it would be possible that it might have some kind of tendency associated.

As a tool for supporting specialists, the Augmented Dickey-Fuller Test, was applied. When considering a p-value > 0.05 and accepting the null hypothesis (H0), we concluded that the series has a unit root and is non-stationary. On the other hand, for a p-value $<= 0.05$ we reject the null hypothesis (H0), i.e. indicating that the series has no unit root and is stationary. Thus, it is possible to have a support to assess whether the key technology is becoming popular or not. In Fig. 6, it is possible to see an example of monitoring of Biomass term. The x-axis represents the months. The y-axis represents the count of documents in which the term was identified (Table 6).

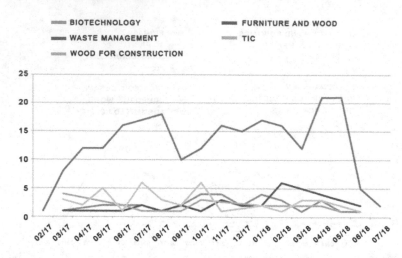

Fig. 5. Total of documents that mention the ontology-defined classes, per month.

Table 5. Number of documents in which the ontology-defined classes are present.

Classes	Number of documents
Furniture and wood	254
Biotechnology	31
Waste Management	31
Information and Communication Technologies	31
Wood for construction	14

Table 6. Augmented Dickey-Fuller Test result for the monitored term "Biomass".

General output	Critical values
Variation in Y axis: +65.22%	1%: −4.223
ADF Test: −2.453971	5%: −3.
p-value found: 0.127079	10%: −2.730
p-value > 0.05: It accepts the null hypothesis (H0); the data has a unit root and is non-stationary	

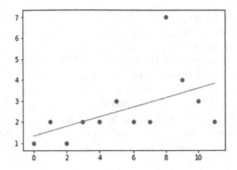

Fig. 6. Number of documents with 'Biomass' term (x: months; y: total of docs).

5   Final Considerations

This paper proposed an automated method for the identification of key technologies in specialized information sources. The proposed method is based on the Technological Surveillance methodology of Sánchez Torres and Palop Marro (2002) [22], and comprises 5 steps: Planning, Collecting, Organization, Intelligence, and Communication. A prototype was developed in order to validate this proposal. Key technologies of the Furniture and Wood sector were identified and quantified from 2,918 publications collected from four specialized Web portals, suggested by Furniture and Wood specialists.

The proposed method for identifying the use of existing technologies in Big Data scenarios, by means of the structured reports, graphs and tables generated by it, is the answer to our research question. That is, now we can automatically identify a set of technologies of interest with increasing popularization in Web portals.

As for future works, we suggest the creation of a sub-step of the Collection phase, which is capable of recommending new possible key technologies to be monitored based on text mining techniques. Although this feature would make the process richer, it would not rule out the human intervention. Another possible evolution would be to extract geographic data from the collected data, in order to check the dispersion of technologies on the map, as well as their intersection with data come from other sources of information.

References

1. Abe, H., Tsumoto, S.: Detecting temporal trends of technical phrases by using importance indices and linear regression. In: Rauch, J., Raś, Z.W., Berka, P., Elomaa, T. (eds.) ISMIS 2009. LNCS (LNAI), vol. 5722, pp. 251–260. Springer, Heidelberg (2009). https://doi.org/10.1007/978-3-642-04125-9_28

2. ABNT. Brazilian Association of Technical Standards. Guidelines for research, development and innovation management systems (R&D&I) [In Portuguese: Associação Brasileira de Normas Técnicas. Diretrizes para sistemas de gestão da pesquisa, do desenvolvimento e da inovação (PDI).]. https://www.abntcatalogo.com.br/norma.aspx?ID=088796. Accessed 28 Oct 2020

3. AENOR. Spanish Association for Standardization and Certification. Spanish Experimental Standard Rule UNE 166006 R&D&I: Technological Surveillance System [In Spanish: Asociación Española de Normalización y Certificación. Norma Española Experimental UNE 166006 Gestión de la I +D +i: Sistema de Vigilancia Tecnológica]. https://www.aenor.com/normas-y-libros/buscador-de-normas/une?c=N0059973. Accessed 28 Oct 2020

4. Salgado Batista, D., Guzmán Sánchez, M. V., Carrillo Calvet, H.: Establishment of a scientific-technological surveillance system. Technol. Forecasting [Original title in Spanish: Establecimiento de un sistema de vigilancia científico-tecnológica]. ACIMED 11(6) (2003)

5. Buckland, M.K.: Information as thing. JASIS 42(5), 351–360 (1991)

6. Dean, J., Ghemawat, S.: MapReduce: simplified data processing on large clusters. Commun. ACM 51(1), 107–113 (2008)

7. Ena, O., Mikova, N., Saritas, O., Sokolova, A.: A methodology for technology trend monitoring: the case of semantic technologies. Scientometrics 108(3), 1013–1041 (2016). https://doi.org/10.1007/s11192-016-2024-0

8. Flick, U.: Introduction to Qualitative Research [Original title in Portuguese: Introdução à Pesquisa Qualitativa]. Artmed, Porto Alegre (2009)

9. Geum, Y., Jeon, J., Seol, H.: Identifying technological opportunities using the novelty detection technique: a case of laser technology in semiconductor manufacturing. Technol. Anal. Strategic Manag. 25(1), 1–22 (2013)

10. Gil, A.C.: How to design research projects [Original title in Portuguese: Como elaborar projetos de pesquisa]. Atlas, Barueri (2010)

11. Hakim, A.R., Djatna, T.: Extraction of multi-dimensional research knowledge model from scientific articles for technology monitoring. In: 3rd International Conference on Adaptive and Intelligent Agroindustry (ICAIA), pp. 300–305. IEEE (2015)

12. Hjørland, B., Albrechtsen, H.: Toward a new horizon in information science: domain analysis. J. Am. Soc. Inf. Sci. 46(6), 400–425 (1995)

13. Zikopoulos, P., Eaton, C.: Understanding Big Data: Analytics for Enterprise Class Hadoop and Streaming Data. McGraw-Hill Osborne Media, New York City (2011)

14. Muñoz Durán, J., Marín Martínez, M., Vallejo Triano, J.: Technological surveillance in R&D&I project management: resources and tools [Original title in Spanish: "La vigilancia tecnológica en la gestión de proyectos de I +D +i: recursos y herramientas"]. El profesional de la información 15(5), 411–419 (2006)

15. Laurila, J., et al.: The mobile data challenge: Big data for mobile computing research. Nokia Research Center (2012). https://www.academia.edu/20092648/The_mobile_data_challenge_Big_data_for_mobile_computing_research. Accessed 27 Oct 2020

16. León, A.M., Castellanos, O.F., Vargas, F.A.: Assessment, selection and relevance of software tools used in technological surveillance [Original title in Spanish: Valoración, selección y pertinencia de herramientas de software utilizadas en vigilancia tecnológica]. Ingeniería e investigación 26(1), 92–102 (2006)

17. Momeni, A., Rost, K.: Identification and monitoring of possible disruptive technologies by patent-development paths and topic modeling. Technol. Forecast. Soc. Change 104, 16–29 (2016)

18. Nam, S., Kim, K.: Monitoring newly adopted technologies using keyword based analysis of cited patents. IEEE Access **5**, 23086–23091 (2017)
19. Neighbors, J. M.: Software construction using components. University of California, Irvine (1980). https://escholarship.org/content/qt5687j6g6/qt5687j6g6.pdf. Accessed 28 Oct 2020
20. OVTT. Virtual Observatory for Technology Transfer. Technology Surveillance Concept [In Portuguese: Observatório Virtual de Transferência de Tecnologia. Conceito De Vigilância Tecnológica]. https://pt.ovtt.org/vigilancia-tecnologica-conceitos. Accessed 28 Oct 2020
21. Palop, F., Vicente, J.M.: Technological Surveillance and Competitive Intelligence. Its potential for Spanish companies [Original title in Spanish: Vigilancia tecnológica e inteligencia competitiva: su potencial para la empresa española]. Cotec, Madrid (1999)
22. Sánchez Torres, J.M., Palop Marro, F.: Software tools for the practice in the company of Technological Surveillance and Competitive Intelligence [Original title in Spanish: "Herramientas de software para la práctica en la empresa de la Vigilancia Tecnológica e Inteligencia Competitiva"], Evaluación Comparativa, 1ª Edición. TRIZ, España (2002)
23. Park, H., Kim, E., Bae, K.J., Hahn, H., Sung, T.E., Kwon, H.C.: Detection and analysis of trend topics for global scientific literature using feature selection based on gini-index. In: IEEE 23rd International Conference on Tools with Artificial Intelligence, pp. 965–969. IEEE (2011)
24. PDIC2022. Industry Development Program of the Brazilian State of Santa Catarina for 2022: competitiveness with sustainability [In Portuguese: Programa de Desenvolvimento da Indústria Catarinense 2022: competitividade com sustentabilidade]. http://www4.fiescnet.com.br/homepdic. Accessed 28 Oct 2020
25. Sharma, S., Gupta, P.: The anatomy of web crawlers. In: International Conference on Computing, Communication Automation, Greater Noida, pp. 849–853. IEEE (2015)
26. Shiryaev, A.P., Dorofeev, A.V., Fedorov, A.R., Gagarina, L.G., Zaycev, V.V.: LDA models for finding trends in technical knowledge domain. In: IEEE Conference of Russian Young Researchers in Electrical and Electronic Engineering (EIConRus), pp. 551–554. IEEE (2017)
27. Smiraglia, R.P.: The Elements of Knowledge Organization. Springer, Cham (2014). https://doi.org/10.1007/978-3-319-09357-4
28. Tomaél, M.I.: Information sources on the internet [Original title in Portuguese: Fontes de informação na internet]. EDUEL, Londrina (2008)
29. Wei, Y.M., Kang, J.N., Yu, B.Y., Liao, H., Du, Y.F.: A dynamic forward-citation full path model for technology monitoring: an empirical study from shale gas industry. Appl. Energy **205**, 769–780 (2017)
30. White, T.: Hadoop: The Definitive Guide. O'Reilly, Beijing (2015)

Feature Importance Investigation for Estimating Covid-19 Infection by Random Forest Algorithm

André Vinícius Gonçalves[1,5]([⊠]) [iD], Ione Jayce Ceola Schneider[2] [iD],
Fernanda Vargas Amaral[3] [iD], Leandro Pereira Garcia[4] [iD],
and Gustavo Medeiros de Araújo[5]

[1] Federal Institute of Northern Minas Gerais, Minas Gerais, Brazil
[2] Federal University of Santa Catarina, Araranguá, Brazil
ione.schneider@ufsc.br
[3] University of Malaga, Malaga, Spain
[4] Florianópolis Municipal Health Department, Florianópolis, Brazil
[5] PGCIN, Federal University of Santa Catarina, Florianópolis, Brazil
gustavo.araujo@ufsc.br

Abstract. The present work raises an investigation about the feature importance to estimate the COVID-19 infection, using Machine Learning approach. Our work analyzed 175 features, using the Permutation Importance method, to assess the importance and list the twenty most relevant ones that represent the probability of infection of the disease. Among all features, the most important were: i) the period comprised between the date of notification and symptom onset stand out, ii) the rate of confirmed in the territory of health units in the last 14 days, iii) the rate of discarded and removed from the health territory, iv) the age, v) variables of the traffic flow and vi) symptoms features as fever, cough and sore throat. The model was validated and reached an accuracy average of 78.19%, whereas the sensitivity and specificity achieved 83.05% and the 75.50% respectively in the infection estimate. Therefore, the proposed investigation represents an alternative to guide authorities in understanding aspects related to the disease.

Keywords: Feature importance · Feature engineering · Machine learning · Prediction model · COVID-19

1 Introduction

In December 2019, a new coronavirus, called SARS-CoV-2, was recognized in the city of Wuhan, China, and spread quickly to other countries in the world. In January 2020, the World Health Organization declared a Public Health Emergency of International Importance, and in March, the COVID-19 pandemic. At

E. Bisset Álvarez (Ed.): DIONE 2021, LNICST 378, pp. 272–285, 2021.
https://doi.org/10.1007/978-3-030-77417-2_20

the beginning of October 2020 there are already more than 33 million confirmed cases and more than 1 million deaths from the disease [14].

When infecting the human body, there is a period of latency, followed by an infectious period. During this period, the infected person can transmit to others through coughing and sneezing. The virus mainly affects the respiratory tract and the first symptoms appear after the incubation period. The main symptoms include fever, cough and fatigue, which appear on average after 11 days of contamination. Other symptoms, such as mucus production, headache, hemoptysis, diarrhea, dyspnoea, lymphopenia can also appear. The main clinical diagnosis is pneumonia [2,13,20,24]. Furthermore, the risk of symptomatic infection increases with age. Thus, older individuals are more likely to have symptomatic infection and worse outcomes [2].

Laboratory diagnosis is an important tool for diagnosis, as well as for follow-up, evaluation and evolution of the case. The recommended diagnostic test is the real-time polymerase chain reaction (RT-PCR) of nasal and oropharyngeal swab samples. Other serological tests can be used to detect immune responses, such as class M (IgM) and class G (IgG). However, it is important to use resources rationally in conducting diagnostic tests [26].

Towards the rational use of the infection spread of detection capabilities, artificial intelligence techniques have been used to predict the diagnosis of COVID-19. The algorithms are managing to predict the stage of COVID-19 by means of several features such as age, comorbidities, symptoms, diagnosis and outcome [7].

In order to create a model, we developed a investigation to assess the main features that can determine Covid-19 infection. To conduct our research, we collected and analysed data from the public health system in the capital of Santa Catarina, a state in southern Brazil. The set of features are composed of several variables from symptoms to demographic data.

Furthermore, we modeled a machine learning algorithm to estimate the infection of an individual. In our work, we conduct several experiments with 175 features to label the 20th most important features that represent the high Covid-19 infection likelihood.

1.1 Contributions

Among the contributions of our work, we can highlight:

1. The verification of the high importance of the features of confirmed, discarded, and removed by region of health, as well as the features of symptoms (fever, cough, and sore throat), all along the time of the notification date.
2. An intensive feature importance investigation results in findings that also highlighted the importance of traffic load, which reflects the people's isolation level.
3. The accuracy of the predictive model with an average of 78.19% of correctness in determining whether the individual is infected with Covid-19.

The remainder of this paper is structured as follows: In Sect. 2, we describe the more relevant related works on the effort to determine the Covid-19 infection; Sect. 3 introduces the methodology applied to feature engineering; Sect. 4 detail the experimental assessments; Sect. 5 outline the discussion about results and finally, in Sect. 6, we present our final remarks and future work.

2 Related Work

COVID-19 had a significant impact on the life and economy of several countries [10]. In addition to collapsing economies, the moral values of nations have been strongly affected by the pandemic [21]. All the impact, economic and social, motivated the Pan American Health Organization to seek to better understand the signs and symptoms of Sars-cov-2, in order to disseminate this knowledge. From now, the challenge of the pandemic is to find the best model that elucidates the initial growth trajectory and the epidemiological characteristics of the new coronavirus [19]. In this sense, the application of Forecasting models has been useful to deal with the dynamic behavior of this virus [22].

Sars-cov-2 is a respiratory virus transmitted through droplets of saliva, sneezing or by close contact. In their study, [25] described 69 cases of COVID-19 in China, where it was identified that 15% of individuals had fever, cough and dyspnoea. However, a survey conducted in the United States, showed that 50% of patients affected by this virus did not have a fever, however cough and dyspnea were reported by 88% of people with the virus [3]. Still, in other studies, reports of symptoms were difficult to measure objectively, such as anosmia (loss of smell), hyposmia (decreased smell) and ageusia (loss of taste) [11].

In addition, infected individuals may never develop symptoms, others may have mild symptoms or develop moderate to severe Sars-cov-2 disease [15]. In order to understand the symptoms that best represent the pandemic, researchers around the world try to understand the behavior of the virus [11]. A group of researchers from Spain found five patterns of skin infection that may be associated with COVID-19. These patterns were repeated in patients with varying demographic characteristics, in different periods and with different severities of the disease. Among these patterns are maculopapular rashes (47% of cases), vesicles or pustules (19% of cases), hives (19% of cases) and other vesicular rashes (9% of cases) and livedo or necrosis (in 6% of cases) [8].

A preliminary analysis by the World Health Organization (WHO) shows that in relation to gender, there is a relatively uniform distribution of infections between women and men (47% versus 51 respectively), however, it seems that men have a higher rate mortality rate (58%) in relation to women [15]. Nevertheless, due to the need to know the outbreak of COVID-19, some studies are being carried out considering exogenous factors such as the social environment, climatic variables, pollution and population density [22]. Other studies point to the role of room temperature in the survival and transmission of viruses. According to the WHO, several environmental factors can influence the spread of communicable diseases that can cause epidemics. The underlying theory is that the number of cases and the spread of previous infectious viruses demonstrate seasonal patterns, affected by the climate, and therefore Covid-19 is likely to be similar in this respect [12].

Therefore, the prediction of a pandemic can be made based on several parameters, such as the impact of environmental factors, incubation period, impact of quarantine, age, sex and many more. The difficulty in predicting the number of cases of a pandemic is the fact that the number of cases to be studied does not match the total infected population. [17]. Considering the importance of knowing this difficult epidemiological scenario in a short-term horizon [22], Forecasting models have had a positive impact to mitigate the pandemic [21].

Forecasting techniques assess past situations, which allows for better predictions about the situation that will occur in the future [21]. These models allow managers to develop strategic planning and carry out decision making in the most assertive manner possible [12]. Since in addition to the concern with public health, the danger of the pandemic is also with the supply of food, medicine, and other supply chains needed by the population. Statistical analyzes such as forecasting allow governments not to focus only on underlying decision-making methods such as personal judgment [17].

To understand the nature of the coronavirus and predict its spread, the analysis models must be trained on a large volume of the data set. The ideal amount of the data set plays an extremely important role in training the task and affects the performance of the proposed algorithms [12]. The forecasting of COVID-19 cases is a challenging task, since Forecasting models are impacted by the effect of a small data set [22].

3   Methodology

The main goal of our work is to analyze which features most contribute to the diagnosis of suspected cases of COVID-19 using the classification technique with Machine Learning. The methodology steps that can be seen in Fig. 1 are: 1) data selection and extraction, 2) data pre-processing and feature engineering, 3) hyperparameterization and feature selection and 4) model validation. Each step is detailed further next.

3.1   Data Selection and Extraction

In the first step, the database used has been set corresponding to 1,930 reported cases of COVID-19 in the period between 02/15/2020 and 05/25/2020. The database was extracted from the Health Department of Florianópolis, capital of the State of Santa Catarina in southern Brazil and is available[1] to be analysed.

According to the [9], the database comes from three sources: 1) anonymized data on suspected and confirmed cases resident in Florianópolis; 2) demographic data of the 49 health regions that make up the municipality; and 3) data on the mobility represented by the traffic flow in the municipality.

---

[1] https://github.com/avgandre/covid_florianopolis/tree/master/dados/Dione.

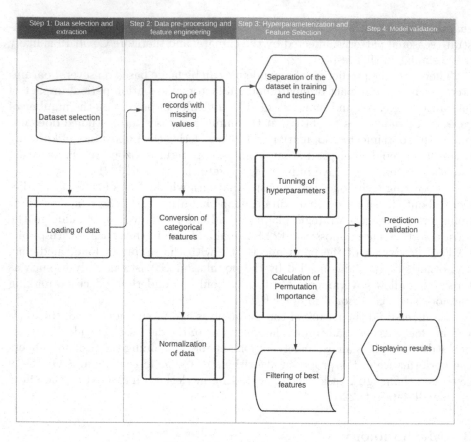

Fig. 1. Methodology flow

The database contains individual data on the diagnosis (confirmed or discarded), sex, age (in years), and age groups (under 10 years old, 10 years old under 20 years old, 20 years old under 40 years old) years old, 40 years old to under 60 years old, 60 years old to under 80 years old, 80 years old or more), skin color (white and not), date of onset of symptoms, in addition to the following clinical data of symptoms of the disease: pain throat, dyspnoea, fever and cough.

There is also data on health regions in the city of Florianópolis. There are 49 territories and 104 sub-territories that correspond to regional divisions of the city.

Furthermore, the database contains the following demographic data for health territories the total number of inhabitants and by sex; the number of persons aged 1 year, 2 years and so on up to 100 years or more; the number of people by skin color (white, black, yellow, Brazilian, indigenous and ignored); the number of people by years of schooling (from 1 to 17 years completed or more, in addition to literate, non-literate, literate through youth and adult literacy programs and with uninformed schooling); the total income per household, the

Table 1. Conversion of categorical features

Categorical feature	Factors	Method	New features
Race/Color	White; Yellow; Black; Parda; Unknown	One-hot-encoding	Race_Col0 to Race_Col4
Age range	<10; $10 <= years < 20$; 20 to 80 step 20; >80	One-hot-encoding	Age_range0 to Age_range5
Screening method	3 methods	One-hot-encoding	Screening_method0 to Screening_method2
Fever	Yes; No	One-hot-encoding	Fever0 to Fever1
Cough	Yes; No	One-hot-encoding	Cough0 to Cough1
Sore throat	Yes; No	One-hot-encoding	Sore_throat0 to Sore_throat1
Dyspnea	Yes; No	One-hot-encoding	Dyspnea0 to Dyspnea1
Health territory	48 regions	FeatureHasher	Health_territory0 to Health_territory19
Health subTerritory	104 subregions	FeatureHasher	Health_subterritory0 to Health_subterritory19

average income of the households, the total income of the heads of households, the average income of the heads of households, the total income per person and the average income per person, the proportion of males, persons with 60 years of age or more, of people with non-white skin and of people with 10 years or less of education, as possible indicators of vulnerability.

Regarding mobility features, the database provides data on the average daily traffic on four major avenues in the city. The time window for calculating the average considers it starts on the day of symptom detection until the thirteenth day before, that is, it is a window delayed in time.

3.2 Data Pre-processing and Feature Engineering

Initially, all records with the value 'Missing' in the attributes of symptoms (Sore Throat, Dyspnea, Fever and Cough) and Diagnosis were removed. Then, the categorical attributes were converted to numerical ones, using the One-hot-encoding technique for Race/Color, Age group, Screening Method and symptoms, and Feature Hashing [23] for Territory and Subterritory. The Table 1 has the conversion process result detailed:

Another procedure performed was the creation of new attributes. As suggested by [9], the number of infected people (with a positive diagnosis and up to 14 days after the onset of symptoms) in each health territory was calculated.

Moreover, according to the principle of the SIR model of epidemiology [4], it was proposed to include the number of people discarded (with a negative diagnosis) and the number of people removed (with a positive diagnosis and more than 14 days after the onset of symptoms).

Furthermore, it was included the rate of people infected by the number of inhabitants of their respective health territory, as well as the rate of discarded and removed rate. Finally, the data were normalized, transforming them to values within the range [0, 1] and, thus, establishing a common scale.

3.3   Hyperparameterization and Feature Selection

The database was divided into training and test basis, 70% for training and 30% for testing. As there is an imbalance in the amount of data between the discarded and confirmed cases, the first being a larger amount, the sample was balanced using the Undersampling technique.

In the training stage, cross-validation was adopted as a way to assess the model's generalization. According to [18], the technique consists of dividing the database into k folds, one of which is selected at a time to be the test set and the other k−1 are used as a training set. The test is repeated until each of the k folds is used as a test set. In the end, the accuracy is given by the average of the accuracy obtained for each of the k folds.

Hyperparametrization was performed using a random combination of parameters with 10 iterations in each tuning process. Accuracy was chosen as the maximization score.

After defining the parameters of the algorithm, the feature selection was performed considering the values of permutation importance as a criterion for assessing the degree of importance [1]. The criterion used was to select only those features with a value greater than zero. In this way, the features with values above this threshold remained in the model and the rest were removed from the database.

3.4   Model Validation

In the last step of the process, with the algorithm trained and configured with the best parameters that fit the model, the algorithm was validated with the test base to assess its prediction capacity.

Steps 3 and 4 were repeated 100 times and the results for each were stored. Then the data were used to calculate the mean and stantard devation of evaluation metrics and permutation feature importances.

The experiments were initially tested with three algorithms: Random Forest and Support Vector Machine (SVM). The equipment used to carry out the experiments had: i) Intel (R) Xeon (R) Gold 6126 CPU @ 2.60 GHz CPU with

12 CPUs, ii) 32.0 GB of RAM, iii) 250 GB of hard disk and iv) Linux Ubuntu 16.04. The entire implementation was developed in the Python programming language, version 3.7.

4   Experiment Assessments

We carried out experiments to analyze the evaluation metrics that measure the accuracy, in addition to ascertaining the features that had the most contribution to the performance.

The specific parameters of the Random Forest and SVM are presented in Table 2 and Table 3, as well as the possible value ranges. Through them, the best configuration is adjusted by means of a random search of hyperparameters.

Table 2. Random forest hyperparameters

Parameter	Value
criterion	[entropy, gini]
n_estimators	[5...100]
max_depth	[None, 1...5]
min_samples_split	[2...5]
min_samples_leaf	[1...5]
min_weight_fraction_leaf	[0.0...0.5]
max_features	['auto', 0.1...0.5]
Bootstrap	False, True

Table 3. SVM hyperparameters

Parameter	Value
C	[0.025...1.0]
degree	[3...25]
gamma	['scale', 'auto', 0.1...2.0]
shrinking	[True, False]
probability	[False, True]
decision_function_shape	[ovo, ovr]

And the parameters described in Table 4 relate to the general settings of the environment.

In the experiments, the metrics used in the analysis of the proposed model to assess performance were accuracy, sensitivity and specificity. The data samples were obtained by running the algorithm repeatedly and they are presented below in the form of average and standard deviation.

Table 4. General settings

Parameter	Value
Execution amount	100
Folds	10
Training/Test	70/30
RandomizedSearchCV interactions	10
Features selection threshold	Above zero (>0)

Among the two algorithms used, Random Forest was the one with the highest efficiency, as shown in Table 5. Therefore, it was extensively explored in this study.

Table 5. Metrics from accuracy score

Metric	Random forest		SVM	
	Training (M ± SD)	Test (M ± SD)	Training (M ± SD)	Test (M ± SD)
Accuracy	0,80682 ± 0,02279	0,79755 ± 0,02198	0,76733 ± 0,02285	0,75328 ± 0,02542
Sensibility	0,83802 ± 0,03916	0,84141 ± 0,04320	0,80397 ± 0,05217	0,80015 ± 0,05508
Specificity	0,77561 ± 0,02779	0,77332 ± 0,03545	0,73069 ± 0,05110	0,72740 ± 0,04980

The Fig. 2 shows the comparison between the evaluation metrics of the two algorithms: Random Forest and SVM. The data presented are only from the Test Set.

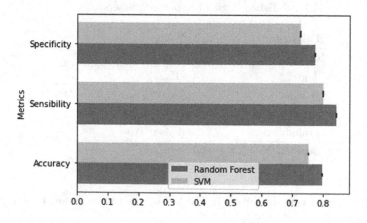

Fig. 2. Metrics from test set

The Random Forest algorithm performed better compared to SVM. It had an accuracy of 0.80682 ± 0.02279 (mean $\pm$ standard deviation) on the training set and 0.79755 ± 0.02198 on the test set. The sensitivity was 0.83802 ± 0.03916 and 0.84141 ± 0.04320 in each of the two bases, respectively. The specificity was 0.77561 ± 0.02779 in the training base and 0.77332 ± 0.03545 in the test base.

The main features selected by Random Forest with their respective Permutation Importance percentages are shown in Table 6. The results presented below are in the form of average and standard deviation for the set of executions.

Table 6. Features permutation importance of accuracy score

Position	Feature	Permutation importance (M $\pm$ SD)
1[a]	Notification date	0,05412 $\pm$ 0,03538
2[a]	Confirmed rate territory 14 days	0,03016 $\pm$ 0,01456
3[a]	Fever1	0,02783 $\pm$ 0,01964
4[a]	Fever0	0,02638 $\pm$ 0,01992
5[a]	Cough0	0,01575 $\pm$ 0,01256
6[a]	Cough1	0,01505 $\pm$ 0,01299
7[a]	Confirmed territory 14 days	0,01100 $\pm$ 0,00677
8[a]	Symptoms start date	0,00454 $\pm$ 0,00488
9[a]	Discarded rate territory	0,00247 $\pm$ 0,00303
10[a]	Sore throat 1	0,00166 $\pm$ 0,00226
11[a]	Removed rate territory	0,00140 $\pm$ 0,00195
12[a]	Sore throat 0	0,00140 $\pm$ 0,00202
13[a]	Age	0,00123 $\pm$ 0,00165
14[a]	Discarded territory	0,00115 $\pm$ 0,00161
15[a]	Age above 91	0,00108 $\pm$ 0,00253
16[a]	Removed territory	0,00105 $\pm$ 0,00164
17[a]	traffic mean	0,00098 $\pm$ 0,00142
18[a]	traffic mean lag2	0,00088 $\pm$ 0,00125
19[a]	traffic mean lag1	0,00084 $\pm$ 0,00125
20[a]	Screening method 2	0,00081 $\pm$ 0,00122

For a better visual understanding of the features importances, the Fig. 3 is shown with the values of each variable.

Lastly, the response time of the algorithm had an average result of 17.49 and standard deviation of 4.87 s, considering the training step that involved the hyperparameter tuning process and feature selection, in addition to the test step that consisted of the model validation.

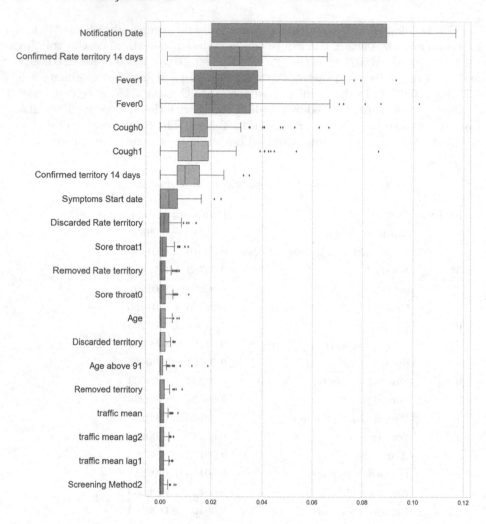

Fig. 3. Features permutation importance of accuracy score

5   Discussion

After nearly a year of its discovery, COVID-19 is a disease that arouses much interest because of its great impact on humankind. Wherefore, this work proposed to investigate the relevance of a set of variables in the diagnostic prediction of the disease.

The investigation started with the acquisition of a preliminary database with 175 features, which after going through pre-processing, increased to 228 due to the techniques of coding categorical variables. Then, the model was processed and analyzed by the Permutation Feature Importance method to assess the impact of each feature on the accuracy metric.

The most important feature was the Notification Date. All symptom features appeared among the twenty most significant, with the exception of dyspnea. This fact corroborates with the researches that investigate the symptoms and indicate fever, cough and sore throat among some of the most common ones [5, 6, 16].

Mobility features also meaningful. There were fourteen features to represent traffic on the city's four major avenues. Thus, four of them are among the most important features, listed in Table 6

The feature engineering process carried out in the pre-processing step resulted in the creation of new health territory variables (Confirmed_territory_14days, Removed_territory and Discarded_territory, Confirmed_rate_territory_14days, the Rate_Discarded_Territory and the Rate_Removed_Territory) contributed significantly to the model performance.

However, the Confirmed_territory_14days, the Rate_Discarded_Territory and the Rate_Removed_Territory features were even more expressive and were among the ten most important. In this way, it is noticed that the factor "health region" was very important in the correct classification of the algorithm's diagnosis.

Finally, the results of the model were somewhat satisfactory. Coping with the research developed in [9], there is an improvement of approximately 15% in the accuracy measured in the final test stage. This matter may be associated with the addition of new variables that were not present in the previous work, including symptoms and those confirmed, discarded and removed related to the health territory.

6   Final Remarks and Future Work

The present work shows a investigation about the feature importance in a prediction diagnostic model for cases of COVID-19, using the classification technique with Machine Learning.

These classification approaches are fundamental for monitoring the number of virus reproductions and for making decisions in the face of the pandemic. The advantage of them is to produce quick responses and relatively low cost compared to laboratory diagnosis.

The methodology section emphasized the hyperparametrization and feature selection techniques, as the research aimed to investigate two aspects: the features that best contributed to the performance of the model and the results of the hit rates in the validation of the test step.

The Random Forest performed better compared to SVM. So it was explored in more detail. In the first investigation step, the Permutation Importance method was used to assess the impact of the features on the results. Among the 228 variables that make up the database, the most relevant are: the date of notification and onset of symptoms, the rate of confirmed in the territory of the last 14 days, the rate of discarded and removed from the territory, age, flow variables traffic, in addition to the attributes of fever, cough and sore throat.

Regarding the second stage of the investigation, the metrics showed consistent results. Accuracy had a mean of 78.19%, whereas sensitivity reached 83.05% and specificity 75.50% of cases.

Therefore, the research conducted has shown that there is a feasible alternative in the process of underdiagnosis COVID-19 disease, considering the most relevant characteristics for the determination of infection.

In near future, we pretend to extend the model including new features, such as climatic conditions, as [22] suggests that could have an impact on people behavior. Another possibility is to reduce the granularity of the health territory to sub-territories in the calculation of confirmed, discarded and removed and, later, to analyze the behavior of the algorithm.

References

1. Altmann, A., Toloşi, L., Sander, O., Lengauer, T.: Permutation importance: a corrected feature importance measure. Bioinformatics **26**(10), 1340–1347 (2010)
2. Ashour, H.M., Elkhatib, W.F., Rahman, M., Elshabrawy, H.A., et al.: Insights into the recent 2019 novel coronavirus (SARS-CoV-2) in light of past human coronavirus outbreaks. Pathogens **9**(3), 186 (2020)
3. Bhatraju, P.K., Ghassemieh, B.J., Nichols, M., Kim, R., Jerome, K.R., Nalla, A.K., Greninger, A.L., Pipavath, S., Wurfel, M.M., Evans, L., et al.: Covid-19 in critically ill patients in the seattle region-case series. New Engl. J. Med. **382**(21), 2012–2022 (2020)
4. Brauer, F.: The Kermack-McKendrick epidemic model revisited. Math. Biosci. **198**(2), 119–131 (2005)
5. Burke, R.M., et al.: Symptom profiles of a convenience sample of patients with covid-19–United States, January–April 2020. Morb. Mortal. Wkly Rep. **69**(28), 904 (2020)
6. Carfì, A., Bernabei, R., Landi, F., et al.: Persistent symptoms in patients after acute covid-19. JAMA **324**(6), 603–605 (2020)
7. Chisari, E., Krueger, C.A., Barnes, C.L., Van Onsem, S., Walter, W.L., Parvizi, J.: Prevention of infection and disruption of the pathogen transfer chain in elective surgery. J. Arthroplasty **35**(7), S28–S31 (2020)
8. Galván Casas, C., et al.: Classification of the cutaneous manifestations of covid-19: a rapid prospective nationwide consensus study in Spain with 375 cases. Br. J. Dermatol. **183**(1), 71–77 (2020)
9. Garcia, L.P., et al.: Estimating underdiagnosis of covid-19 with nowcasting and machine learning: experience from Brazil. medRxiv (2020)
10. He, S., Tang, S., Rong, L.: A discrete stochastic model of the covid-19 outbreak: forecast and control. Math. Biosci. Eng. **17**, 2792–2804 (2020)
11. Iser, B.P.M., Sliva, I., Raymundo, V.T., Poleto, M.B., Schuelter-Trevisol, F., Bobinski, F.: Definição de caso suspeito da covid-19: uma revisão narrativa dos sinais e sintomas mais frequentes entre os casos confirmados. Epidemiologia e Serviços de Saúde **29** (2020)
12. Malki, Z., Atlam, E.S., Hassanien, A.E., Dagnew, G., Elhosseini, M.A., Gad, I.: Association between weather data and covid-19 pandemic predicting mortality rate: machine learning approaches. Chaos, Solitons Fractals **138**, 110137 (2020)
13. Meo, S., et al.: Novel coronavirus 2019-nCoV: prevalence, biological and clinical characteristics comparison with SARS-CoV and MERS-CoV. Eur. Rev. Med. Pharmacol. Sci. **24**(4), 2012–2019 (2020)
14. World Health Organization: WHO coronavirus disease (COVID-19) dashboard (2020). https://covid19.who.int/

15. Organization, W.H., et al.: Diagnostic testing for sars-cov-2: interim guidance, 11 September 2020. World Health Organization, Technical report (2020)
16. Pan, L., et al.: Clinical characteristics of covid-19 patients with digestive symptoms in Hubei, china: a descriptive, cross-sectional, multicenter study. Am. J. Gastroenterol. **115** (2020)
17. Petropoulos, F., Makridakis, S.: Forecasting the novel coronavirus covid-19. PLoS ONE **15**(3), e0231236 (2020)
18. Rodriguez, J.D., Perez, A., Lozano, J.A.: Sensitivity analysis of k-fold cross validation in prediction error estimation. IEEE Trans. Pattern Anal. Mach. Intell. **32**(3), 569–575 (2009)
19. Roosa, K., et al.: Real-time forecasts of the covid-19 epidemic in china from February 5th to February 24th, 2020. Infect. Dis. Model. **5**, 256–263 (2020)
20. Russell, C.D., Millar, J.E., Baillie, J.K.: Clinical evidence does not support corticosteroid treatment for 2019-NCoV lung injury. Lancet **395**(10223), 473–475 (2020)
21. Shinde, G.R., Kalamkar, A.B., Mahalle, P.N., Dey, N., Chaki, J., Hassanien, A.E.: Forecasting models for coronavirus disease (covid-19): a survey of the state-of-the-art. SN Comput. Sci. **1**(4), 1–15 (2020)
22. da Silva, R.G., Ribeiro, M.H.D.M., Mariani, V.C., dos Santos Coelho, L.: Forecasting Brazilian and American covid-19 cases based on artificial intelligence coupled with climatic exogenous variables. Chaos, Solitons Fractals **139**, 110027 (2020)
23. Soheily-Khah, S., Wu, Y.: A novel feature engineering framework in digital advertising platform **10**, 21 (2019). https://doi.org/10.5121/ijaia.2019.10403
24. Vannabouathong, C., et al.: Novel coronavirus covid-19: current evidence and evolving strategies. J. Bone Joint Surg. Am. **102**(9), 734 (2020)
25. Wang, Z., Yang, B., Li, Q., Wen, L., Zhang, R.: Clinical features of 69 cases with coronavirus disease 2019 in Wuhan, china. Clin. Infect. Dis. **71**(15), 769–777 (2020)
26. Xavier, A.R., Silva, J.S., Almeida, J.P.C., Conceição, J.F.F., Lacerda, G.S., Kanaan, S.: Covid-19: clinical and laboratory manifestations in novel coronavirus infection. Jornal Brasileiro de Patologia e Medicina Laboratorial **56** (2020)

Neural Weak Supervision Model for Search of Specialists in Scientific Data Repository

Sergio Jose de Sousa[1]([⊠]) [iD], Thiago Magela Rodrigues Dias[1] [iD], and Adilson Luiz Pinto[2] [iD]

[1] Departamento de Modelagem Matemática e Computacional, Centro Federal de Educação Tecnológica de Minas Gerais (CEFET-MG), Belo Horizonte, MG, Brazil
thiagomagela@cefetmg.br
[2] Departamento de Ciência da Informação, Universidade Federal de Santa Catarina (UFSC), Florianópolis, SC, Brazil
adilson.pinto@ufsc.br

Abstract. With the growing volume of data produced today, it is clear that more and more users are using different types of systems, such as, for example, professional and academic data storage systems. Given the large amount of stored data, the difficulty of finding candidates with appropriate profiles for a particular activity is noteworthy. In this context, to try to solve this problem comes the expertise retrieval, a branch of information retrieval, which consists of, given a query, documents are recovered and used as indirect units of information for the candidates and some aggregation techniques are used in these documents to generate a score to the candidate. There are several models and techniques to work with this problem, some have been tested extensively but the search for specialists in the academic field with neural models has a smaller amount of research, this fact is due to the complexity of these models and the need for large volumes of data with judgments of relevance or labeled for your training. Therefore, this work proposes a technique of expansion and generation of weak supervised data where the relevance judgments are created with heuristic techniques, making it possible to use models that require large volumes of data. In addition, is proposed a technique of deep auto-encoder to select negative documents and finally a ranking model based on recurrent neural networks and that was able to overcome all the baselines compared.

Keywords: Expertise retrieval · Deep learning · Weak supervision.

E. Bisset Álvarez (Ed.): DIONE 2021, LNICST 378, pp. 286–296, 2021.
https://doi.org/10.1007/978-3-030-77417-2_21

1 Introduction

The challenges of finding the required information increases with the volume of data. Therefore, support tools such as recommendation and information retrieval systems are essential when searching, whether searching for websites, shopping website recommendations, searching for specialists, among other possibilities.

Search for specialists has existed since before the invention of the computer and denotes the need to find someone with some specific knowledge. In the field of psychology, in [4] it is said that the superior performance of specialists is acquired through experience, repetitions and structuring of long-term activities. For [15] specialists are people with knowledge or who have mastered detailed skills in specific areas. This task is challenging because it is necessary to evaluate what the person has already done, worked and produced in order to find out what their specialties are and who stands out among various options.

The sub-domain expertise retrieval in academic environment is one of the areas that has been receiving more and more attention despite having less related research [2], which may be related to the complexity of the problem or the difficulty of obtaining relevant data on the topic. In the literature it is possible to find some bases such as in [1,3,23] where it is possible to observe expert datasets with a focus on only one topic, not being generalist, centralized in just one organization or in only one area. Some platforms for sharing and storing projects and academic works that stand out, for example Academia[1], DBLP[2], Google Scholar[3], Plataforma Lattes[4], Microsoft Academic Search[5], ResearchGate[6].

Specifically, the *Plataforma Lattes* developed and maintained by the *Conselho Nacional de Desenvolvimento Científico e Tecnológico* (CNPq) appears as an important source of academic knowledge. In [6] the authors demonstrated how relevant the data contained in the platform are for the understanding Brazilian research and science, including personal, professional and academic information, such as scientific and technological production. Therefore, the *Plataforma Lattes* is an expressive source of high quality information from individuals [13] and is a good source of data to feed models and techniques in general.

For any Information Retrieval (IR) system, the relevance of the items returned given a query is extremely important. These measures may vary according to the robustness, sensitivity and efficiency that the system is expected to demonstrate [18]. These attributes help us to compare and select traditional and neural models that recently gained prominence.

The first artificial neural network proposed in [21] with a very simple format called perceptron that today is basically a neuron unit of current network, it works like a linear binary classifier that tries to find a better separation of

---

[1] Academia: https://www.academia.edu/.

[2] DBLP: https://dblp.uni-trier.de/.

[3] Google Scholar: https://scholar.google.com/.

[4] Plataforma Lattes: http://lattes.cnpq.br/.

[5] Microsoft Academic Search: https://academic.microsoft.com/.

[6] ResearchGate: https://www.researchgate.net/.

the data. One of the biggest problems of this model is the inability to separate nonlinear data as a xor distribution, to be solved we must increase the depth and width of the network, adding neurons and more layers. This creates other problems, overfitting, when the model decorates training data and fails to generalize validation, another major problem is the need for an advanced hardware architecture that was expensive at the time. But with the reduction of hardware prices like GPUs, the advancement in techniques that reduce overfitting like dropout and maturity and specialties of the architectures provided a great advance and highlight in deep neural network.

Neural networks have brought great improvements in the areas that uses unstructured data like of computer vision, natural language processing and speech recognition [14], unlike traditional techniques, these networks benefit from large amounts of data, having an ability to learn contexts and relationships that are difficult to identify with handcraft methods. More recently, some attempts have been made to propose and adapt these techniques for information retrieval.

In these neural models applied to IR, a major problem stands out, the need for large amounts of labeled data. These labels consist of judgment of relevance, that is, a set of triples containing a query, a document and the score of this relationship. Data labeling is expensive and can take a long time, which makes it impossible to apply these models in many cases.

Motivated by the new neural models and the need for data with judgments of relevance, the present work presents an alternative model with weak supervision where the labels are generated by heuristics, trying to answer the following questions:

Q1: Can weak supervision obtain better results than traditional technique used to generate the judgments of relevance applying to problems of search for specialists?

Q2: Can the generation of negative documents for queries through a deep auto-encoder surpass the standard of selecting random documents?

Q3: Can the Dual Embedding LSTM model surpass the models in the literature?

Therefore, in this work, a technique is proposed to generate pseudo-judgments of relevance to the documents, considering that for a given query it is also necessary samples of relevant documents and samples of non-relevant documents. The response of $Q1$ is positive when they combine the technique of weak supervision and the deep autoencoder to select the negative documents. The answer from $Q2$ is also positive, the deep autoencoder provides a selection of documents that negatively represent a query more effectively than random selection.

To select samples of positive documents for the queries, a strategy is proposed to generate relevance judgments using classical information retrieval techniques based on language model with Bayesian smoothing using Dirichlet [25] distribution; for samples of negative documents is proposed a deep autoencoder [9], calculating the most distant candidates from each consultation and extracting their documents. To carry out the reclassification of documents, a dual neural architecture with recurring layers is proposed, in order to recalculate the sequence and scores of the documents that ultimately compose the score of each candidate, answering positively the question $Q3$.

In the next section, it will be presented the Related Works (Sect. 2) to this research, following the applied Methodology (Sect. 3), presenting the entire proposed framework, right after there is a detailed description about the used dataset in Sect. 4 and finally presented the Final Results (Sect. 5) and Conclusions (Sect. 6).

2   Related Works

In [2] an extensive literature review is described that highlights the advances in models and algorithms for searching and ranking specialists, summarizing and establishing the relationships of these approaches. More recently, [11], a survey was conducted that selected 96 articles consisting of 57 journals, 34 conferences and three book chapters, analyzing the domain of expert search, knowledge sources, methods and databases. There was a growing trend in the amount of IR research for searches models by academy experts.

The work [16] use the database LExR the authors proposed a technique based on information theory where the document-author association is given by a probability, that is, a non-Boolean model, and two alternatives normalization schemes which measures how discriminative a particular document-author association is in view of the other associations involved in each author's document. The approach surpassed the proposed baselines.

In [5] the authors proposed a weak supervised model of deep neural network. The tests were carried out on two data-sets, one for news and other with general data from the internet. For the queries, query logs from the service provider AOL[7] were used. Pseudo-judgments of relevance are generated using BM25 [20] with 1000 documents with the highest score being selected for each query and 1000 other negative documents sampled at random. In the experiments three network architectures are proposed, one point-wise and two pair-wise, three representations of the input data, being a dense representation with several calculated attributes, a sparse representation with bag-of-words and finally an embedding vector learned during training. The combination of the pair-wise architecture with representation of embedding as an input surpassed the baseline technique.

The tutorial proposed by [18] presents basic concepts and intuitions behind the neural models applied in IR, reviewing recent neural network architectures, pointing out their positive and negative sides and finally discussing possible future directions.

Part of the hyper-parameters and architecture of the Dual Embedding LSTM model used in this work were inspired by [19] which proposes a duet architecture where the network has a part that learns local representation and second part that learns representation distributed from the sets of queries and documents.

---

[7] AOL: https://www.aol.com/.

3   Methodology

In this section, the framework for generating the necessary data is presented, such as an extensive list of queries, a list of documents with positive and negative relationships given a query (Sect. 3.1). Next the Dual Embedding LSTM model is described (Sect. 3.2).

3.1   Weak Supervision

Weak supervision is usually a term given to refer to models that have noisy labels but it also refers to models that use labels generated by heuristic techniques [10]. As neural models are greedy with labeled data, it is necessary to obtain pseudo-judgments of relevance, and for correct learning it is necessary for each query a set of documents with a positive relationship and a set of negative documents, that is, results that would be incorrect for a given query. The following are techniques for extracting, given a query, obtaining positive and negative documents together with their labels or relevance judgments. That is,

$$S = f(q, d),$$

where S means the pseudo-judgment score, q the query and d document, with documents varying between relevant $+d$ and no relevant $-d$.

Positive Documents. To select positive documents together with a pseudo-label, the technique based on Language Model with smoothing and distribution of Dirichlet [25] was used as a calculation of similarity between documents and queries. Thus, for each query, up to 20 documents with the best scores are selected.

The Bayesian smoothing with Dirichlet distribution can be seen in Eq. 1, where it tries to estimate the smoothed probability of finding the term i given the probabilities model of the document j ($P^s(i|\theta_{d_j})$) where $tf_{i,j}$ is the frequency of term in the document, F_i the frequency of term i in all documents of corpus divided by the sum of the frequencies of all terms in all documents. Sum up these values and divide by the sum of the frequencies of all the terms of the document j and these terms are weighted by the constant λ ranging from 0 to 1, the closer to 1 the more smoothed the language model becomes.

$$P^s(i|\theta_{d_j}) = \frac{tf_{i,j} + \lambda \frac{F_i}{\sum_i F_i}}{\sum_i tf_{i,j} + \lambda} \tag{1}$$

Negative Documents. To find the most distant documents of each query a deep auto-encoder is proposed in this work. To learn a representation of the candidates in a reduced and latent dimension space [22], the queries are also transformed in this new dimensional space and with that, the cosine similarity between the query and each candidate is calculated, thus selection the most distant candidates and we extract their documents as negative samples equaling the number of positive and negative documents for each query.

In Fig. 1 it is possible to see the used architecture, having a total of 6 layers, the first 3 being the encoder. The idea is to train the model to reconstruct the data in order to obtain a compact and latent representation. To summary, this process goes through the steps:

1. Documents are grouped by author, generating a bag-of-words vector for each candidate.
2. This vector is used in the training of the auto-encoder.
3. Reconstruction error calculated by the cosine similarity between the input vector and the vector returned by the auto-encoder output.
4. Once trained, the encoder is used to transform all candidates and queries into a reduced dimensional space.
5. Cosine similarity is applied between each query with each candidate.
6. From the most distant candidates, their documents are extracted.

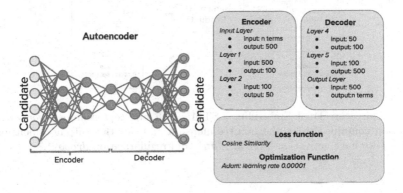

Fig. 1. Deep autoencoder architecture. The first half being called the encoder and the second decoder, next to it you can see the configuration of the proposed neural network.

3.2 Dual Embedding LSTM

As can be seen in Fig. 2, the Dual Embedding LSTM architecture has two input nodes, the first for queries and the second for documents. The representation of the input data is done through one-hot-encode, the terms are indexed by a value Long-Int, both queries and documents go through a layer of type embedding in which one-hot is converted into 100 dimensions that are used as input for two layers followed by two LSTM [7], a recurring layer capable of memorizing important information and forgetting the less. Then a fully connected layer and finally the query data and documents are aggregated through the Hadamard Product, ending with 3 fully connected layers, this architecture was inspired by [19].

The idea of this architecture is to try to merge a distributed and local model in a single architecture through the LSTM layers, memorizing the relationship that each term has to each other. As an optimization function, Adam [12] was used with a learning rate of $5E-5$ and L2 regularization. The loss function used was MSE calculated trying to predict whether a query is given and a document is positive or negative.

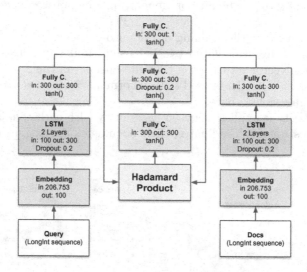

Fig. 2. Architecture Dual Embedding LSTM: Each box represents a layer with the information contained in *in* the input dimension and *out* the output dimension, in some cases we have activation layer like *tanh* and regularization *dropout*.

4   Data Characteristics

Training and evaluation use data from LExR collection [17], a public collection extracted from the *Plataforma Lattes* containing metadata from 10,942,014 publications among 206,697 candidates with title, keywords and some cases the summary. This set also includes 235 queries suggested by 513 experts who made judgments of relevance to themselves and some others, reaching 1,635 relevant judgments in all.

The keywords describe clearly and objectively the main issues that permeate the documents [24]. In this way, to generate the queries for our model, we extracted the keywords of all articles, totaling 1,876,279 valid queries. With these queries,we use the configuration mentioned in Subsect. 3.1 applied to a search server Elasticsearch [8], using the LM Dirichlet similarity function, reducing all lowercase terms and removing stop words in Portuguese and English. We then extracted up 20 documents for each query, totaling 20,641,359 triple of document, query and score.

For negative documents we use the technique of Subsect. 3.1. For each of the 1,876,279 queries, the most distant candidates were found after the reduction of dimension using the encoder, finally, we extracted the documents of the candidates until totaling the same amount of positive documents for each query.

2,456,446 terms were extracted from the documents, so we removed the terms with less than 20 uses due to the reduced size of the data and to collaborate with the model that requires a lot of training data, and may not create a good representation of the terms with few samples. After the reduction we have 206,753 terms that will be used in indexing queries and documents that will serve as input in the Dual Embedding LSTM model.

A summary of pre-processed data:

- 206.753 Terms
- 1.876.279 Queries
- 8.428.270 Documents used
- 41.282.718 Triples of Query + Document + Score (including half positive and half negative)

With data generated, the next step includes indexing documents and queries, transforming each term into its previously determined index. The data are separated into 80% for training and other 20% for validation and the 1,635 relevance judgements of LExR to perform the model test.

The training was carried out with size 300 batches, that is, 300 triples of query, document and score are inserted in the model. The architecture was developed in Pytorch and we used the Google Colab platform to train the model for 5 epochs, taking about 402 min to train and validate in each iteration.

5   Results

Applying the proposed model to the transformed data, we obtained the results from Table 1, which shows the evolution of the model according to each epoch, the accuracy increases until epoch 3, after which the model suffers from *overfitting* when the model starts to decorate the data and stops generalizing to unseen data . To calculate the *nDCG@10* we performed the test set queries in the model present Subsect. 3.1 returning 2000 documents for each query , that are finally re-ranked using the Dual Embedding LSTM. This new score is grouped by candidate and added, generating a final score for each. It is then compared with the *LExR* template as seen in Table 1.

The performance comparison between the proposals can be seen in Table 2 where we have the best models trained with negative documents randomly sampled and selected from the deep autoencoder as well as the generated baselines by LM Dirichlet and presented by [16].

Table 1. Evolution of the model by epoch

	Random		Autoencoder	
Epoch	nDCG@10	Accuracy	nDCG@10	Accuracy
1	0.159	0.725	0.167	0.6973
2	0.170	0.761	0.169	0.8737
3	0.169	0.782	**0.184**	**0.8894**
4	0.168	0.797	0.176	0.8894
5	0.166	0.808	0.174	0.8892

Table 2. Performance between different models.

Method	nDCG@10
Inf. Theoretic ρKL [16]	0.135
Inf. Theoretic ρH - ψDC [16]	0.146
Inf. Theoretic ρH - ψSDC [16]	0.164
LM Dirichlet	0.178
Dual Emb. LSTM + Random	**0.170**
Dual Emb. LSTM + Autoencoder	**0.184**

6   Conclusions

An improvement over baselines can be seen, indicating that it is possible to train a weak supervised model and obtain good results. The methodology of selecting documents with negative relevance with deep autoencoder for queries combined with the weak supervision generated by the language model for selecting documents with positive relevance and the Dual Embedding LSTM model to reclassify candidates surpassed the other techniques.

The next steps include verifying the model's performance using the new embedding language-agnostic BERT [8] to select new documents with positive and negative correlation given a query. Other works include the elaboration of a pair-wise architecture where the model receives two documents and the query, returning positive if the first document is more relevant, testing variations on the Dual Embedding LSTM and applying other statistical techniques to evaluate the models.

References

1. Balog, K., Bogers, T., Azzopardi, L., De Rijke, M., Van Den Bosch, A.: Broad expertise retrieval in sparse data environments. In: Proceedings of the 30th Annual International ACM SIGIR Conference on Research and Development in Information Retrieval, pp. 551–558. ACM (2007)

---

[8] LaBSE Language-agnostic BERT Sentence Embedding: https://dblp.uni-trier.de/.

2. Balog, K., et al.: Expertise retrieval. Found. Trends® Inf. Retrieval **6**(2–3), 127–256 (2012)
3. Berendsen, R., De Rijke, M., Balog, K., Bogers, T., Van Den Bosch, A.: On the assessment of expertise profiles. J. Am. Soc. Inf. Sci. Technol. **64**(10), 2024–2044 (2013)
4. Chi, M.T., Glaser, R., Farr, M.J.: The Nature of Expertise. Psychology Press, London (2014)
5. Dehghani, M., Zamani, H., Severyn, A., Kamps, J., Croft, W.B.: Neural ranking models with weak supervision. In: Proceedings of the 40th International ACM SIGIR Conference on Research and Development in Information Retrieval, pp. 65–74. ACM (2017)
6. Dias, T.M.R., Moita, G.F.: A method for the identification of collaboration in large scientific databases. Em Questão **21**(2), 140–161 (2015)
7. Gers, F.A., Schmidhuber, J., Cummins, F.: Learning to forget: continual prediction with LSTM. In: 9th International Conference on Artificial Neural Networks (ICANN1999). IET (1999)
8. Gormley, C., Tong, Z.: Elasticsearch: the Definitive Guide: a Distributed Real-time Search and Analytics Engine. O'Reilly Media, Inc., Sebastopol (2015)
9. Hinton, G.E., Salakhutdinov, R.R.: Reducing the dimensionality of data with neural networks. Science **313**(5786), 504–507 (2006)
10. Hoffmann, R., Zhang, C., Ling, X., Zettlemoyer, L., Weld, D.S.: Knowledge-based weak supervision for information extraction of overlapping relations. In: Proceedings of the 49th Annual Meeting of the Association for Computational Linguistics: Human Language Technologies, vol. 1. pp. 541–550. Association for Computational Linguistics (2011)
11. Husain, O., Salim, N., Alias, R.A., Abdelsalam, S., Hassan, A.: Expert finding systems: a systematic review. Appl. Sci. **9**(20), 4250 (2019)
12. Kingma, D.P., Ba, J.: Adam: a method for stochastic optimization. arXiv preprint arXiv:1412.6980 (2014)
13. Lane, J.: Let's make science metrics more scientific. Nature **464**(7288), 488 (2010)
14. LeCun, Y., Bengio, Y., Hinton, G.: Deep Learn. Nat. **521**(7553), 436–444 (2015)
15. Lin, S., Hong, W., Wang, D., Li, T.: A survey on expert finding techniques. J. Intell. Inf. Syst. **49**(2), 255–279 (2017)
16. Mangaravite, V., Santos, R.L.: On information-theoretic document-person associations for expert search in academia. In: Proceedings of the 39th International ACM SIGIR Conference on Research and Development in Information Retrieval, pp. 925–928. ACM (2016)
17. Mangaravite, V., Santos, R.L., Ribeiro, I.S., Gonçalves, M.A., Laender, A.H.: The lexr collection for expertise retrieval in academia. In: Proceedings of the 39th International ACM SIGIR conference on Research and Development in Information Retrieval. pp. 721–724. ACM (2016)
18. Mitra, B., Craswell, N., et al.: An introduction to neural information retrieval. Found. Trends® Inf. Retrieval **13**(1), 1–126 (2018)
19. Mitra, B., Diaz, F., Craswell, N.: Learning to match using local and distributed representations of text for web search. In: Proceedings of the 26th International Conference on World Wide Web, pp. 1291–1299. International World Wide Web Conferences Steering Committee (2017)
20. Robertson, S., et al.: The probabilistic relevance framework: Bm25 and beyond. Found. Trends® iInf. Retrieval **3**(4), 333–389 (2009)
21. Rosenblatt, F.: The Perceptron, a Perceiving and Recognizing Automaton Project Para. Cornell Aeronautical Laboratory, New York (1957)

22. Salakhutdinov, R., Hinton, G.: Semant. Hash. RBM **500**(3), 500 (2007)
23. Tang, J., Zhang, J., Yao, L., Li, J., Zhang, L., Su, Z.: ArnetMiner: extraction and mining of academic social networks. In: Proceedings of the 14th ACM SIGKDD International Conference on Knowledge Discovery and Data Mining, pp. 990–998. ACM (2008)
24. Yi, S., Choi, J.: The organization of scientific knowledge: the structural characteristics of keyword networks. Scientometrics **90**(3), 1015–1026 (2012)
25. Zhai, C., Lafferty, J.: A study of smoothing methods for language models applied to ad hoc information retrieval. In: ACM SIGIR Forum. vol. 51, pp. 268–276. ACM (2017)

Propensity to Use an Aerial Data Collection Device in Agricultural Research

Jacquelin Teresa Camperos-Reyes[1]([✉]) [iD], Fábio Mosso Moreira[4] [iD],
Fernando de Assis Rodrigues[2] [iD], and Ricardo César Gonçalves Sant'Ana[3] [iD]

[1] Faculty of Philosophy and Sciences, UNESP, Marília, SP, Brazil
[2] Institute of Applied Social Sciences, UFPA, Belém, PA, Brazil
deassis@ufpa.br
[3] Faculty of Sciences and Engineering, UNESP, Tupã, SP, Brazil
ricardo.santana@unesp.br
[4] Faculty of Sciences and Technology, UNESP, Presidente Prudente, SP, Brazil
fabio.moreira@unesp.br

Abstract. The access to information as a success factor in areas of human activity gains relevance with the intensive adoption of ICT. An important sector such as agriculture cannot be left out of this phenomenon. The results obtained in agricultural research show an increasing dependence on the intensive use of data, with a bigger volume and variety, origin in different environments, and through different technologies. The objective of this study is to categorize the possibilities of using Aerial Data Collection Device described in agriculture academic studies, to understand how these devices are being used in research carried out by Brazilian universities. The research is limited to the results of thesis and dissertations between the years 2015 and 2017, considering the graduate programs of the USP, UNICAMP, and UNESP. Five categories were defined regarding the propensity for the use of Aerial Data Collection Devices: Soil Diagnosis, Plant Diagnosis, Management of Grazing Areas, Crop Management, and Hydrographic Monitoring. It was identified that more research is needed to reflect on how science can help its application in this strategic productive sector.

Keywords: Research data · Data collection · Aerial data collection device · Drone agricultural research

1 Introduction

Access to information as a key success factor in all areas of human activity gains prominence with the intensive adoption of Information and Communication Technologies (ICTs). One of many characteristics identified in the economically developed countries in the 21st century is the intensive use of technologies addressed to the digital informational environment, specifically computers and mobile devices connected to the internet, which caused an effect of incorporation of several digital services in professional and entertainment activities [35]. This fact implies in a scenario characterized by the increase

E. Bisset Álvarez (Ed.): DIONE 2021, LNICST 378, pp. 297–312, 2021.
https://doi.org/10.1007/978-3-030-77417-2_22

in the volume of information generated and shared, the speed of dissemination of that content, and the variety of informational objects present in the accessible data flows in the digital environment [28].

The education, research, and outreach projects activities are also impregnated by this effect: technological development has allowed academics to carry out new practices, considering the possibilities of collecting, processing, and recovering large amounts of research data. The science area has provided these new options by increasing the number of data generated and collected from technological devices such as microscopes, telescopes, satellites, among others [36]. Those new options currently transform the phenomena of nature into electrical signals, standardized and machine-understandable, with little or no human intervention.

Although science increasingly relies on actions of data collection and processing, it should be noted that part of the technological body was not designed to deal with characteristics such as scalability and heterogeneity of data, which ends lacking the quality of the information that will support the scientific research process. This problem is associated with the requirement to consolidate e-science, which is strongly dependent on the fall of barriers to open access to research data, and grounded on an infrastructure based on the internet and digital devices [19].

One of the most relevant research domains in Brazil is the agricultural area, that places the country as the most productive in Latin America, and fifth in the world ranking [40]. Agriculture includes a set of activities associated with nourishment, environmental management, and human culture, and may be diversified concerning the techniques used and in terms of production systems and social organization [33]. Activities such as the management of soil fertility and plant varieties, the adaptation of animals to the characteristics of the local environment, and the management of environmental resources are inserted in the practices carried out by farmers [42].

Agricultural practices passed by transformations during the 19th and 20th centuries, influenced by the development of technologies during and after the Second World War which resulted in the mechanization of activities performed by farmers and the development of new chemical and biological products [33, 42, 47].

The agricultural modernization process in Brazil was influenced by the international model and the local incentive for scientific and technological production research. This incentive was supported by the public sector since the beginning through the establishment of departments, centers, institutes, and public companies, especially through the research performance performed by the Brazilian Agricultural Research Company [17, 21].

Article 12 of law No. 8.171/1991 [8], which provides the Brazilian Agricultural Policy, establish that agricultural research is the practice of research generated or adapted from the biological knowledge of the integration of the various ecosystems to increase productivity and generating technologies focused on animal and plant health, as well as the preservation of health and the environment.

The research of development technologies and innovation for agriculture collects and generates a variety of data, ranging from results of practical experiments in greenhouses and laboratories to data generated from the use of electronic remote sensors [23], in which the data is regularly generated in diverse data types, with a high expense, and through experiments that sometimes consume a lot of time, implying an urge for optimization, from data integration and reuse processes [15].

With the occurrence of different techniques for data collection in agricultural research, this investigation is constructed in the context of the aerial survey - a set of operations to obtain land, air, and sea data from embedded technologies for data collection and transmission using the technological resources from aerial platform and control station [1].

This research addresses the Unmanned Aerial Vehicles (UAVs) used like an instrument applied in aerial data collection activities within studies in the agricultural area. It starts with the following questions: I) What is the propensity to use Aerial Data Collection Device (ADCD) in agricultural area research as a resource for data collection? ii) What results may be generated with the use of this type of resource? Considering those questions, the objective of the research is to categorize the possibilities of using ADCD for data collection in agricultural research, paying attention to how this device is being used in research performed by higher education institutions.

The Information Science comprises the area of knowledge that is interested in research related to studies on research data, addressing subjects such as i) digital repositories and curation of research data [39]; ii) use and reuse of research data [16], and; iii) research data management [18]. Thus, it is considered that the Information Science perception may (and it should) also be applied in the context of agricultural research data, focusing on the use of technological devices in data collection and processing activities.

Based on the Information Science perspective, the research scope is oriented towards the information flows and the phases of the data life cycle, with special attention to the data collection phase [37], in a sense that the technical resource involved becomes only the instrument through which the data collection is carried out.

2 Theorical Framework

The use of UAVs may replace the use of manned aircraft, bringing as main benefits i) the reduction of operational cost (cost of an hour of flight, maintenance, among others); ii) greater application flexibility in obtaining data; iii) dispatch of transport and storage due to its small size; iv) convenience in takeoff and landing operations, and; v) technical capacity for shooting and collecting photographic records with a resolution similar to manned aircraft. In addition, the UAV foregrounds other aerial data collection platforms for having the ability to perform low-altitude flights and, consequently, suffer fewer impacts from climatic factors [34, 49].

In 2010, the Brazilian Department of Air Space Control defined UAV as an air vehicle designed to operate without a pilot on board, with a payload onboard and not used for recreational purposes. This definition includes all three-axis controllable airplanes, helicopters, and airships, therefore excluding traditional balloons and model airplanes [11].

For DECEA, a department of the Brazilian Ministry of Defense, the term UAV is considered obsolete by the international aeronautical community because the main aviation organizations no longer use the term "vehicle" and due to the requirement for an existing system to carry out the flights [9]. In this sense, the Brazilian Ministry of

Defense replaced the acronym UAV with RPA, referring to the term Remotely Piloted Aircraft in the English language [9]. In this sense, an RPA is an unmanned aircraft piloted from a remote pilot station [9].

Another term that refers to this aerial platform is DRONE, a term from the English language meaning a low continuous humming sound, observed in scientific publications and commercial communications. The term DRONE is associated with a more generic and informal concept has no technical protection or legal definition in Brazil, and it can be used to describe any unmanned flying object, even for recreational use [10].

Considering the presence of more than one designation for these instruments (UAV, RPA, and even DRONE) emerges the requirement to use in this study the definition of Aerial Data Collection Device (ADCD) to refer to the set determined by the aerial platform and the control platform, through embedded technologies, has the ability to carry out aerial data collection.

Operations with ADCD are restricted to airspace, involving autonomous, semi-autonomous, or remotely operated aircraft, authorized and coordinated by the ANAC through Brazilian Special Civil Aviation Regulation No. 94/2017. This regulation is an instrument complementary to the standards to operate this type of device, established by DECEA and by the Brazilian National Telecommunications Agency (ANATEL) [2].

Initially, the ADCDs were designed for military purposes in the 1950s and 1960s. These devices were developed to facilitate the gathering of information on signs of hostilities, obtaining enemy reconnaissance photographs [6].

Over time the use of ADCDs changed to a more flexible application to other scenarios, such as surveillance actions; cartographic and oceanographic studies; search and rescue missions in difficult areas; road traffic control; foreign borders control; monitoring of polluting gas emissions, and; wildfire monitoring [41].

It was possible to follow that the ADCD for data collection in agricultural scientific research is an ongoing discussion. Besides, there is an open debate about the recognition of the importance of regulation about the use of this kind of technological resource in agricultural research by the Commission of Constitution, Justice, and Citizenship (CCJC) from the Brazilian Federal Senate of Brazil: Bill - PLS No. 698/2015. The bill seeks to update Law No. 8,171/1991 and includes, among the purposes, the prioritization of the use of this type of technological tool [8, 30].

A Brazilian pioneering initiative to use an ADCD in agriculture was made by the Radio Assisted and Autonomous Reconnaissance Aircraft Project (ARARA) developed in association with the Institute of Mathematical and Computing Sciences (USP), EMBRAPA, and AGX Technology. The ARARA project had the goal to replace the use of conventional aircraft to gather data and images for monitoring areas subjected to environmental issues [22].

According to [41], since the use of ADCD in agricultural scientific research is a recent phenomenon, some Brazilian universities use it as a resource to expand the amount of data available in the development of scientific studies, in addition to being a key element in the capture of resources before development agencies.

3 Methodological Procedures

This research starts from a systematic collection of research in the agricultural area in which the ADCD are used or referenced in data collection. The research universe is the results of studies in thesis or dissertations formats, with a sample restricted to publications of the Postgraduate Programs (PGP) of the University of São Paulo (USP), State University of Campinas (UNICAMP), and São Paulo State University (UNESP), published in the years between 2015 and 2017, available in the respective institutional online repositories.

The selection criteria were defined on the concept that these documents (thesis and dissertations) need to correspond to the results of research that are in experimental status with a single and defined topic, developed in the scope of a PGP. Thus, it is possible to observe the characteristics of the use of these devices in each educational institution, as well as the thematic areas that each PGP research is investigating. It was considered to establish the sample on those three state universities of São Paulo by the excellent performance that they present in the scientific environment such as the Web of Science [13], and because the state of São Paulo has the highest number of ADCD records enabled for operation, representing 35.20% of the total records of this type of aircraft in Brazil [3].

The repositories of Theses and Dissertations of the three educational institutions were used as a source for obtaining the studies to be analyzed. The procedure employed was to access documents through the advanced search interfaces available to retrieve the document set. The search filters were configured to retrieved only documents published between 2015 and 2017, and types of documents retrieved only composed on thesis or dissertations forms. The search expressions were used were: "Non-Manned Air Vehicle", "Drone", and "Remotely Piloted Aircraft", concatenated with the terms "Agriculture", "Agricultural", "Livestock" and "Environmental". All those expressions were written in Portuguese, respectively: "Veículo Aéreo Não Tripulado", "Drone" and "Aeronave Remotamente Pilotada", concatenated with the terms "Agricultura", "Agrícola", "Pecuária" and "Ambiental".

After recovering the scientific production in those repositories, it was made a classification of the results from the analysis of fragments of the texts that cited at least one of the terms: "UAV", "RPA" or "DRONE". This procedure sought to filter, by reading the title and the abstract, the existence of adherence to the subject analyzed.

The Content Analysis [5] was applied for the treatment of the corpus, which consists of a set of communication analysis techniques intended to obtain indicators (quantitative or not) by systematic procedures and objectives of content description in the messages, that allow the inference of knowledge related to the conditions of production and reception (inferred variables) of these messages.

It was chosen the technique of category analysis, which implies the selection of qualitative criteria of choice and categories for the association of the content present in the analyzed messages; in this case, textual domain messages [5]. The definition of the categories was performed after the analysis of the corpus and was based on principles of homogeneity and mutual exclusion, conceptually based on the possibilities of data collection with ADCD in agricultural research.

4   Results and Discussion

The results obtained from the sample analysis allow observing a partial view of how Brazilian higher educational institutions use ADCD in agricultural research. After applying the methodological strategy, it retrieved 460 thesis and dissertations from the institutional repositories: 11 from USP, 6 from UNICAMP, and 443 from UNESP. A total of 442 thesis and dissertations were discarded during the classification of results because it was verified that has no thematic adherence to ADCD use to collect data in agricultural research. In other words, they did not contemplate topics related to Agriculture. The discard pile was formed by 8 documents from USP, 3 documents from UNICAMP, and 431 documents from UNESP. After discarding, the analyzed corpus was formed by 18 documents, which are synthesized in Appendix A. The corpus consists of 12 documents from UNESP, 3 documents from USP, and 3 documents from UNICAMP (Appendix A).

From the content analyzed were defined five categories to represent the types of use provided by the ADCDs in data collection within agricultural research. The Table 1 presents the five defined categories: Soil Diagnosis (SD), Plant Diagnosis (PD), Management of Grazing Areas (MGA), Crop Management (CM), and Hydrographic Monitoring (HM). It can be observed the definitions established for each class; the consulted works that are associated with each one; as well as the attributes of the data collected in each investigation.

Table 1. Categorization of ADCD use in agricultural research.

Category	Category description	Analyzed document	ADCD data collected during the research
Soil Diagnosis (SD)	The analytical process to qualify the soil based on characteristics such as fertility, typology, and structure.	Ávila [4]	Soil type mapping.
		Tagliarini [44]	Estimation of soil losses by erosion.
Plant Diagnosis (PD)	The analytical process to determine crop Phyto-physiognomy and identify the presence of pests and threats.	Moriya [31]	Spectral characterization of healthy and diseased sugarcane, to identify the most suitable wavelengths for detecting diseases.
		Martins [26]	Coffee mapping: healthy, the initial state of infection or severely infected.
		Borges [7]	Plant ecophysiological information.
		Vasconcelos [48]	On the effects of applying doses of phosphorus and potassium on growth in *Khaya Senegalensis* plants in the implantation phase.
Management of Grazing Areas (MGA)	The systematic practice of planning, strategies, and control of factors related to organization and the welfare of livestock in grazing space.	Teixeira [45]	Management and rotation of animals in grazing areas.

(continued)

Table 1. (*continued*)

Category	Category description	Analyzed document	ADCD data collected during the research
Crop Management (CM)	Activities of planning, strategies, and control of factors related to the development of the crop, such as quality of the vegetation cover, the arrangement of plants in the area, and the application of techniques to improve productivity.	Maldonado Júnior 24]	Counting of fruits of green oranges to estimate quantities of fruits present in the fruit trees.
		Miyoshi [29]	Spectral characterization of vegetation species at the foliar and crown level to contribute with information that can be used for forest monitoring.
		Souza [43]	Fault mapping in sugarcane planting lines; Represent variability of the productivity field.
		Criado [14]	Monitoring the recomposition of reforested areas.
		Torres [46]	Monitoring of forestry activities, estimation of forest volume and biomass, and biodiversity mapping.
		Santos [38]	Mapping of sugarcane areas.
		Martello [25]	Height and productivity estimation; evaluation of sugarcane fields.
		Niemann [32]	Information related to the surface as plant canopies.
		Chiacchio [12]	Mapping of rural cultures and properties.
Hydrographic Monitoring (HM)	The state of surveillance of availability of water resources and the control of irrigation systems.	Marton [27]	Environmental monitoring; Water frames.
		Garcia [20]	Environmental diagnosis of the physical environment of the hydrographic basin.

The stacked bar graphic element of Fig. 1 summarizes the incidence of studies for each category differentiated by the institution where the research was carried out.

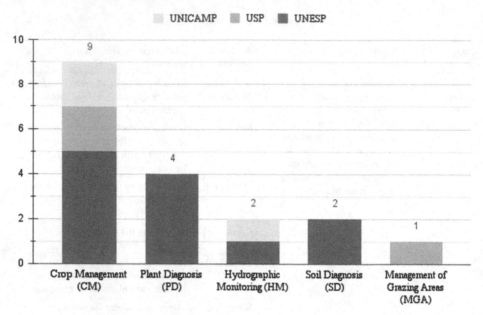

Fig. 1. Incidence of research by category and by institutions.

The category in which most of the documents were associated was Crop Management, which includes the types of use of ADCD related to obtaining information on productivity in the harvest, monitoring of environmental and productive areas, and monitoring of forests and planting failures. A total of 9 documents were associated with this category, 5 related to UNESP, carried out in the Cartographic Sciences and Geography PGP (2 research each), and Agronomy. The 2 investigations carried out at USP correspond to one in the area of Sciences and the other in the area of Agricultural Systems Engineering, and 2 studies were developed at UNICAMP, one in the area of Agricultural Engineering and the other in the area of Geography.

The category Plant Diagnosis includes plant diagnostic studies at a micro level, related to the chemical and biological properties of each plant that make up the tillage, such as measurement of macronutrients, abiotic changes, control of fruit ripening, and detection of hydric and or nutritional stress. A total of 4 documents of the corpus were associated with this category, being the second most representative category in the analyzed context. All the studies associated with this category were carried out at UNESP, two of them in the area of Cartographic Sciences, one in Agronomy and the other in Ecology.

The category defined as Hydrographic Monitoring comprises ADCDs to support strategies for the use of water resources, mapping and geo-referencing of hydrographic basins, and monitoring of the degradation of water areas. Two documents of the corpus were associated with this category, one being developed in the Mechanical Engineering PGP of UNICAMP, and another in the UNESP at the Agronomy PGP.

The category Soil Diagnosis presents ADCD applications for data collection that assist the analysis of soil moisture and temperature, monitoring of erosive processes, and measurement of hydric and nutritional deficiencies in plantations. Two documents were associated with this category, both carried out in UNESP, one in the area of Agronomy and the other in Geography.

Studies related to the observation of the movement of cattle in grazing areas and the elaboration of rotation strategies were classified in the category Management of Grazing Areas, in which only 1 research was associated with this category carried out in the PGP of Sciences at USP.

Regarding the representativeness of the categories concerning the sample of documents analyzed, research related to the use of ADCD for Crop Management (50%) predominated, followed by the use of ADCD in Plant Diagnostics (22%). The categories Soil Diagnosis and Hydrographic Monitoring obtained the same percentage of representativeness (11% each) and as a less representative category the use of the device for Management of Grazing Areas (6%).

Besides, it was possible to establish a perception of the use of ADCD by higher educational institutions and by areas of knowledge that work with agricultural research. In general, there was a predominance of the use of ADCD in the areas of Agronomy, Geography, Cartographic Sciences, Engineering (Agricultural, Agricultural Systems or Mechanics), and Sciences.

In summary, the ADCD use applied to Soil Diagnosis is more representative in the areas of Agronomy and Geography; while for Plant Diagnosis, the use of this type of device predominates in areas such as Cartographic Science, Agronomy, and Ecology.

Regarding the Management of Grazing Areas category, the only associated document was the result of an investigation carried out in the area of general science. The investigations that used ADCD to obtain data were predominant in the areas of Geography, Cartographic Sciences, and Engineering (Agricultural or Agricultural Systems), followed by Agronomy and Sciences. The Hydrographic Monitoring included the use of this type of device in research carried out in programs in the areas of Agronomy and Mechanical Engineering.

An outstanding fact is that there is a higher frequency of use of ADCD at UNESP by researchers from the campus of the Faculty of Sciences and Technologies, located in the municipality of Presidente Prudente. The existence of an agreement between this campus and the Finnish Geodesy Institute (FGI), called Unmanned Airborne Vehicle-Based 4D Remote Sensing for Mapping Rain Forest Biodiversity and its Change in Brazil (UAV_4D_BIO), which involves mapping and monitoring of the biodiversity and the use of ADCD for obtaining images and geospatial data in forest mapping.

Regarding how each investigation used or mentioned the use of ADCD, it was observed that 50% of the research used ADCD to obtain their data, including the description of the elements involved in the procedure for using the device. A total of 11% of the research also used the resource for data collection but did not describe its mode of use, only presenting the data gathered. A part corresponding to 39% of the total of analyzed research did not use ADCD for data collection. However, they referred to its use as a possibility of improvement in the data collection carried out, or as an example to replace the method applied in their studies.

5 Conclusions

Through the analysis of thesis and dissertations that used or mentioned the use of ADCD as a data collection device, it was possible to establish categories that could facilitate the understanding of the possibilities and implications of the use of this type of device in agricultural research. These results may contribute to researchers who present the need to collect data and could project trends of applying ADCD to obtain data with a higher level of quality and economic benefit compared to other existing resources.

Information Science area contributed to this research because it allowed us to glimpse the information flows related to the use of ADCD in the collection and processing of data in agricultural research since it was necessary to understand the main attributes of the data treated in each document analyzed to establish the definition of the categories. In addition, the area contributed with its interdisciplinary vision, allowing to discussion data collection activities from a modern and innovative device, but with a perspective directed to the content that is generated, and not to the technical aspects of the machine.

The results obtained determined that the use of ADCD for data collection in agricultural research is already a reality and that it can be applied both in Nature Sciences (e.g. Agronomy and Biology) and Math-Sciences (e.g. Engineering and Cartography) and Humanity Sciences (e.g. Geography) or any specialization that address research within the agricultural context.

In the analyzed sample, it was verified that the propensity is towards studies that focus on improvements in Crop Management, although considering the geographical location and the predominance of the agricultural sector, it is possible to extrapolate to other regions that are eager for developments technologies that optimize processes paramount to them.

It should be noted that because it is an instrument that recently became part of research data collection procedures, it is necessary to warn about aspects involved in its application in the scientific area, mainly in the description of its use procedure when publishing the studies. In this sense, other developments are glimpsed that can also be covered by Information Science, such as the singularity related to curating the data obtained through ADCD, and what are the essential requirements for a repository to support the availability of the data collected with this type of device.

Acknowledgments. This work was developed with the support of the Coordination for the Improvement of Higher Education Personnel - Brazil (CAPES) - Financing Code 001.

Appendix A. Synthesis of the Analyzed Corpus

Year	Advisor	Author	Title	University	PGP	Location (Municipality)
2015	IMAI, N. N.	MORIYA, E. A. S.	Identification of spectral bands for the detection of healthy and sick sugarcane cultures using a hyperspectral camera embedded in UAV	UNESP	Cartographic Sciences	Presidente Prudente
2015	PANCHER, A. A.	ÁVILA, M. R.	Scenarios of urban expansion and legislation and reflections on tree and shrub cover in the city of Americana-SP	UNESP	Geography	Rio Claro
2016	BARBOSA, J. C.	MALDONADO JUNIOR, W.	Estimation of the number of green fruits in orange trees using digital images	UNESP	Agronomy	Jaboticabal
2016	FIORAVANTI A. R.	MARTON, A. S.	Linear control of the trajectory of the robotic airship with quadruple propulsion	UNICAMP	Mechanical Engineering	Campinas
2016	GALO, M. de L. B. T.	MARTINS, G. D.	Inference of nematode infection levels in coffee culture from remote sensing data acquired in multi-scale	UNESP	Cartographic Sciences	Presidente Prudente
2016	IMAI, N. N.	MIYOSHI, G. T.	Spectral characterization of ir land Atlantic Forest species at leaf and crown level	UNESP	Cartographic Sciences	Presidente Prudente
2016	LAMPARELLI, R. A. C.	SOUZA, C. H. W.	Acquisition of information at sugarcane field level using data from an Unmanned Aerial Vehicle (UAV) under different methodologies	UNICAMP	Agricultural Engineering	Campinas

(continued)

(continued)

Year	Advisor	Author	Title	University	PGP	Location (Municipality)
2016	MORELLATO, L. P. C.	BORGES, B. D.	A new perspective to understand functional connectivity by integrating landscape and Phenology	UNESP	Ecology	Rio Claro
2016	PIROLI, E. L.	CRIADO, R. C.	Change in land use and land cover in municipalities in Paranapanema from 1984 to 2014	UNESP	Geography	Presidente Prudente
2016	TECH, A. R. B.	TEIXEIRA, B. E.	Use of unmanned aerial vehicle with fixed-wing to monitor and collect images of animals and environments in rural properties	USP	Sciences	Pirassununga
2016	TOMMASELLI, A. M. G.	TORRES, F. M.	Assembly and Evaluation of a Laser Scanning System embedded in UAV	UNESP	Sciences	Presidente Prudente
2016	VASCONCELOS, S. T.	VASCONCELOS, R. T.	Phosphate and potassium fertilization in the implantation of *Khaya senegalensis A.Juss*	UNESP	Agronomy	Jaboticabal
2017	CAMPOS, S.	TAGLIARINI, F. S. N.	Geoprocessing techniques applied to quantify soil losses in a hydrographic basin	UNESP	Agronomy	Botucatu
2017	CAMPOS, S.	GARCIA, Y. M.	Environmental diagnosis of the Ribeirão Pederneiras River Basin - Pederneiras / SP	UNESP	Agronomy	Botucatu

(continued)

(*continued*)

Year	Advisor	Author	Title	University	PGP	Location (Municipality)
2017	CASTILLO, R. A.	SANTOS, H. F.	Regional competitiveness of the sugar-energy sector in the mesoregion of Minas Gerais / Alto Paranaíba: Globalized scientific agriculture and socio-environmental implications in the city of Uberaba / MG	UNICAMP	Geography	Campinas
2017	FIORIO, P. R.	MARTELLO, M.	Estimation of height and productivity of sugarcane using images obtained by remotely piloted aircraft	USP	Agricultural Systems Engineering	Piracicaba
2017	SILVA, T. S. F.	NIEMANN, R. S.	Comparison of filtering methods and generation of digital terrain models from images obtained by an unmanned aerial vehicle	UNESP	Geography	Rio Claro
2017	TECH, A. R. B.	CHIACCHIO, S. S. R.	An unmanned aerial vehicle with rotary-wing in the activity of mapping and image collection in precision agriculture and animal monitoring	USP	Sciences	Pirassununga

References

1. Agência Nacional de Aviação Civil. Resolução nº 377, de 15 de março de 2016. Regulamenta a outorga de serviços aéreos públicos para empresas brasileiras e dá outras providências (2016). https://www.anac.gov.br/assuntos/legislacao/legislacao-1/resolucoes/resolucoes-2016/resolucao-no-377-15-03-2016
2. Agência Nacional de Aviação Civil. RBAC-E no 94/2017. Requisitos gerais para Aeronaves Não-Tripuladas de Uso Civil (2017). https://www.anac.gov.br
3. Agência Nacional de Aviação Civil. Quantidade de Cadastros (2018). http://www.anac.gov.br/assuntos/paginas-tematicas/drones/quantidade-de-cadastros
4. Ávila, M.R.: de. Cenários da expansão urbana e da legislação e os reflexos na cobertura vegetal arbórea e arbustiva na cidade de Americana-SP. (Master's thesis). Universidade Estadual Paulista (2015). http://hdl.handle.net/11449/138523
5. Bardin, L.: Análise de conteúdo 1. ed. Almedina, Portugal (2011)
6. Blom, J.D.: Unmanned Aerial Systems: A Historical Perspective. Combat Studies Institute Press, Fort Leavenworth (2010)
7. Borges, B.D.: Uma nova perspectiva para entender a conectividade funcional integrando paisagem e fenologia. (Master's thesis). Universidade Estadual Paulista (2016). http://hdl.handle.net/11449/143954
8. Brazil: Lei No 8.171, de 17 de Janeiro de 1991. Dispõe sobre a política agrícola. Portal do Planalto (1991). http://www.planalto.gov.br/ccivil_03/LEIS/L8171.htm
9. Brazil: Ministério da Defesa. Comando da Aeronáutica Departamento de Controle do Espaço Aéreo. Sistemas de Aeronaves Remotamente Pilotadas e o Acesso ao Espaço Aéreo Brazileiro. ICA 100-40 (2015a). https://www.decea.gov.br/static/uploads/2015/12/Instrucao-do-Comando-da-Aeronautica-ICA-100-40.pdf
10. Brazil: Departamento de Controle do Espaço Aéreo Força Aérea Brasileira. Voos de VANT (drones). Entenda melhor! Portal do DECEA (2015b). https://www.decea.gov.br/?i=midia-e-informacao&p=pg_noticia&materia=autorizacoes-para-voos-de-vant-entenda-melhor
11. Brazil: Ministério da Defesa. Departamento de controle do espaço aéreo. Veículos Aéreos não Tripulados. AIC-N 21 (2010). http://web.archive.org/web/20100826122344/https://publicacoes.decea.gov.br/?i=publicacao&id=3499
12. Chiacchio, S.S.R.: Veículo aéreo não tripulado de asa rotativa na atividade de mapeamento e coleta de imagem na agricultura de precisão e no monitoramento de animais. (Master's thesis). Universidade de São Paulo (2017). http://www.teses.usp.br/teses/disponiveis/74/74134/tde-24042017-104340/pt-br.php
13. Coordenação de Aperfeiçoamento de Pessoal de Nível Superior. Research in Brazil: A report for CAPES by Clarivate Analytics (2018). https://www.gov.br/capes/pt-br/centrais-de-conteudo/17012018-capes-incitesreport-final-pdf
14. Criado, R.C.: Mudanças no uso e na cobertura da terra em municípios do Pontal do Paranapanema de 1984 a 2014. (Doctoral thesis). Universidade Estadual Paulista (2016). http://hdl.handle.net/11449/148629
15. Ćwiek-Kupczynska, H., et al.: Measures for interoperability of phenotipic data: minimum information requirements and formatting. Plant Meth. **12**(44) (2016)
16. Dias, G.A., dos Anjos, R.L., de Araújo, D.G.: A gestão dos dados de pesquisa no âmbito da comunidade dos pesquisadores vinculados aos programas de pós-graduação brasileiros na área da Ciência da Informação: desvendando as práticas e percepções associadas ao uso e reuso de dados. Liinc em Revista **15**(2), 5–31 (2019). http://revista.ibict.br/liinc/article/view/4683/4327
17. Empresa Brasileira de Pesquisa Agropecuária. Quem somos? Portal EMBRAPA (2018). https://www.embrapa.br/quem-somos

18. Estevão, J.S.B., Arns, E.M., Strauhs, F.: do R. Gestão de dados de pesquisa: uma prática para abrir a caixa preta da pesquisa científica. Revista Digital de Biblioteconomia e Ciência da Informação 17(1), 1–26 (2019). https://periodicos.sbu.unicamp.br/ojs/index.php/rdbci/art icle/view/8656239/21458

19. Fox, P., Hendler, J.: eScience semântica: o significado codificado na próxima geração de ciência digitalmente aprimorada. In: Hey, T., Transley, S., Tolle, K. (eds.) O quarto paradigma: descobertas científicas na era da eScience. Oficina de Textos, São Paulo (2011)

20. Garcia, Y.M.: Diagnóstico ambiental da bacia hidrográfica do ribeirão Pederneiras – Pederneiras/SP. (Tesis doctoral). Universidade Estadual Paulista (2017). http://hdl.handle.net/11449/15088

21. Ichikawa, E.Y.: O Estado no apoio à pesquisa agrícola: uma visão histórica. RAP 34(3), 89–101 (2000). http://bibliotecadigital.fgv.br/ojs/index.php/rap/article/view/6282/4873

22. Jorge, L. A. de C.; Inamasu, R. Y. Uso de veículos aéreos não tripulados (VANT) em agricultura de precisão. In: A. C. de C. Bernardi et al. (Eds.). Agricultura de Precisão: Resultados de um Novo Olhar 1. ed. 109–133, EMBRAPA, Brazilia (2014)

23. Leonelli, S., Davey, R.P., Arnaud, E., Parry, G., Bastow, R.: Data management and best practice for plant science. Nat. Plants 3(17086), 1–4 (2017). https://doi.org/10.1038/nplants.2017.86

24. Maldonado, Júnior, W.: Estimativa do número de frutos verdes em laranjeiras com o uso de imagens digitais. (Doctoral thesis). Universidade Estadual Paulista (2016). http://hdl.handle.net/11449/136455

25. Martello, M.: Estimativa da altura e produtividade da cana-de-açúcar utilizando imagens obtidas por aeronave remotamente pilotada. (Master's thesis). Universidade de São Paulo (2017). http://www.teses.usp.br/teses/disponiveis/11/11152/tde-16102017-170204/pt-br.php

26. Martins, G.D.: Inferência dos níveis de infecção por Nematoides na cultura cafeeira a partir de dados de sensoriamento remoto adquiridos em multiescala. (Doctoral thesis). Universidade Estadual Paulista (2016). http://hdl.handle.net/11449/148760

27. Marton, A.S.: Controle linear de trajetória de dirigível robótico com propulsão quádrupla. (Doctoral thesis). Universidade Estadual de Campinas (2016). http://repositorio.unicamp.br/jspui/handle/REPOSIP/305467

28. Mayer-Schönberger, V., Cukier, K.: Big Data: A Revolution That Will Transform How We Live, Work, and Think. Houghton Mifflin Harcourt, Boston (2013)

29. Miyoshi, G.T.: Caracterização espectral de espécies de Mata Atlântica de Interior em nível foliar e de copa. (Master's thesis). Universidade Estadual Paulista (2016). http://hdl.handle.net/11449/13641

30. Morais, W.: Projeto de Lei do Senado no 698 de 2015. Altera a Lei no 8.171, de 17 de janeiro de 1991, que dispõe sobre política agrícola, para incluir entre as finalidades da pesquisa agrícola no Brazil o apoio ao uso de Veículos Aéreos Não Tripulados (VANTs) (2015). https://www25.senado.leg.br/web/atividade/materias/-/materia/123755?o=d

31. Moriya, É.A.S.: Identificação de bandas espectrais para detecção de cultura de cana-de-açúcar sadia e doente utilizando câmara hiperespectral embarcada em VANT. (Doctoral thesis). Universidade Estadual Paulista (2015). http://hdl.handle.net/11449/133961

32. Niemann, R.S.: Comparação de métodos de filtragem e geração de modelos digitais de terreno a partir de imagens obtidas por veículo aéreo não-tripulado. (Master's thesis). Universidade Estadual Paulista (2017). http://hdl.handle.net/11449/152635

33. Offutt, S.: What is agriculture? Trabajo presentado en Conference on Agricultural and Environmental Statistical Applications (CAESAR 2001). Conference on Agricultural and Environmental Statistical Applications, ISTAT, Rome (2002)

34. De Oliveira, et al.: Potencialidades da utilização de drones na agricultura de precisão. Braz. J. Dev. 6(9), 64140–64149 (2020)

35. De Rodrigues, F.A., Sant'Ana, R.C.G.: Use of taxonomy of privacy to identify activities found in social network's terms of use. Knowl. Organ. **43**(4), 285–295 (2016)
36. Sales, L.F., Cavalcanti, M.T.: Seleção e avaliação de coleções de dados digitais de pesquisa: uma possível abordagem metodológica. Informação & Tecnologia (ITEC) **2**(2), 88–105 (2015)
37. Sant'Ana, R.C.G.: Ciclo de vida dos dados: uma perspectiva a partir da ciência da informação. Informação & Informação **21**(2), 116 (2016)
38. Dos Santos, H.F.: Competitividade regional do setor sucroenergético na mesorregião Triângulo Mineiro/Alto Paranaíba: agricultura científica globalizada e implicações socioambientais no município de Uberaba – MG. (Doctoral thesis). Universidade Estadual de Campinas (2017). http://repositorio.unicamp.br/jspui/handle/REPOSIP/324346
39. Sayão, L.F., Sales, L.F.: Curadoria digital e dados de pesquisa. Atoz **5**(2), 67–71 (2016). https://revistas.ufpr.br/atoz/article/view/49708/3016
40. SCIMAGOjr (2020). http://www.scimagojr.com/countryrank.php?area=1100
41. Da Silva, G.G., et al.: Veículos aéreos não tripulados com visão computacional na agricultura: aplicações, desafios e perspectivas. In: 2o Seminário Internacional de Integração e Desenvolvimento Regional, Universidade Católica Dom Bosco, Ponta Porã (2014)
42. Soglio, F.D.: A agricultura moderna e o mito da produtividade. In: Soglio, F.D., Kubo, R.R. (eds.). Desenvolvimento, agricultura e sustentabilidade, vol. 1, pp. 11–38, Editora UFRGS, Porto Alegre (2016)
43. De Souza, C.H.W.: Aquisição de informações em nível de campo da cana-de-açúcar utilizando dados de um veículo aéreo não tripulado (VANT) sob diferentes metodologias. (Doctoral thesis). Universidade Estadual de Campinas (2016). http://repositorio.unicamp.br/jspui/handle/REPOSIP/330772
44. de Tagliarini, F.S N.: Técnicas de geoprocessamento aplicadas na quantificação de perdas de solo em bacia hidrográfica. (Tesis de maestría inédita). Universidade Estadual Paulista (2017). http://hdl.handle.net/11449/150268
45. Teixeira, B.E.: Utilização de veículo aéreo não tripulado de asa fixa no monitoramento e coleta de imagem de animais e ambientes em propriedades rurais. (Master's thesis). Universidade de São Paulo (2016). http://www.teses.usp.br/teses/disponiveis/74/74134/tde-06042016-133819/pt-br.php
46. Torres, F.M.: Montagem e avaliação de um sistema de varredura a LASER embarcado em VANT. (Master's thesis). Universidade Estadual Paulista (2016). http://hdl.handle.net/11449/138900
47. Troian, A., Klein, Â.L., Dalcin, D.: Relato de caso: novidades e inovações na agricultura familiar: debates e discussões da produção de tecnologias. Revista Brasileira de Agropecuária Sustentável (RBAS), **1**(1), 6–17 (2011). https://periodicos.ufv.br/ojs/rbas/article/view/2604
48. De Vasconcelos, R.T.: Adubação fosfatada e potássica na implantação de Khaya senegalensis A.Juss. (Doctoral thesis). Universidade Estadual Paulista (2016). http://hdl.handle.net/11449/14501
49. Zhang, C., Kovacs, J.M.: The application of small unmanned aerial systems for precision agriculture: a review. Prec. Agric. **13**(6), 693–712 (2012). https://link.springer.com/article/10.1007/s11119-012-9274-5

A Model for Analysis of Environmental Accidents Based on Fuzzy Logic
Case Study: Exxon Valdez Oil Spill

Ana Claudia Golzio(✉) and Mirelys Puerta-Díaz

São Paulo State University, Av. Hygino Muzzi Filho, 737, Mirante, Marília,
SP CEP 17525-900, Brazil

Abstract. This research aims to present a fuzzy-logic-based conceptual model
for environmental accidents analysis , to reveal corporate social responsibility
initiatives by companies responsible for the disasters. We studied one of the biggest
environmental man-made disasters in history, the one that occurred on March 24,
1989 in Prince William Sound, Alaska when the oil tanker Exxon Valdez spilled
10.8 million gallons of American crude oil. The data was collected from the online
database of the newspaper The New York Times for the timespan 03/24/1989–
09/01/2017 . As a central point of the research, we investigate ethical issues based
on the mapping of an ethical vocabulary carried out in the corpus of the analyzed
documents. The results show that the proposed model can be replicated, after some
adjustments, to verify actions in accordance with the principles of corporate social
responsibility for other environmental accidents.

Keywords: Fuzzy Rules-Based System · Corporate social responsibility · Data
mining

1 Introduction

Environmental accidents can result in a series of damages to the population, leaving
irreparable marks on the planet. In this context, the agent who caused the accident
(directly or indirectly) may be asked by several sources to take measures to mitigate
the damage. These sources could be legal, that is, agents could be legally penalized
and forced to take corrective measures, or, due to some type of social pressure, agents
could be led to reverse damages and generate benefits, favoring the quality of life of
the affected population and the preservation of nature. The latter characterizes what is
known as corporate social responsibility (CSR).

The second half of the 20th century witnessed a long debate on corporate social
responsibility and, in recent decades, a renewed interest has arisen with the proposal of
new concepts and interdisciplinary relations to study this topic [2, 3, 6, 7].

The echo in the interest of the scientific community on the topic is linked to the
drastic increase in criticism of the business system. In the last decade, national and
international companies have faced increasing demands due to transparency problems

E. Bisset Álvarez (Ed.): DIONE 2021, LNICST 378, pp. 313–327, 2021.
https://doi.org/10.1007/978-3-030-77417-2_23

and the occurrence of accounting, legal scandals and reputation failures or corporate collapses, such as the famous cases of Enron, WorlCom and Tyco in the United States.

Critical public opinion has led companies that operate in the environment to continually improve their social, environmental and ethical performance. Increased media interest and public pressure also reinforced stakeholder awareness [12]. In this scenario, corporate social responsibility goes beyond being seen as a potential competitive advantage for companies and it starts to be considered as a real strategic need [11].

The growing interest in CSR was accompanied by the substantial publication of CSR articles in journals on management, focusing on ethical and environmental issues [13]. In practice, CSR theories have four dimensions: profits, political performance, social demands and ethical values [3]. With regard to the latter dimension, theories are based on the ethical responsibilities of companies with society.

CSR aims to raise ethical and environmental issues within and outside organizations, insisting that organizations adopt socially responsible and sustainable attitudes, aiming at protecting the environment and the ethical regulation of the conduct of their members.

Awareness and attribution of the reasons for CSR are essential aspects that need to be evaluated, as they represent the main drivers of the type of assessment that society will make of the company and the crisis situation [5].

Although there has been a parallel growth in scientific publications that analyze this phenomenon, there are few critical approaches based on methods of data mining and fuzzy logic to identify the assignation and recognition of corporate social responsibility. Fuzzy logic is the logic underlying approximate, rather than exact, reasoning modes [15] and, therefore, it can be considered adequate for analyses involving subjective properties or inaccurate attributes, absent in the research that is traditionally conducted on the topic of interest to CSR.

This research aims to fill this gap that exists in the traditional literature, by examining the capacity of corporate response when facing environmental accidents based on the processing of massive databases on public opinion. Considering issues about environmental disasters, we believe it is interesting to find out if there was an initiative, over time, to assume corporate social responsibility by companies that caused major environmental disasters.

The main objective of this work is to propose a model, which gathers analyses of an ethical vocabulary and a Fuzzy Rules-Based System, to predict possible ethical and social responsibility responses given by companies that have caused environmental disasters reported in the media.

Our research is based on a case study and the model is applied to the environmental accident that occurred on March 24, 1989, in which the oil tanker Exxon Valdez, one of the ships belonging to the Exxon Mobile Company, ran aground in the north of Prince William Sound, in Alaska, spilling 42 million liters of oil and contaminating 1,990 km of coastline.

We believe that the conceptual model proposed here can be used as a basis for understanding the dynamics of other environmental accidents, as well as identifying ethical actions for assigning or recognizing corporate social responsibility in companies based on pressure by the public opinion.

This article is organized into four sections. In the first section the methodological issues are addressed. The second section presents the details of the model developed to deal with the problem and the pre-processing of the collected data, describing in detail each of its phases, as well as its application in the studied environmental accident. In the next section, the questions related to the validation of the model are described. In the final section, the results obtained in the previous sections and the scope of application of the model are discussed.

2 Methodology

The proposed research is descriptive and combines qualitative and quantitative methodological approaches based on theoretical assumptions of corporate social responsibility with special emphasis on the dimension of ethical values proposed by [3].

Data Collection
As the main aim of our research is to propose a conceptual model based on fuzzy logic, our data was collected from the online news database of the newspaper The New York Times (NYT). The choice of this source of data is justified by the prestige, influence and extensive international academic reference to the New York Times newspaper, which was also portrayed in the scientific literature. For instance, in [4], the authors explain their preference of this data source over other newspapers, an aspect that also fits the purposes of our research, stating that one of the advantages of its use is the free access to the abstracts.

In addition, our research is part of the Project *Understanding Language and Opinion Dynamics using Big Data*, which specifically focuses on the study of the dynamics of opinion and the evolution of language based on data emitted by this kind of media.

The search and retrieval of data were conducted using the *Article Search API* to obtain news available on the developer platform (https://developer.nytimes.com/docs/articlesearch-product/1/routes/articlesearch.json/get). The use of the API's function *get /articlesearch.json* allowed filtering the search by date of publication of the news and keywords. With the application of filters it was possible to reduce the existing computational cost in handling large amounts of data thus making subsequent processes manipulable through current hardware and software.

The considered parameters were the following ones:

- begin_date: 19890324;
- end_date: 20170901
- facet_fields: null
- Query: exxon
- Sort[1]: null

The retrieved news included the word 'exxon' in the title field, keywords and lead paragraph, the choice of this term allows to apply a more comprehensive search of the

---

[1] The parameter allows retrieving by relevance criteria or by the most recent publication data or by the oldest published on the NYT platform.

news about the company responsible for the accident and thus guarantees the retrieval of the entire universe of data in the temporary window under analysis. The considered temporary window is viable to study the dynamics of public opinion about the event, since the collection allowed to retrieve a representative data set from the date of the accident until the year 2017, including a total of 28 years of media coverage.

Following these search criteria, a total of 2001 news items were retrieved and incorporated into the file named 'DB Exxon 1989–2017.txt' (data set available at: https://doi.org/10.6084/m9.figshare.13555661 as Attachment 1). Other criteria on the characteristics of the data records of the retrieved news are included in the 'Data collection, filtering and pre-processing' section.

Methods and Techniques

In this work, we apply fuzzy system techniques, specifically designing a fuzzy rules-based system, to meet the research objective. Basically, fuzzy logic was applied as an attempt to make rational decisions in an environment of inaccurate information [16]. The research was initiated through the presentation of a conceptual model that includes in one of its phases the application of a fuzzy model in massive databases. Alternatively, the conceptual model based on fuzzy logic uses inference rules that allow adaptation to various contexts, especially those that involve the treatment of some degree of uncertainty present in the data analysis.

An important phase of the present work is based on the identification of the presence of an ethical vocabulary for the assignation and recognition of corporate social responsibility. The authors of the present research developed a categorization of this vocabulary into three main groups, described below:

a) Concepts of assignation of responsibility,
b) Concepts of recognition of responsibility and
c) Neutral concepts (whose role is defined due to the proximity of the terms of assignation or recognition, that is, they can launch both functions depending on the context).

A description of how this vocabulary was mapped will be further explored in the next section, following the criteria (a)–(c) of categorization.

3   Development and Application of the Model

In the construction phase of the model, we developed a workflow that represents the model and is divided into four phases, shown in the Fig. 1 below.

We analyzed in detail the viability of each stage of the model, considering that the ability to abstract and understand the dynamics of public opinion in relation to companies that caused major environmental disasters is an important point that should guide the methodology. The database used to verify the hypothesis was composed of news collected from online newspapers, because, although they do not directly reflect public opinion, they indicate the way the accident affected the social environment.

Each stage of the model will be described below, indicating the application in the environmental accident caused by the ship Exxon Valdez concurrently.

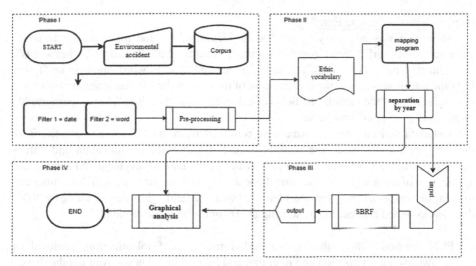

Fig. 1. Environmental accident analysis model (Source: elaborated by the authors).

Phase 1 - Data Collection, Filtering and Pre-processing

Collection: The environmental accident to be analyzed and the news database used in the analysis.

Collection Application:

Environmental accident to be analyzed: Oil spill by the tanker Exxon Valdez.

News database: Online corpus of the newspaper The New York Times (NYT).

Filtering: The model uses two types of filters, one by time period and the other by keyword. The purpose of applying the filters is to delimit the universe of analysis and ensure that the analysis is conducted on the largest number of news related to this environmental accident.

Filtration Application:

Filter by period: Only news published in the period from March 24, 1989, to September 1, 2017, were selected.

Keyword filter: Only news items containing the word ("exxon") were selected.

After applying the filters, a total of 2001 news were retrieved from the NYT online database. Each news item received an identification number (ID) from 0 to 2000.

Pre-processing[2]: the metadata was filtered and some Natural Language Processing (NLP) methods were also applied.

Metadata filtering: the collected data were used to map the ethical vocabulary and to prepare the fuzzy system.

---

[2] The scripts applied in the processing were considered those published in the methodology published on the website https://www.machinelearningplus.com/nlp/topic-modeling-gensim-python/. These were optimized and adapted according to the context of the research data and interests.

Application of Metadata Filtering:

At first, it was necessary to remove irrelevant values and metadata for research purposes ("web_url", "print_page", "source", "multimídia", "headline", "keywords", "document_type", "news_desk", "type_of_material", "_id", "word_count", "uri"), that is, some information about the organization of the data in the file that is not necessary for the application of the model. The fields considered were: "headline", "lead_paragraph", "pub_date", "snippet" and "abstract".

Metadata and unnecessary data were removed from the 2001 news selected from the NYT online database. The records were left only with the metadata and values of the fields that are relevant for the processing of the natural language and subsequent application of the mapping of the ethical vocabulary and the fuzzy system. It is important to mention that each news item was identified with the field "_id" (denoted here by "ID") that was assigned consecutively, starting at "0" until the value of "2000".

1. PLN methods: this sub-phase included methods of tokenization, removal of stopwords, lemmatization and n_grams and bigram in the news from the database.

   The PLN methods were applied to the 2001 news retrieved and stored in the file 'DB Exxon 1989–2017.txt' (Attachment 1[3]).
   Description of natural language processing at the morphological level:

- Tokenization: this technique was applied to transform a set of documents into a matrix, where each document is transformed into a list of words, and each generated word list will occupy a line in the matrix, using the Gensim library. The following symbols were removed: punctuation marks, isolated letters (a, e, o), numbers and special symbols present in the analyzed corpus.
- Removal of stopwords: the stopwords of common words in the language without seemingly relevant meaning, such as "I", "that", "for", were removed using the NLTK library in the English version.
- Lemmatization, as part of the NLP at the morphological level, this technique is applied to transform the word in its "root" form.
- n_grams: bigram was applied to identify those terms that appear together frequently, such as 'social network', etc.

After filtering and pre-processing, the data were used in the following phase of the model.

Phase 2 - Mapping the Ethical Vocabulary

In the second phase, the mapping of the ethical vocabulary was done in the filtered and pre-processed database. Attachment 2 (available at https://doi.org/10.6084/m9.figshare. 13582898) shows the deployment of the terms of the ethical vocabulary following the categorization described in the methodology:

a) **Group 1** - Concepts of assignation of responsibility (23 concepts),
b) **Group 2** - Concepts of recognition of responsibility (9 concepts) and

---

[3] Data set available at: https://doi.org/10.6084/m9.figshare.13555661.

c) **Group 3** - Neutral concepts (whose role is defined due to the proximity of the terms of assignation or recognition, that is, they can perform both functions depending on the context) (6 concepts).

This phase was developed in four stages:

First stage: Identification of the news, in the filtered and pre-processed database, which contains the words of the ethical vocabulary.
Second step: Selection and ordering of only the news IDs that contain the words from the ethical vocabulary.
Third stage: The selected news that contained the terms from the ethical vocabulary were related to their respective terms of the ethical vocabulary and years of publication. With that, we present a table that contains the ID of the news, the year in which it was published and the words from the ethical vocabulary that are present in the news.
Fourth stage: The table developed in the third stage was divided by year, indicating the occurrence of words from the ethical vocabulary for each of the three groups of concepts (neutral, assignation and recognition).

Ethical Vocabulary Mapping Application:

First and Second Stage: In these stages we made a mapping and located those news (from the clean and pre-processed database) that contain words from the ethical vocabulary. This mapping was divided into three parts:

1. We identify the *Porter stem* of each word in the ethical vocabulary, using the website: https://morphadorner.northwestern.edu/morphadorner/lemmatizer/example/.
2. We searched the *Porter stem* of the words in the clean and pre processed database using a searcher developed in Python [10] (Attachment 3 available at https://doi.org/10.6084/m9.figshare.13567817). In some cases, where the *Porter stem* changed some letters of the word in relation to the original word, we searched both versions, for example, in "duty", the *Porter stem* turn back "duti", so we seek "duty" and "duti". The same happened with "penalty", we searched "penalty" and "penalti" and with "culpability", we seek "culpability" and "culpabl". Our search program uses the resource *wildcard* to search for variations of the word in the ethical vocabulary. A *wildcard* is a symbol used to replace or represent one or more characters in the word. In particular, the asterisk "*" is a *wildcard* is normally used at the end of a root word, when we want to search for variable endings of a root word. In our case we use the *Porter stem* of the word as the root of the search.
3. The search program, developed in Python, returns as a response a list with the ID of the news item that contains the searched word. We organized these data, relating the words of the ethical vocabulary searched with the ID of the news in which they appear. It is important to note that our search does not take into account the frequency of the word in the news, we only mark whether the word occurs in the news or not.

Third and Fourth Stage: In the third stage we related the data from the second stage to the year in which the news was published, that is, we showed the relationship between the ID of the news, the year in which it was published and the words of the ethical vocabulary present in the news (where "1" indicates the presence of the word in the news and "0" indicates the absence). In the fourth step, we separated this information year by year (from 1989 to 2017) and calculated the occurrence of words from the ethical vocabulary for the group of neutral concepts, for the group of assignation concepts and for the group of recognition concepts. The table in Attachment 4 available at https://doi.org/10.6084/m9.figshare.13567838 illustrates how the calculation was made.

In Attachment 4^4, for example, we calculated that in the year 2000, we recorded 12 news items that have some words from the ethical vocabulary. Of these 12 news items, we have 276 possibilities of occurrences of words in the vocabulary belonging to the group of responsibility assignation, 108 possibilities of occurrences of words belonging to the group of responsibility recognition and 72 possibilities of occurrences of words from the group of neutral concepts. For each group of concepts, we counted the respective occurrences of the words. Thus, we have 4 occurrences of assignation concepts among the 276 possible occurrences for this group, that is 1%, we have 7 occurrences of recognition concepts among the 108 possible for this group, that is 6% and we have 4 occurrences of neutral concepts. among the 72 possible occurrences for this group, which also means 6%.

The same calculation was made for the other years analyzed by the proposed model. The data from all 4 stages of mapping the ethical vocabulary were organized in tables (Attachment 4 (See footnote 4) and Attachment 5^5), and used in the next stages of application of the model. The data from the mapping of the ethical vocabulary were used to construct the variables of the fuzzy system, described below .

Phase 3 - Fuzzy Rules-Based System

Lotfi Zadeh, in the 1960s [14], suggested an alternative set theory, less rigid than usual. This theory was called the fuzzy set theory.

In this theory, proposed by Zadeh, the change from pertinence to non-pertinence is gradual and not abrupt. Thus, in the fuzzy sets, for each element of the discourse universe, we have a corresponding degree of relevance in the fuzzy set that is given by a real number between 0 and 1. With the fuzzy sets, the possibility of interpreting non quantitative and vague, also increasing the need to seek mechanisms for inferences from these data. One of these mechanisms was the Fuzzy Rules-Based System^6 (FRBS).

A FRBS is a system that uses fuzzy logic to produce outputs from fuzzy inputs. In these systems the linguistic variables^7 they play a fundamental role, since the linguistic terms, translated by fuzzy sets, are used through a rule base to obtain a fuzzy inference relation in which an output is produced for each system input.

---

^{4} Available at https://doi.org/10.6084/m9.figshare.13567838.

^{5} Available at https://doi.org/10.6084/m9.figshare.13567871.

^{6} Due to its multidisciplinary nature, the fuzzy rules-based system is known by several other names, such as "fuzzy rules-based inference system", "expert fuzzy system", "fuzzy model" and "controller logical fuzzy".

^{7} A *linguistic variable* is a variable whose values are fuzzy set names. For example, the temperature of a given process can be a linguistic variable assuming low, medium, high, etc. These values can be described by fuzzy sets.

In short[8], a FRBS consists of four components connected, they are:

1) Fuzzification module;
2) Rule base module;
3) Inference module;
4) Defuzzification module.

Application of FRBS:
The FRBS, called "Mars" (Model for social responsibility analysis or *Modelo de análise de responsabilidade social* in Portuguese), was developed using Mamdani's inference method and implemented using Matlab software, through the "Fuzzy Logic Designer" toolbox.

Fuzzification and Defuzzification
The FRBS has three input variables, they are: "assignation", denoted by "a" (which refers to the occurrences of concepts of assignation of responsibility), "recognition", denoted by "r" (which refers to occurrences of concepts of recognition of responsibility) and "neutral", denoted by "n" (which refers to occurrences of neutral concepts). These data are obtained in the previous phase by mapping the ethical vocabulary, based on the tables in Attachment 4 (See footnote 4) and Attachment 5 (See footnote 5), and were used here in the construction of the system variables (Table 1).

Table 1. Definition of membership functions and characteristics of input variables.

Definition of the membership functions of the input variables: neutral (n), recognition (r) and assignation (a)			
Universe of discourse: [0; 12]			
Notation	Type	Delimiters	Terms
x1	Trapezoidal	[0 0 3 5]	Low
x2	Trapezoidal	[3 5 7 9]	Medium
x3	Trapezoidal	[7 9 12 12]	High

We present the membership function (trapezoidal type) of each one of these input variables. As they are identical in the three cases: neutral (n), recognition (r) and assignation (a). The values of the input variables are mapped by their respective membership functions that were constructed based on the occurrences (measured in percentage) of the words of the ethical vocabulary in the corpus.

The input variables were classified as "low", "medium" and "high", according to an equidistant partition of the speech universe, taking into account the results of the mapping of the ethical vocabulary. The universe of discourse was defined considering the maximum value obtained in the table in Attachment 5 (See footnote 5), in relation to the three variables (recognition, assignation and neutral).

---

[8] For more information about a FRBS, see [1].

We have in Mars an output variable called "indicator_SR", denoted by "i". We present the membership functions (of the triangular type[9]) of the output variable in Table 2 below:

Table 2. Definition of the membership functions and the characteristics of the output variable.

Definition of the output variable membership function: indicator_SR (i)			
Notation	Type	Delimiters	Terms
z1	triangular	$[-1 \, -1 \, -0,5]$	assigned
z2	triangular	$[-0,5 \, 0 \, 0,5]$	no_evidence
z3	triangular	$[0,5 \, 1 \, 1]$	recognized

The universe of discourse of the output variable was defined considering the 3 values: $-1, 1, 0$. They refer, respectively, to the idea that "assigned" is a negative measure for the company responsible for the environmental disaster, "recognized" is a positive measure and 0 means that a value could not be assigned. The output variable was classified as "assigned", "no_evidence" and "recognized", according to the results of the mapping of the ethical vocabulary and from an analysis in the database. The intervals of the triangular functions were constructed considering the idea of "around", that is, we classified as "assigned" the values around -1, as "recognized" the values around 1 and "no_evidence" the values around 0.

The defuzzification[10] method used was the LOM (last of maximums) because it showed a better adherence to our proposal to approach the problem.

Rules and Inference
The rules were built after a careful analysis of the database and the results of the mapping of the ethical vocabulary, taking into account the following premises:

1. If $(r > a)$ and $(n \, \delta \, r)$, then (no_evidence)
2. If $(r > a)$ and $(n > r)$, then (recognized)
3. If $(r < a)$ and $(n \, \delta \, r)$, then (no_evidence)
4. If $(r < a)$ and $(n > r)$, then (assigned)
5. If $(r = a)$, then (no_evidence)

We need at least 27 rules so that all combinations of terms for the input variables were attended, that is, the number of rules we built was $3 \char`^ 3 = 27$ rules.

The fuzzy rules used in Mars are defined in the table in Attachment 6[11].

---

[9] Triangular-type functions were chosen, as it was the one that presented the best result, according to what was expected in the model.

[10] For more information on other defuzzification methods, see [1].

[11] Available at https://doi.org/10.6084/m9.figshare.13582988.

The inference was made based on the Mamdani method, for more details on the Mandani method, we recommend the book by Barros and Bassanezi [1]. The Mars code developed on Matlab can be found in Attachment 7^{12}.

Phase 5 - Graphical Analysis of the Results

The results obtained in the previous phases were interpreted graphically. Based on the data from the mapping of the ethical vocabulary, it was possible to analyze the behavior of the ethical vocabulary over time, as indicated in Attachment 8^{13}.

The figure in Attachment 8 (See footnote 13) indicates the dynamics, over time, of the ethical vocabulary found in the analyzed corpus. We identified that the year 1991 was the year with the highest peaks of terms of assignation or recognition of responsibility (present in the NYT news about the ecological disaster). In 1991, there was a greater presence of words such as *damage, accusation* and *guilt* in the news, which leads us to infer that NYT's journalistic discourse in 1991 had a predominance of the responsibility assignation vocabulary. This inference is also justified based on the annual analysis of the mapping of the ethical vocabulary in the news, since the year 1991 appears in the table in Attachment 5 (See footnote 5) (in the column of concepts of assignation of responsibility) with the second highest value of the average 0.048007 (5%) in relation to other years. This behavior in relation to the concepts of assignation even increased in 1992 to 0.058385 (6%).

Coincidentally, in the year 1991 the NYT reported that a judge accepted the Exxon Pact and the Exxon Valdez spill lawsuits ended with a $900 million payment by Exxon over 10 years to a trust fund administered by three state officials and three federal officials [9]. In addition, the Exxon company was responsible for an additional payment of US $ 100 million, necessary for the recovery of the affected area.

It is interesting to note that the term *compensation* (concept of assignation of responsibility) reached a peak in 2006, which may be related to the proximity of the deadline of 15 years to request a new financial compensation to the civil agreement of Exxon Corporation with the Department of Justice and the state of Alaska, valued at 900 million dollars (federal and state lawyers requested 92 million more for the recovery of the affected ecosystem).

In 2010, the debate about this environmental disaster started again after the Deepwater Horizon oil spill, which was an industrial disaster that started on April 20, 2010 in the Gulf of Mexico, in the Macondo Prospectus, operated by the BP company. This year registered a higher average and percentage value in neutral concepts (those that can play both the assignation and recognition role depending on the context) 0.116667 (12%).

This new oil spill was echoed in the international press with an emphasis on the NYT, as new information on the consequences of the 1989 Exxon spill was published; an example of this was the news about Exxon's new precautionary culture, which allowed the company to socialize opinions on the safety issues linked to the environmental disaster at Prince William Sound. On the other hand, that same year new information emerges about the difficulties that hinder the progress of the cleaning due to the lack of research and rules that hinder new tests in the affected area [8].

---

[12] Available at https://doi.org/10.6084/m9.figshare.13567883.
[13] Available at https://doi.org/10.6084/m9.figshare.13583042.

Other inferences can be made considering the crossing of the data obtained from the mapping of the ethical vocabulary over time, thus facilitating new discoveries in the NYT journalistic discourse on the phenomenon under study.

The developed FRBS was applied, considering the data provided in the table in Attachment 5 (See footnote 5). We show, in Attachment 9[14] an example of application of Mars in Matlab.

The example in Attachment 9 (See footnote 14) shows that in 1991, when we have a percentage of 5% of neutral concepts, 3% of recognition concepts and 5% of assigned concepts, Mars returns the value of Indicator_SR as -1. This means that, in 1991, the NYT news about Exxon addressed issues related to the assignation of responsibility to the company in relation to the environmental disaster caused by Exxon Valdez. We observed that the FRBS result agrees with the analysis made previously from the results obtained in the mapping of the ethical vocabulary.

The results of applying Mars in all years collected in the corpus are described in the table in Attachment 10[15].

In Attachment 11[16] we graphically show the result of the application of Mars in the corpus, the diameter of the circles reflects the occurrence of concepts from the ethical vocabulary in the year (column "News with concepts from the ethical vocabulary (%)" in Attachment 5 (See footnote 5)) and the colors green, red and purple represent, respectively, the classification made: "no_evidence, "assigned" and "recognized".

We note, both in the table in Attachment 10 (See footnote 15) and in the graph (Attachment 11 (See footnote 16)) that Mars shows that in the years 1991, 1992, 1994, 1995, 2005, 2006, 2008 and 2012 the news published about Exxon by the NYT, highlights, in the majority, a certain assignation of social responsibility to the company due to the Exxon Valdez accident. While it was only in 1999 that we had some prominence in the news published about the recognition, by the company, of its responsibility in relation to the oil spill in Alaska. We also note that in most of the years analyzed in the corpus, we do not have enough evidence to conclude the content of the news published by the NYT about Exxon.

4   Model Validation

A scientific model must be reliable, that is, it must be possible to be replicated and validated. The validity of the model must be external and internal. Internal validity concerns the analysis of the causality relationship in the model independently of subjective factors, that is, we must answer the question: Can we obtain the same type of relationship between the elements by other means, in addition to those used in a specific experiment? External validity is related to the generalizability of the model.

In the proposed model, reliability and external validity are guaranteed, considering that the proposed model is described in detail and can be easily replicated for analysis of other environmental disasters, after some adjustments of the data, mainly with regard

---

[14] Available at https://doi.org/10.6084/m9.figshare.13583081.

[15] Available at https://doi.org/10.6084/m9.figshare.13583123.

[16] Available at https://doi.org/10.6084/m9.figshare.13583216.

to the input variables of Mars. The internal validity will be evaluated by comparing the results obtained in Mars with a manual classification of the specialist (expert) list (in: "no_evidence", "assigned" and "recognized"), of 30% (10 random years) of the 29 years analyzed in the corpus. The individual classification of each article in the 10-year period can be found in Attachment 12[17]. We present the results of the validation in Table 3 below:

Table 3. Most frequent responses annually from the expert assessment on the nature of the news.

Year	Number of news with ethical vocabulary	Expert assessment	Results in the Mars	Comparison
1989	41	no_evidence	no_evidence	equal
1990	42	no_evidence	no_evidence	equal
1991	48	assigned	assigned	equal
1994	23	assigned	assigned	equal
1996	8	no_evidence	no_evidence	equal
1999	15	no_evidence	recognized	different
2000	12	no_evidence	no_evidence	equal
2008	9	assigned	assigned	equal
2014	5	no_evidence	no_evidence	equal
2017	7	no_evidence	no_evidence	equal

Although the error rate seems high (10%) when analyzed year by year, we observed that 30% of the years analyzed were tested. We have a total of 518 news items with some term from the ethical vocabulary (according to Attachment 4 (See footnote 4)) and in total we manually classify 210 news items (referring to the 10 years tested), which represents 40% of the result of mapping the ethical vocabulary. The year 1999 presents only 15 news items and the application of the model presented an error in the classification of 4 of them. From these data, we can infer that the model had an error in 4 out of 210 news classified manually, reducing the error rate to less than 2%. Considering the large number of tested news and the complexity involved in the analysis in question, we consider that the model performs well in the classification of news.

5   Analysis of Results and Some Comments

After completing the application of the model, here we present a brief analysis of the obtained results as well as a discussion of the scope of application and the limitations of the proposed model.

According to our model, in the years 1991, 1992, 1994, 1995, 2005, 2006, 2008 and 2012 the news published about Exxon brings information on topics related to the assignation of social responsibility to Exxon for the environmental disaster in Alaska.

---

[17] Available at https://doi.org/10.6084/m9.figshare.13583555.

Regarding the events that occurred in those years, we highlight that in the year 1991 a federal judge accepted a package of US$ 1 billion in criminal and civil agreements to close the state and federal lawsuits against Exxon Corporation. Subsequently, in 1992, transcripts of telephone conversations were released among employees of the oil industry during the response to the environmental disaster. A federal jury in 1994 imposed a penalty of $ 5 billion in punitive damages to 34,000 fishermen and other Alaskans, but the American public opinion found the sums imposed on Exxon to be totally inadequate to prevent it from being penalized in the future. In 1995, Exxon announced a donation of US $ 5 million for the protection of the tiger habitat in the wild, which, in our opinion, is a strategy to manipulate the public opinion, since the company was accused of not being fast enough during the oil spill. Lawrence G. Rawl, who led the review of Exxon's operations in 1980 and shaped the company's response to the Exxon Valdez oil spill, died at the age of 76 in 2005. A year later, the federal appeals panel reduced the punitive damages from $ 4.5 billion against Exxon Mobil to $ 2.5 billion; arguing that the company's negligent conduct was unintentional. In 2008, the United States Supreme Court further reduced the claim to $ 500 million. In 2012, the book "Private Empire" was published, revealing the true extent of the ExxonMobil corporation's power in American and foreign politics.

Through the facts that occurred, we realized that the classification obtained through our model as "assigned" is related to the relevant events in the same period.

We also highlight that, during the reading of the database, for the manual classification of the specialist in relation to the corpus, in the 210 analyzed news, only 21 of them were classified manually as "recognized", which means that only 10% of the news who have some term in the ethical vocabulary indicate the theme of recognizing social responsibility. The news classified manually as "no_evidence" was 118, that is, more than half of the classified news did not have enough evidence to be classified as dealing with the topic of assignation or recognition of social responsibility, and the news classified as "assigned" was 71 of the 210 tested, which represents 34% of them.

The results of the annual analysis of the model show that 3% of the years that were analyzed indicate themes related to the recognition of social responsibility by Exxon; 28% refer to themes surrounding the assignation of responsibility for causing the accident to Exxon, and 69% do not have enough evidence to indicate whether the most popular texts about the accident of the year point to assignation or recognition of social responsibility.

Despite comparing different magnitudes, we noticed that the results are similar in both types of analysis (by year and by news). This result shows that, despite the complexity of an analysis of this kind in which the model is applied considering isolated words in the text and not the semantic aspect, the proposed model proved to be adequate to develop this type of analysis. We also highlight the advantage of the high potential for adapting this model to the study of assigned and recognized social responsibility in relation to other environmental accidents.

Acknowledgments. We would like to thank professors Mariana Claudia Broens, Jose Arthur Quilici Gonzalez and Maria Eunice Quilici Gonzalez for the helpful supervision and indications during the fundamental stages of this research. This paper was developed with the collaboration of the team of project "Understanding opinion and language dynamics using massive data", financed by The São Paulo Research Foundation - FAPESP (process number: 2016/50256-0), to whom we thank the opportunity to conduct this research. Ana Claudia Golzio also received support FAPESP (process number: 2019/08442-9) during the development of this paper, and Mirelys Puerta-Díaz received funding from Coordination for the Improvement of Higher Education Personnel (CAPES) - Financial Code 001.

References

1. Barros, L.C., Bassanezi, R.C.: Tópicos de Lógica Fuzzy e Biomatemática. Unicamp/IMECC (2015)
2. Campbell, J.L.: Why would corporations behave in socially responsible ways? An institutional theory of corporate social responsibility. Acad. Manag. Rev. **32**(3), 946–967 (2007)
3. Garriga, E., Melé, D.: Corporate social responsibility theories: mapping the territory. J. Bus. Ethics **53**(1–2), 51–71 (2004). https://doi.org/10.1023/B:BUSI.0000039399.90587.34
4. Hicks, D., Wang, J.: The New York times as a resource for mode 2. Sci. Technol. Hum. Values **38**(6), 851–877 (2013). https://doi.org/10.1177/0162243913497806
5. Janssen, C., Sen, S., Bhattacharya, C.B.: Corporate crises in the age of corporate social responsibility. Bus. Horiz. **58**(2), 183–192 (2015). https://doi.org/10.1016/j.bushor.2014.11.002
6. Matten, D., Moon, J.: "Implicit" and "explicit" CSR: a conceptual framework for a comparative understanding of corporate social responsibility. Acad. Manag. Rev. **33**(2), 404–424 (2008)
7. McWilliams, A., Siegel, D.: Corporate social responsibility: a theory of the firm perspective. Acad. Manag. Rev. **26**(1), 117–127 (2001)
8. NYT: Since Exxon Valdez, Little Has Changed in Cleaning Oil Spills. Print edition, p. A23, 25 June 2010. https://www.nytimes.com/2010/06/25/us/25clean.html?searchResultPosition=5
9. NYT: Judge Accepts Exxon Pact, Ending Suits on Valdez Spill. Print edition, 9 October 1991. https://www.nytimes.com/1991/10/09/us/judge-accepts-exxon-pact-ending-suits-on-valdez-spill.html
10. Python Software Foundation: Gensim 3.8.3: project description (2020). https://pypi.org/project/gensim/
11. Story, J., Neves, P.: When corporate social responsibility (CSR) increases performance: exploring the role of intrinsic and extrinsic CSR attribution. Bus. Ethics A Eur. Rev. **24**(2), 111–124 (2015). https://doi.org/10.1111/beer.12084
12. Våland, T., Heide, M.: Corporate social responsiveness: exploring the dynamics of "bad episodes." Eur. Manag. J. **23**(5), 495–506 (2005). https://doi.org/10.1016/j.emj.2005.09.005
13. Visser, W.: Corporate social responsibility in developing countries. In: The Oxford Handbook of Corporate Social Responsibility (2008). https://doi.org/10.1093/oxfordhb/9780199211593.003.0021
14. Zadeh, L.A.: Fuzzy sets. Inf. Control **8**, 338–353 (1965)
15. Zadeh, L.A.: Fuzzy logic. Computer **21**(4), 83–93 (1988). https://doi.org/10.1109/2.53
16. Zadeh, L.A.: Is there a need for fuzzy logic? Inf. Sci. **178**(13), 2751–2779 (2008). https://doi.org/10.1016/j.ins.2008.02.012

Luminiferous Funeral

Journeying in Delusional Pavilions

Sarah Vollmer[1,2(✉)] and Racelar Ho[1,2]

[1] IVAS GROUP, Toronto, Canada
racelar@yorku.ca
[2] York University, Toronto, ON, Canada
vollmer@yorku.ca
http://www.ivas.studio/

Abstract. In response to the growing climate crisis, *Luminiferous Funeral* is an interdisciplinary Virtual Reality game-art work with a physical sensory perception installation. This work explores the invisible erosion of climate change and environmental breakdown by offering audiences an opportunity to dialogue with nature and seeks to focus participants on the inner communication with oneself about the essential nature of life and death. The relationship to nature is harnessed by our open-source framework in which we seek collaborative interactivity from others - encouraging them to journey within their local *nature space* and document their phenomenological relationship with the environment through sound clips, sketches, video, photographs, and other forms of digital media. Through communication with corresponding environmental and climate scientists, and by combining this user-centric data input with known local climate and weather models, the playable game-art is continuously evolving - downloadable game patches periodically transform a player's virtual world. With a Zen inspired ideology, our cloud-based Artificial Intelligence systems employ Natural Language Processing on texts describing Eastern and Western philosophies of nature, power, fear and love, space and environment - crafting responses into poetic expressions, and physical interpretations of, this ongoing accumulation of climate content used to create the downloadable game.

Keywords: Virtual reality · NLP · AI · Game design · Social issues · Climate issues · Interactivity · Collaborative content creation · XR

1 Introduction

1.1 Inspiration

A blended medium formed by immutability and mutability of digits and strings with specific syntax rules, *digital technology* was once placed as the hope

---

Supported by Graham Wakefield and the Alice Lab, York University.

E. Bisset Álvarez (Ed.): DIONE 2021, LNICST 378, pp. 328–347, 2021.
https://doi.org/10.1007/978-3-030-77417-2_24

of a Utopia of neutrality. Since digital technology has been acting as a sort of *new-age cold-war weapon* with tools of propagating political ideologies, the negative impression of **Technological Phobia** in the general public has been growing since the middle of last century. Such an implicit association with war and politics, its existence seems like a kind of **Original Sin** and accompanies biased impressions. The popularity and the growth of Internet technologies further amplifies the level of panic of this phobia. People are afraid that the possibility of being monitored and threatened by prerogatives or a 'higher special power' would become a permanent 'normal'. Again, while facing uncertain and abstract objects that are difficult to sense and understand directly, people would arguably rather panic and consider them as a threat and present a propensity of negative affective responses. In particular, the missing **right to discourse** regarding the use and adoption of digital technologies further expands this anxiety - people consider that their life may spiral out of control and without visible and perceivable precursors or traces.

Arguably, the World-Wide-Web (WWW) extends the boundary of physical geography and enriches the global narrative amplified to, and multiplied by, the world, where clusters in the various relationships with different cultures are loud, pervasive, and omnipresent. This connected web flattens the spatial morphometric of the world; it shortens the cost of communication and explodes the complexity of power relations among the many different societies. Under the everyday impact of globalisation, the power relationships on the WWW are striking that capital - political and ideological exchanges from different organisations that run on this virtual world. The one who **holds the right to discourse** of digital technologies is empowered to control the standard of defining society, the world and even people themselves, where the gain of this relevant converted result - tangible and intangible capital, is made uniquely accessible. This is similar to a **Gestalt Shift** in that the closed-loop of input/output strengthens the hegemony of such purposeful ideology or so-called 'common sense'; people are seduced into attempts at self-adjusting and self-acquiring power relations to their surroundings (space, nature, station) without recognising that they have been held captive within the grip of a permeating perceptive altering mind trap [10,21,30,31].

Overall, the so-called **Digital Divide** or the democratic claim of technologies is more about the consideration of **cultural prejudice** and the justice of **space (geography of power)**. Instead of suggesting that the public directs their hatred, anger and phobia toward machines, it might be more suitable to say that they are instead suffering the loss of being deprived the **right to self-distribute** the *space* in their power relations within the society. The proposition of the open-source movement provides unique possibilities to create new spatial relations and allows individuals to re-examine the importance of the multiplicity of space, thus regaining their allocation rights to the **Grand Narrative** of the world. This is one of the original aspirations of our current research program. We wish to demonstrate the multiplicity and simultaneity of space with a perceptible form to the public through a methodology of cross-cutting space occupation and allocation to therefore understand the means in which one can regain this **right to discourse**.

As part of our research-creation program - *Luminiferous Funeral—Journeying in Delusional Pavilions* is a Mixed Reality game project focusing on experimenting with relationships between physical space, virtual (digital) space and imaginative space; it is also a disciplinary game accessible through the WWW with two forms - In-Browser and Virtual Reality, to form a playful means of interacting with *space*.

1.2   Context

Philosophy. Aesthetic experience is the primacy of understanding and theorising the nature of aesthetics and art. Martin Heidegger [15], a well-known philosopher of phenomenology, hermeneutics, and existentialism, deems that aesthetic experience is a dominant manner of interpreting the nature of art. This proposition that over-stresses the subjectiveness of experiencers insofar catalyses a hypothesis regarding the end of art. To ignore the presence of the subjectivity and entity of experiencers (audiences and artists) deliberately in the experiential process of appreciating and creating aesthetic objects herein resembles a way of rescuing art. However, the essential fact of art attests that art is created by and interdependently exists with human beings. So, the significant purpose of contemporary aesthetics is to determine and seek for a shareable, common base that allows the experiencers and the objects (artworks) to coexist in the coordinate systems of nature. In other words, it ties the body and the object together in a gentle way, then pauses human intervention on the renovation and reconstruction to the nature of instruments and the passion of the soul. In brief, the subjectiveness in the process of aesthetic experience presents as corporeality (Leiblichkeit) - somatic perception, experiencing art through embodied bodies, the physical self.

The concept of *atmosphere* and *climate* is not obscure to understand, especially for the discourse of East Asian culture and aesthetics. In ancient Chinese aesthetic theory, it is expressed as a fusion of Spirit and Rhythm - vitality, harmonious manner and aliveness. It also frequently occurs in the cultural discourse of Western art to emphasise an aesthetic property that is beyond rational explanation but firmly emotion- and perception- related. Therefore, atmosphere in the discourse of contemporary art is generally used to indicate and describe something vague and diffuse, challenging to express. As to what aesthetic discourse and its relevant theories in Western and Eastern worlds suggest, the concept and the reality of atmosphere is a thing that is distinct from daily life - it is a catalyst of fusing a subject and an object and an intermediary status of expressing and representing this fusion.

Sensory Perception. Let us now assert that one's existence precedes their essence. To illustrate this concept we reference medical studies on patients who have severe brain injuries - those characterised by their current *consciousness*

- persistent vegetative state (PVS)[1]. Patients who rest on hospital beds, no matter how they look, no matter who they are, are nearly the same. Most of them cannot avoid a tragic destiny that they pass away slowly; eventually, their families cannot help but must turn off all machines that maintain their life. It is noteworthy that the symptom of PVS is an *unresponsive wakefulness* - it is different from patients who are presenting with a coma. This means that a PVS patient's body is still functional - the problem is that the neurons in their brain have a disturbance of communication with each other. Since human brains are skilful at self-restoration, there is a chance that in patients with a low level of consciousness, awareness can be aroused when a certain nerve gets stimulated.

"*Awakening of PVS patients*" is an interdisciplinary art project initiated by artists Chao Wu and Weilun Xia in 2014, that combines medicine, science, psychology, the science of religion and sociology, music and art to explore the potentiality of individuals' self-consciousness. This project has awakened more than 300 PVS patients and created over 50 shared documentary libraries. Similarly, we contextualise this excitement of sensory perception by acknowledging that the brain, the connections within, can be activated by direct impression. By presenting a carefully crafted virtual (curated) space, we then activate the largest organ - skin, through participatory tactile and haptic impressions of substantiated environmental information and climate data. Thus, it could be suggested that we hope to *awaken* this *persistent vegetative state* of *climate illiteracy*.

Virtual Reality.   Immersion, from the Latin word *immergere* (in(to) + dip/plunge), is a kind of sensory perception phenomenon representing a feeling when one's body exists transgressing the physical dimension. As a particular mental state, experiencers journey between the physical world and the virtual world, where their capacity to measure boundaries of time and space gradually becomes vague. There are two dimensions of the meaning in relation to virtual reality: the emerging Virtual Reality (VR) technology, and the other, led by one's self-consciousness, are the internal, emotional, and panoramic transgressive experiences. Human beings strive to transcend limitations of the representation and expression of word, time and space - transforming and virtualising abstract and concrete information into a novel dimension. VR technology increases this common vision by delivering a more tangible way to amplify the sensory perception while penetrating the boundary between physical and imaginative worlds - bridging one's internal perception with communal experiences.

In Venice Biennale 2019, during the interview about her work "*Endodrome*", artist and musician Dominique Gonzalez-Foerster commented, "As opposed to thinking of VR as a tool for escape or for constructing an artificial world, it is more exciting for me to envision it as a kind of organic and mental space in which abstraction and consciousness can be questioned" [14]. "*Endodrome*" was

---

[1] The authors acknowledge the pejorative connotation of describing one as *vegetative*, and although alternative vocabulary has been established, we retain this terminology to connect to the art project "*Awakening of PVS patients*".

the first VR work by Gonzalez-Foerster as well as the first VR work exhibited in Venice Biennale history. It stages an immersive environment to evoke, stimulate, and amplify spectators' memories, emotional feelings, and imaginations through multiple abstract coloured shapes floating and wandering around the VR environment (digital world). To Gonzalez-Foerster, VR technology is not only a tool and a medium for re-morphing and re-mapping a world but also an intermedia to exchange internal and external conscious experiences; VR explores further immersive possibilities by making present one's body and relating oneself to others (things, living beings, and natural environments).

Where is the very beginning of Virtual Reality on any given technological and anthropological perspective? It is *human beings* themselves. The eager desire to explore and create new spatial panoramic layers is derived from the very initial impulse of imagination and curiosity. From the traditional Panoramic Theatre to Happening Art, Gesamtkunstwerk (Total Art), and Digital Media Art, human beings continuously endeavour to extend and transgress their self-awareness throughout substantial and mental (imaginative) space. In functionalism and conventional biology, human beings negotiate and communicate with the world's stirring stimulation through the nervous system. VR technology and the anthropological and psychological phenomenon of virtual reality both trigger and are mutually triggered by each other, immersing one's subjective consciousness along an excitable journey into a strange new world. As Murray suggests, "Immersion is a metaphorical term derived from the physical experience of being submerged in water... the sensation of being surrounded by a completely other reality, as different as water is from the air, that takes over all of our attention, our whole perceptual apparatus" [22].

Climate Perspective. Regarding the issue of the perceptual atmosphere in aesthetic experience, conventional aesthetic theories are uncertain and argumentative. On one hand, they indicate that it is an experience with exceptionally spiritual perceptions: fascination (high arousal and concentration), synaesthesia (high cognitive participation), and resonance with an aesthetic object. Still, there are no agreed upon explicit interpretations of how such a perceptible and conscious process operates. On the other hand, they likewise admit that specific types of art, such as Ecological Art and Climate Art [3,9,17,18,20,28,29,32,39], offer contextualised spaces with multi-dimensional sensorial narratives, allowing people to receive and transmit thoughts intuitively. The interdisciplinary art project *"Awakening of PVS patients"* and the supporting psychological research, relates perception, action and clinical therapies to body-compassion and self-compassion and provides a novel idea of understanding aesthetic experience. In this perspective, experiencers, even those who have a disorder of consciousness, can still recognise the aesthetic atmosphere of artistic works.

Luminiferous Funeral can therefore be thought of and placed within an *art-as-activism* context [1,2,5] where educational aspirations for discourses on atmosphere, climate and environment are decidedly enhanced by integrating methods of sensory perception experiences. In this way, we make the climate crisis *tangible* by integrating the essence of *touch* in a *virtual space* - the perception of seeing with the mind through the body.

2   Concept

Luminiferous Funeral is a long-term and ambitious project that seeks to educate through art-based activism. For this reason, our paper will focus primarily on the game design and collaborative content creation central to the virtual and digital evolution of the game-art work. However, for context, a brief summary of the whole work is provided.

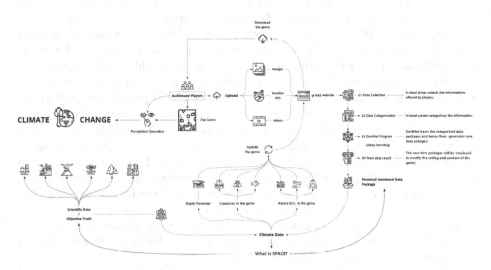

Fig. 1. Concept map of *Luminiferous Funeral*

2.1   Overview of Project Components

In total, this project provides three ways to interact with the digital world we designed: first, audiences can play with the VR (Virtual Reality) game during an installation or through downloading the VR game; second, they can interact with an AR (Augmented Reality) animation app projected on the floor during the installation; third, they can contribute to the evolution of the game content through both a ChatBox app on our website and by submitting digital content (e.g., images, sound clips). The game will also be available In-Browser through a ThreeJS[2] WebXR[3] environment. In keeping with an open-source mindset,

---

[2] threejs.org—a JavaScript framework for building online 3D and Mixed Reality content.

[3] WebXR is a standardised format for adapting the existing WWW framework to include VR and AR content - together they are known as XR.

all designs and code for our customised haptic devices used for interacting in the VR world will be made available such that anyone may re-create them for personal use. Figure 1 is the concept map establishing the relationships between the different aspects of the project.

Education and Learning Theories. Ultimately, we expect to create an accessible and continuously evolving platform for discussion and education on world issues, and in this context, the *climate crisis*. Through public installations and community workshops we aim to facilitate a discourse on polarizing issues; game-play and aesthetic practices of art-as-activism are the inspirational tools behind our participatory design. With respect to a discourse on climate, local and global positions are explored. By partnering with established climate and environmental scientists the project is supported through a fact-based lens upon which the exploratory and fantastical artwork is developed. Through the evolving conditions of the game content, versions that highlight a local anomaly or concern can be tailored through on-site collection of environmental data - the content of which is fed back into the system for a global perspective, effectively rendering multiple occurrences of the VR world that remain playable as separate versions. Overtime, the differences in human choice on behaviour affecting the environment will be made tangible, and therefore, discussable [24–27]. Here we see a pedagogical framework similar to Carina Girvan and Timothy Savage, who popularized the potential for *Communal Constructivism* and *Knowledge Building* as appropriate pedagogies for virtual worlds. In **Luminiferous Funeral**, "...artefacts created by [one individual] are fed back into subsequent iterations... emphasising the use of past learners and their artefacts to influence the learning experience of future [interactants]... artefacts are thus leveraged to extend their *own* knowledge"(emphasis added) [13].

Education, in general, is a means of transmitting and cultivating accumulated cultures, skills, and habits from one to another through appropriate pedagogies and didacticism, producing a permanent change in one's behaviour [7]. In its narrow sense, we can consider two ways of educating - personal and popular, direct and indirect, where popular education occurs beyond formal educational institutions, different from schooling practices. It primarily enlightens and affects the self-awareness extension in individuals on the basis of direct and indirect manners through grassroots social activism and movements. Instead of a given and closed system as in school education, public pedagogy of popular education is dynamic, fluid, and open, embedding itself within multiple overlapping and contested sites of learning [23]. Through public pedagogies of art installations, the general public learns social and political insights by participating in the process of *happenings*. As in Elizabeth Ellsworth's ambitious study in relation to the nature of pedagogy, she suggests converting the concept of knowledge from a *made thing* to a *thing which is in the making* [6]. This derives from a crucial concept of **transitional space** posited by psychologist D. W. Winnicott and describes the progressive transition from an habitual state to that of a novel dimension [40]. To creatively transform established and given knowledge structures into new practices rather than placing new content on an old scaffold, transitional space transgresses between self, other, given and in-the-creating - arranging real and imagined boundaries into open-ended worlds with multiple novel relations. In other words, this learning format is time affording and immersive, allowing individual learners to morph their long-term knowledge schemas through transgressing this transitional space. Scholars, such as Henry A. Giroux who popularised linking public pedagogy with the study of popular culture in broader education, state that public education *as a practice of neoliberalism* can become a force for progressive social change [11]. It is such global and widespread conditions that favour the self-motivation of individuals to didactically reproduce identities, values, and behaviours. More importantly, Giroux encourages strengthening public education's power to resist the hegemonic culture of capitalism and asserts the significance to educators, artists, and cultural workers of creating critical and democratic public discourse. This conceptualises public intellectuals as educators to advance democratic transformation [12].

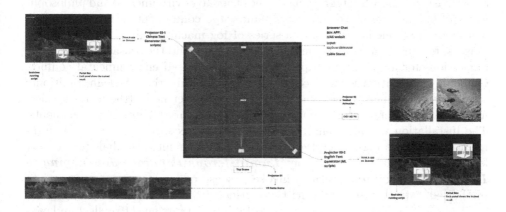

Fig. 2. Exhibition scene - top view of layout and concept plan

Installation. The curation plan mentioned below is structured to let audiences communicate with our project in person (e.g., museum, gallery, public space). We plan to offer an immersive world for audiences to perceive the meaning of *space* under different forms of interactivity. Guests of the exhibit may interact in a variety of ways: actively experiencing the VR world; watching projections of poetic philosophy; uploading content to the cloud AI system; and other participatory actions.

Additionally, by literally cross-cutting the available physical space into separate sections we aim to demonstrate how the meaning of space is generated and to display the difference in understanding of space between Western and East-Asian cultures. These expression will be projected onto the walls as described in Fig. 2. Our audience will not only interact with this curated world through a VR headset but are also encouraged to play with an AR animation projected on the floor, and to communicate with our ChatBox app through a computer provided in the exhibition area (this conversation can also be accessed through a mobile device connected to our website). Figure 3 displays multiple perspectives of the installation plan and the projected curated content.

3   Design

The overall game mechanics are described by an *ecosystem* of game logic as shown in Fig. 4. The traversable VR game-art world component of the Mixed Reality installation is organised around four themes representing the four seasons on Earth. Absurd Utopic environments are contrasted with metaphorical representations of the severity of our nature crisis - beings such as heteromorphic ghost birds, withered toxic plants, fogs and polluted streams will engulf the minimalist-inspired game space.

Figure 5 and Fig. 6 are current screenshots of the original concept game used as the basis for the VR integration and the ethereal environmental and philosophical space. The real-world interactive landscape component of the installation will incorporate elements such as a series of fog machines, flash equipment and sensors, to provide a hybrid space-time atmosphere and the sensation of twisting dimensions for audiences to experience the emphasised environmental conflict. Gameplay navigation is enhanced by our artistic direction through additional sounds with distinct pitches and tones, such as violet noise (the acoustic thermal sound of water), natural-sound recordings, and through interactive elements. The installation will also pull live data and sonic recordings from the environment in real-time permitting the installation to offer intimate dialogues in any location it is exhibited. In this way, *Luminiferous Funeral—Journeying in Delusional Pavilions* seeks to invoke a sense of urgency in participants to further educate themselves during this critical time.

To be sustained in this alternative world, interactants need to collect and capture secret poems and invisible ghosts that are scattered and hidden throughout our afterlife and can only be sensed through the 360° soundscape and hints transmitted by our customised haptics. The poetic expressions are crafted through Natural Language Processing (NLP) AI techniques trained on Eastern and Western philosophies and understandings of nature, power, fear and love; the ghosts are those of the elements of nature that have become extinct or tragically harmed due to climate change. Figure 7 and Fig. 8 are early concept sketches during development of the methods of sensorial integration and kinaesthetic experimentation.

Pulling from the duality found in fundamental theories of physics, nature and existence we envision a sense of parallel worlds enticing one to hear the call of a collapsed time function - to feel the urgency grounded through a futuristic Dystopic death reaching back in time to touch us here and now. As such, participants are forced to interact with invisible elements (ghosts of nature that we are presently losing or have already lost) that can only be felt (unexpected haptic feedback and sensorial experiences) as one explores their new environmental

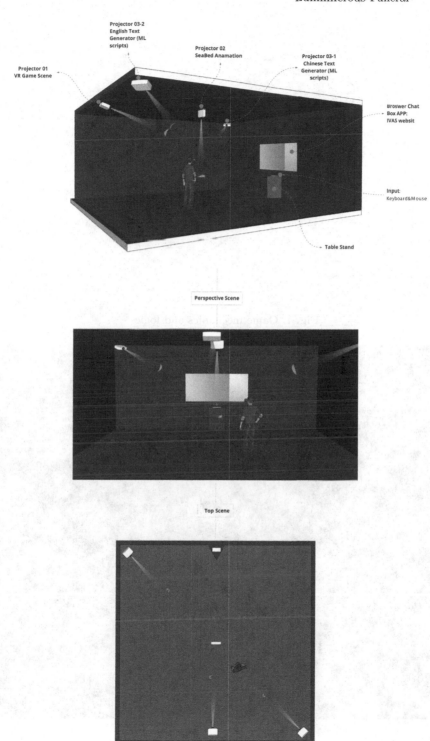

Fig. 3. Exhibition scene - detail perspective view

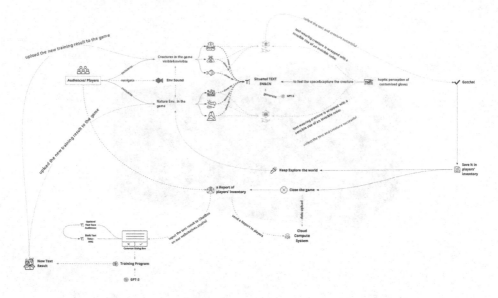

Fig. 4. Game mechanics and logic

Fig. 5. Screenshot of gameplay

reality. Thus, a crucial game element is the constant and mysterious dynamical flow of physical perception transmitted by our custom-built VR glove-like controllers and on-body haptics that materialise and diffuse the sensation between the real and the virtual. In this way, interactants are immersed in and feel the environmental damage in such a way that the visible scenes are only the beginning for one to evoke self-introspection. The intensity of the haptic feedback of the custom wearables increases with time and context to parallel the unseen injuries accruing from current environmental collapse.

Fig. 6. Screenshot of existing game content

3.1   Content Creation Through Collaborative Construction

A core perspective we took when first establishing the parameters of the project was to explore the possibility of leveraging the benefit of a (formally) educated perspective with the self-appeasing instinct of personal experience. Would it be possible to create an art-work that actively engages one on a personal, yet factual basis? Could a simulated, fantastical virtual world provide a connection to real-world experiences? By combining the contribution of established environmental and climate data sets, verification and filtering provided by field experts and the user-centric individual world-views provided by the uploaded digital content, we weave an evolving ecosystem, explorable through curiosity driven sensorial perception.

Features. To illustrate and experiment with the perception of space - which may or may not be affected by different cultural ideologies, we propose to set up two different NLP data packages in which the content is constructed through *imagined relationships* to space [4, 8, 19, 33–38]. Figure 9 identifies a sample categorisation made by the machine learning algorithm trained to identify common everyday items. However, for our project, we expand beyond categorisation and identification and instead utilise the input content (climate models, weather patterns, provided images and sounds from local environments, and, for example, environmental scientists' expert input on eliminating 'garbage' content) to generate custom and curated content within the VR world itself. Not unlike a regularly occurring game patch (update of content), our game will undergo continuous evolution as it grows in supplied content and arterial 'understanding' of *'itself'*.

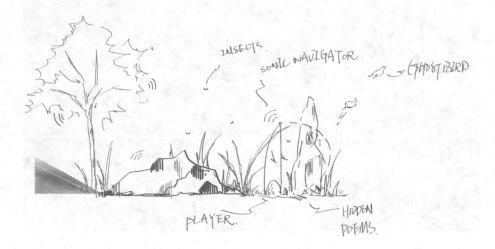

Fig. 7. Concept art of game environment

Figure 10 summarises the AI data input analysis chain. Interaction by audience participants with this program, both through uploading content to our hosting site as well as through the physical kinaesthetic exploration of the populated VR game-art world, would be based on these two pre-trained models. The relationship to language and cultural ideas on nature, environment and power are driven by poetic NLP emergent patterns trained on texts as previously suggested. We chose vital philosophical materials that can be understood to represent features of the two cultures we contrast and compare within this project. The resulting generated poetic structure provided by the algorithms will cause the narrative content to change within the game itself while also impacting the perceptive parameters of the haptic response felt by a participant. Moreover, every updated (AI output) summary data patch would be recorded in our document as our research material and then associated with each successive version of the VR program, forming a complete ecosystem. These updates will be loaded during game start-up sessions as an additional script used to populate the game world. An example of a training data set is shown in Fig. 11.

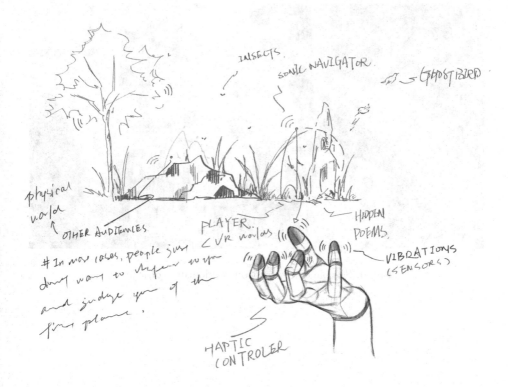

Fig. 8. Concept art of game mechanics

3.2 Interactivity and the Haptic Response

A critical aspect of the curated experience rests on the experiencer's capacity to interact with the VR world through the stimulation of a touch-based discourse. Spatial recognition is transmitted as one explores the world through haptic feedback situated over the entire body - with specific attention positioned at the hands. We consider here a Husserlian [16] perspective - one that distinguishes between spatial things that can be interacted with and temporal experiences that can be lived. The former is experienced through incomplete and one-sided profiles that are presented to the interactant; the later is experienced such that *what it is* and *what it is perceived to be* are the same. In either case, both are existing in a state of essence - this is their true nature. Sensory profiles are both presented and received by means of a mediating technology - one sided profiles are crafted explicitly by the isolation and amplification of individual senses yet the experience of a resulting or reactionary embodied emotional state is (individually) experienced in the same way as it is perceived.

Fig. 9. Sample categorisation of identified objects

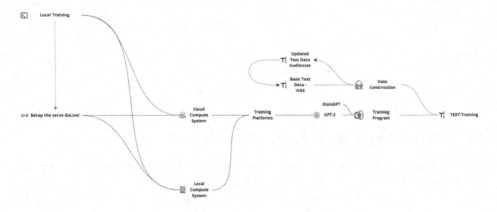

Fig. 10. AI and NLP analysis chain

By forcing a directed and consistent exposure to others via linked technology and content exchange, one is lifted from the confines of their internalised perceptive ideations (Husserl's eidetic intuition) and instead asked to engage in a participatory exchange of a collective presence. Thus, the use of technology that is intentionally designed to pervasively inflict interconnected somatic experiences allows one to exist along, and within, an *action-response continuum* where one is encouraged to gain, and reinforced to retain, a complex awareness of self, others and environment. It is conceivable that the relationships developed between individuals and a particular mediating technology may intentionally create a space where pre-existing assumptions are challenged through sensory fusions.

```
Generating samples...
======== SAMPLE 1 ========
'   , and this is the meaning of the 'child' stage at Z1.1 (and see HS.88, AC26). Likewise, this anal
ysis needs to begin with Zeus and the other guds, and then Plato, and then become essentially appearan
ces. Thus, at H2.40, N writes of eternal recurrence as 'the appearance of the diseased to balance the
gifts' of the good and wicked (and see Z1.14). See also amor fati.

<|endoftext|>

The dithyramb was in ancient Greek literature a choral hymn   to Dionysus, set to music and dance. T
he dithyramb, N believes (following Aristotle) was thus the original form of Greek tragedy (see BT5 an
d 3). Only fragments of dithyrambs survive, along with observations concerning characteristic rhythmic
, formal, and narrative qualities, but without any clear evidence as to how they were performed. N thu
s sees himself and Wagner as reinventing the form (Wagner is the 'dithyrambic dramatist' at UM4.7).
N attempts to recreate the dithyramb as a literary form (albeit without music) in Z. In EH, N draws at
tention to Z2.9 and Z3.4 as his finest examples (see EHZ7). Near the end of his working life, he pre
pared for publication a set of poems with the title Dionysus Dithyrambs.

<|endoftext|>

Hund. A dog is a domesticated, and degraded, beast of prey. Thus, it is both suddenly aggressive, and
contemptible in its display of cowardliness (D135), distress (1883.7.42, Z4.19.8) or anger (1885.38.20
, GM3.14, WCPostscript1). Having been a slave, it has become nothing other than a slave (H3. closing d
ialogue). Thus, the dog is an important analogy for internalization, the opinions of or relationships
to others becoming part of one's identity. The dog's famous loyalty is nothing more than submission,
and it is thus no more capable of love, but for a different reason, than the cat (1882.1.30). This su
bmission is akin to a human's religious feeling (1885.34.141) ~ obedience out of fear, or long disci
pline. The dog is something that is supposed to be wretched, so that its owner can feel important or p
owerful (D369 and see GS312). The sheepdog is an obvious metaphor for those who devote themselves to t
he protection of the herd (ZP9, Z4.7). The firedog in Z (a volcano) is noisy, spectacular, but irrelev
ant (Z2.18, 1883.10.28) ~ and thus like the socialist revolution and similar events. The truly import
ant moments are quiet creating of values. In Z4, Zarathustra tells the story of a traveller accidental
ly stepping on a sleeping dog, making enemies of those who could be friends (Z4.4 ~ but N worked on t
his passage in many notebook entries), the meaning is similar to the story of the adder at Z1.19.

<|endoftext|>

Trauma. Early in his career, N uses the dream to characterize the Apollonian. The key distinction ther
e is not between dreaming and being awake; the dream is employed to understand a certain domain of cul
ture, and particularly of art. The dream is a self-created world, one that is created to be aligned wi
th our instincts of preservation and our need to justify our existence and our values. The dream is no
t simply illusion ~ we can be aware of ourselves dreaming ~ but the validity of the beautiful forms
that we dream are derived from their role in our health and our preservation, and not in their truth s
trictly speaking. This is the account throughout BT and is found again at UM4.4. In H, N tries a gener
al psychology of the dream, particularly the way dreams make sense of stimuli through a retrospective
ascription of causation. But this is precisely how we live our waking lives also (H1.13). At D119, the
analysis changes focus, to dreams as the space in which our drives act or realize themselves; again,
this is not different to waking life. Such ideas were important for Freud.

<|endoftext|>

Later, N explores the idea that not only is waking life not a great deal different from dreams but tha
t, to extent that we insist on it being so, this is a sign of weakness or degeneracy. Waking life is t
he domain in which our instincts or drives are despised, but the values expressed, and developed, in m
y dreams will inevitably find their way into
```

Fig. 11. NLP training data sample text

We therefore consider the integration of custom haptic technology (see Fig. 12) and kinaesthetic experiences that evolve alongside, and are influenced by, the content evolution of the submitted climate data, reflect the suggested importance of *making tangible* these perceivably more abstract and displaced environmental issues. Thus, a series of haptic devices have been, and will con-

Fig. 12. Custom on-body haptics

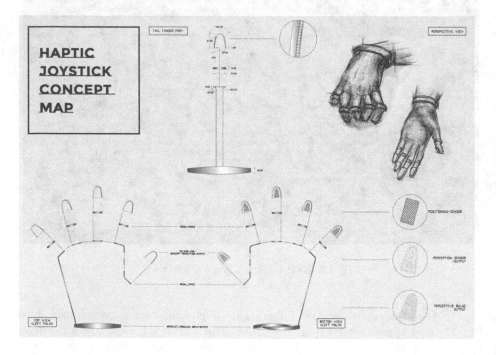

Fig. 13. Design of custom hand haptics

tinue to be, developed to expose multiple body-centric aspects of sensory stimulation. Figure 13 illustrates the concept plan for a set of custom specialty tactile gloves and Fig. 14 is a live shot of testing hand-tracking and haptic integration in VR during the early stages of WebXR development.

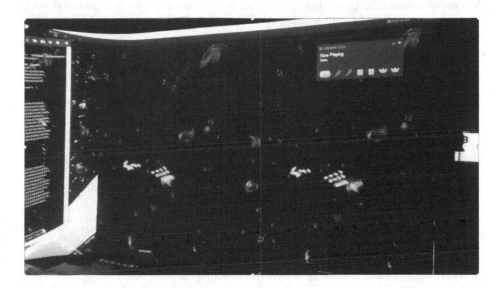

Fig. 14. Demonstration of hand tracking in test VR world

4 Conclusion

Luminiferous Funeral is a game that will evolve to structure itself into a thriving *ecosystem* through the contribution of text, images, videos, sounds and location data by AI, audiences, and artists. Overall, we wish to amplify the perceptions of the multiplicity and simultaneity of space and natural environments for the general public to have a closer look at the world they are living in. Further, we ask those who experience the piece to attune to an *awareness of knowing* to re-attain their *right to discourse* in the **Grand Narrative** of the world and current state of climate change. Moreover, since it is also an open-source project, we wish for the general public to not only make their own set of haptic devices based on our instruction but to also share their creative modifications, allowing others to experience and sense an alternative perspective of perception based on their individual understanding of the world. Finally, since this game-art project is based on the concept of a revolving open-world narrative, we wish for it to be a sort of everyday experience for the general public and provide them with the tools to consciously question their awareness of the world, their space, and the environment.

Acknowledgments. Funding for this project is provided by a VISTA: Vision Science to Application scholarship, a CFREF program, the Centre for Vision Research (CVR) at York University, Canada; a CIBC Student award fund; the Susan Crocker and John Hunkin scholarship in the fine arts; RA funding from Dr. Doug Van Nort (DisPerSion Lab, York University, Canada). Both authors would also love to give many thanks to Dr. Graham Wakefield (Alice Lab, York University, Canada) who has provided both RA funding as well as incredible support as a primary PhD supervisor.

References

1. Anderson, L., Huang, H.C.: The Chalkroom. Interactive Art Installtion (2017)
2. Anderson, L., Huang, H.C.: To The Moon. Virtual Reality Interactive Art Application (2018)
3. Aula, I., Niskanen, M., Salo, J.M.: Ghost Light. Video Documentation of the Installtion (2020)
4. Burnham, D.: The Nietzsche Dictionary. Bloomsbury Academic, London (2014)
5. Davies, C.: Ephémère. Virtual Reality Interactive Art Application (1998)
6. Ellsworth, E.: Places of Learning: Media, Architecture, Pedagogy. Routledge, New York (2004). https://doi.org/10.4324/9780203020920
7. Esu, A., Junaid, A.: Educational development: traditional and contemporary. http://www.onlinenigeria.com/education/. Accessed 30 Feb 2013
8. Young, E.B.: The Deleuze and Guattari Dictionary. Continuum Publishing Corporation, London (2013)
9. Feingold, K.: Sinking Feeling. Interactive Art Installation (2001)
10. Fleischmann, M., Strauss, W.: Staging of the thinking space: from immersion to performative presence, pp. 266–281. Transcript-Verlag (2015)
11. Giroux, H.A.: Public pedagogy as cultural politics: stuart hall and the "crisis" of culture. Cult. Stud. **14**(2), 341–360 (2000). https://doi.org/10.1080/095023800334913
12. Giroux, H.A.: Cultural studies, public pedagogy, and the responsibility of intellectuals. Commun. Crit. Cult. Stud. **1**(1), 59–79 (2004). https://doi.org/10.1080/14791420420000180926
13. Girvan, C., Savage, T.: Identifying an appropriate pedagogy for virtual worlds: a communal constructivism case study. Comput. Educ. **55**(1), 342–349 (2010). https://doi.org/10.1016/j.compedu.2010.01.020
14. Gonzalez-Foerster, D.: Endodrome, VR environment. Lucid Realities Studio, December 2019. https://vimeo.com/379260341
15. Heidegger, M.: Being and time. In: Philosophy's Higher Education, pp. 49-73. Springer, Dordrecht (2005). https://doi.org/10.1007/1-4020-2348-0_3
16. Husserl, E.: The Phenomenology of Internal Time-Consciousness. Indiana University Press, Bloomington (1964). https://doi.org/10.2307/j.ctvh4zhv9
17. Lozano-Hemmer, R.: Airborne Projection-Relational Architecture 20. Interactive Art Installation (2013)
18. Lozano-Hemmer, R.: Cloud Display. Interactive Art Installation (2019)
19. Magee, G.: The Hegel Dictionary. Bloomsbury Academic, London (2010)
20. McRobert, L.: Char Davies' Immersive Virtual Art and the Essence of Spatiality. University of Toronto Press, Toronto (2007). https://doi.org/10.3138/9781442684171
21. Merleau-Ponty, M.: Penomenology of Perception. Routledge/CRC Press, London/New York (2017)

22. Murray, J.: Hamlet on the Holodeck: The Future of Narrative. MIT Press, Cambridge (2017)
23. O'Malley, M.P., Sandlin, J.A., Burdick, J.: Public pedagogy. In: Encyclopedia of Curriculum Studies, pp. 697–700 (2010). https://doi.org/10.4135/9781412958806. n375
24. Seaman, B.: The World Generator/The Engine of Desire - Engine Series. Interactive Art Installation (1996)
25. Seaman, B.: An Engine of Many Senses. Generative Art Installation (2013)
26. Seaman, B.: Luminous Hands. Generative Art Installation (2015)
27. Seaman, B.: Navigating/Negotiating Sound Architectures of the Night. Interactive Art Installation (2016)
28. Shaw, J., Kender, S.: We are like Vapours. Interactive Art Installation (2013)
29. Small, D., White, T.: Jardin Poetique Interactif. Interactive Art Installation (1998)
30. Strauss, W., Fleischmann, M.: Energie Passsagen. Interactive Art Installation (2004)
31. Strauss, W., Fleischmann, M.: The art of the thinking space: a space filled with data. Digit. Creat. **31**, 156–170 (2020). https://doi.org/10.1080/14626268.2020. 1782945
32. Vesna, V.: Bodies, Inc. Interactive Art Installation (1995)
33. Wang, G.: Manual of the Mustard Seed Garden - Full Volume, vol. 1–4. Zhonghua Book Company, Hongkong (1986/1972)
34. Wang, X.: An Architecture Toward Shanshui. Tongji University Press, Shanghai (2015)
35. Wang, X., Qiuye, J.: ARCADIA: Painiting and Garden, vol. I. Tongji University Press, Shanghai (2014)
36. Wang, X., Qiuye, J.: ARCADIA: Illusion and Reality, vol. II. Tongji University Press, Shanghai (2017)
37. Wang, X., Qiuye, J.: A contemporary Chinese garden experiment. Archit. Des. **88**(6), 24–31 (2018). https://doi.org/10.1002/ad.2301
38. Wang, X., Qiuye, J.: ARCADIA: Contemplation and Construction, vol. III. Tongji University Press, Shanghai (2018)
39. Wardrip-Fruin, N., Carrol, J., Coover, R., Greenlee, S., McClain, A., Shine, B.S.: Screen. Interactive Art Installation (2002-present)
40. Winnicott, D.W.: Playing and Reality. Routledge, New York (1989)

Design of Artificial Intelligence Wireless Data Acquisition Platform Based on Embedded Operating System

Amei Zhang$^{(\boxtimes)}$

Xi'an International University, Xi'an 710077, China

Abstract. Advances in microprocessor technology, sensor technology and wireless communication technology have promoted the generation and development of wireless data acquisition systems. The wireless data acquisition system is an ad hoc network system formed by a large number of sensor nodes through wireless communication. Data acquisition is an important means for people to obtain external information, it is an indispensable and important link for preparing a measurement and control system. With the advent of the network era, the traditional data acquisition method has been unable to meet the new production requirements. Based on the embedded operating system, the design of the artificial intelligence wireless data acquisition platform focuses on the characteristics of high compatibility and flexible interface. This paper presents an artificial intelligence wireless data acquisition platform based on embedded operating system. Through this platform, wireless data acquisition and USB interface transmission can be carried out to realize centralized monitoring and management.

Keywords: Wireless communication · Data collection · Embedded Operating System · Artificial Intelligence

1 Introduction

With the rapid development of microelectronic technology and the arrival of the post-PC era, the number of embedded computer systems has far exceeded all kinds of general-purpose computers. Various embedded devices are changing people's daily life in different forms, and at the same time greatly promoting the development of automation and informatization in industry and other fields. In order to be able to feed back the product information in time, people need such a system to realize concise, efficient and real-time data collection and analysis [1]. The traditional data acquisition system has the disadvantages of low efficiency, large error and difficulty in inputting data into the computer, etc. In addition, most of these methods originally used wired network and other communication methods, and their inherent defects greatly limited their use occasions [2]. Traditional data acquisition and transmission systems mostly use single chip microcomputer as the core. Although the implementation is simple and the cost is low, the wired data transmission mode greatly limits its application occasions and cannot be

E. Bisset Álvarez (Ed.): DIONE 2021, LNICST 378, pp. 348–353, 2021.
https://doi.org/10.1007/978-3-030-77417-2_25

applied to some scattered and unattended sites. Therefore, it is necessary to collect data regularly in order to understand the site situation in time [3]. The data acquisition system is a system for real-time acquisition, detection, processing and control of various analog signals generated in the fields of data analysis systems, instrument detection, industrial real-time control, and medical devices.

In the past few years, the CPU has become a low-cost device, and various industrial controls, network equipment, communication equipment, information appliance systems, home medical equipment, and electromechanical equipment have been or are embedded in CPU chips, thus forming an embedded system [4–7]. Before the special computer system has matured, most data acquisition systems use a PC to connect all sensors and actuators, and the control decisions in the system are done by the PC [8]. If a wired transmission method is adopted, it is technically and economically undesirable, and wireless transmission is required for long-distance data transmission. Commonly used data acquisition terminals must transfer data to the computer through keyboard emulation or communication ports, and cannot be used offline. The wireless data collection terminal can make up for the shortcomings of the online data collection terminal, and has good mobility, mobility and flexibility [9–12]. As the application conditions of the data acquisition system become more and more complex, the data acquisition system based on the single-chip computer gradually fails to meet the needs in terms of function, user interface, operation speed and accuracy [13]. The artificial intelligence wireless data collection platform based on the embedded operating system is designed to reflect the characteristics of high compatibility and flexible interface, with multiple data collection methods, high speed and large storage capacity.

2   Overall Structure Design of the System

Embedded system refers to an independent system composed of embedded microprocessor, which has its own operating system and specific functions, is used in specific occasions, and has strict requirements on reliability, cost, volume and power consumption. Due to the huge amount of data brought by human activities, wireless sensor networks will face many security risks in the process of collecting, storing and using these huge amounts of data [14]. As a server, the monitoring center can simultaneously receive data from a plurality of acquisition terminals distributed at far geographical locations, and control and manage all the acquisition terminals.

Table 1. Data format.

Name	Length
Starting mark	2
Data length	4
command word	3
Data section	12
Termination code	3

The data removed by the data preprocessing module should be error data caused by some unpredictable factors, such as error information caused by external interference during data collection or transmission. Data transmission is carried out between the health data acquisition equipment and the data receiving module through an interface. The data format is shown in Table 1.

The monitoring center is a computer running monitoring software. The monitoring software has the functions of displaying data of each terminal, processing and analyzing data, generating alarm signals, and controlling the operation of the monitoring terminal. The software process is shown in Fig. 1.

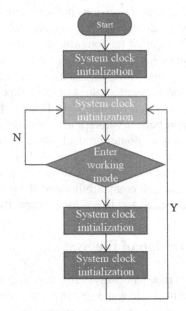

Fig. 1. Software flow.

The data receiving module is mainly responsible for receiving the data information sent by the health data collection device and storing it in the corresponding user data table. If any errors such as reception errors occur during data reception, corresponding error feedback is sent to the health data collection device to ensure the correctness and integrity of the data reception. There are usually three methods for long-distance communication, low-voltage power line carrier, radio stations, and the use of existing mobile communication networks. Because of the interference characteristics, impedance changes and signal attenuation of low-voltage power lines, it is difficult to find a clear analytical or digital model to describe, which is the main technical obstacle. Although the more types of data collection, the better, which makes it easier to analyze energy consumption, the increase in the types of data collection will definitely have requirements on the system cost, the power consumption of the system will also increase, and the stability will also decrease. Product design using a combination of high-performance, low-power microprocessors and wireless communication technology will make it have broad application prospects and a long life cycle.

3 Safety and Reliability Design of Data Acquisition System

The data analysis module is the core of the system, which is used to analyze the current health status of users and predict the development trend of users' health status. We first cluster a large number of preprocessed data to form sample data with small data volume, and then calculate the average value of corresponding types of sample data through weighted average algorithm. The acquisition terminal and the monitoring center adopt C/S mode, and the monitoring center serves as a server, passively waiting for the connection of the acquisition terminal, receiving data and sending control commands. As a client, the acquisition terminal actively connects to the server, sends data to the monitoring center through UDP protocol, and receives control commands from the center [15–19]. Data acquisition and analysis systems have different requirements for data security according to different application occasions. A perfect data acquisition system should have data protection and be able to flexibly cut and expand the system. Before the specific software development, the designer needs to plan which parts of the software functions should be implemented from the macro level, and then select the corresponding solutions according to the actual situation.

The whole wireless data acquisition system is designed with integrated chips. Compared with discrete components, the reliability is improved and the bit error rate is extremely low. The system is divided into sensor module and receiver module. The sensor module comprises an independent power supply part and a triaxial acceleration sensor. Through digital filtering technology and data fusion technology, these first-hand information can be processed quickly and necessary. In order to test it, a small test system is built. The system has two wireless data acquisition cards, one of which continuously sends known data. Data processing and storage preprocesses the collected original data and stores it locally. Any system is not absolutely ideal and reliable. It is necessary to test the reliability and bit error rate of this communication system. As the bit error rate of the system is at a lower level, it is possible to find the bit error phenomenon only by transmitting a large amount of data. For any type of wireless RF transmission chip, its effective transmission distance is a very critical parameter.

4 Conclusion

With the development of microprocessor, embedded technology has been fully developed and widely used in wireless communication, information appliances and industrial control. Embedded-based data transmission system, supported by fast-developing embedded high-performance processors and increasingly powerful embedded operating systems, combined with increasingly perfect wireless network technology, is gradually developing towards multi-function, multi-task, multi-communication mode and high real-time performance, with broad application prospects. Starting from the extensive application of embedded systems in the industrial field, this paper briefly analyzes the current development situation and the problems that will be faced in the remote transmission of data by wired and wireless methods. The adoption of a completely open source and free embedded artificial intelligence operating system not only greatly improves the plasticity and expansibility of the wireless data acquisition and analysis system, but also is very beneficial to the commercialization of products. Embedded systems are widely used in industrial control field due to their high efficiency, stability, configurability and convenient installation. With the rapid development of wireless network communication technology, more and more information devices begin to adopt wireless communication technology. Wireless communication technology eliminates the need to arrange special cables and connectors between devices in offices, homes and on trips.

References

1. Peng, J., Li, D.B., Bao, M.: Design of multi-channel data acquisition system based on RTX51 embedded operating system. Ind. Instrum. Autom. **02**, 31–34 (2015)
2. Yan, J.Y.: Research on energy consumption optimization method of embedded data acquisition system based on machine learning. Modern Electron. Technol. **39**(15), 149–151 (2016)
3. Xing, X.Z.: On the application of embedded computer technology in power systems. Inf. Comput. Theoretical Ed. **343**(21), 111–112 (2015)
4. Filomeno, A.: Application of the western art form based on artificial intelligence. Cultural Commun. Soc. J. **1**(1), 16 (2020)
5. Azram, A.A., Atan, R., Kasim, S.: Use of ontology approach to standardized scientific experimental data representation. Acta Informatica Malays. **2**(2), 10–11 (2018)
6. Hosseinioun, S.: Knowledge grid model in facilitating knowledge sharing among big data community. Acta Informatica Malays. **2**(1), 17–18 (2018)
7. Wei, X.S., Sun, W.R., Wang, L.L., et al.: Design of a universal sensing layer data acquisition and transmission system. Electron. Technol. **07**, 163–166 (2015)
8. Wang, B., Qiu, D., Zhang, Y.N., et al.: Design of wireless transmission system based on CC3200. Inf. Technol. Informatization **007**, 38–40 (2018)
9. Dong, G.T., Zhou, Z.J.: Design of video acquisition system based on embedded Linux. Instrum. Technol. **319**(11), 18–20 (2015)
10. Zhao, L., Wang, W.S.: Parametric architectural design based on optimization algorithm. Eng. Heritage J. **3**(1), 13–17 (2019)
11. Balogh, Z.: Analysis of public data on social networks with IBM Watson. Acta Informatica Malays. **2**(1), 10–11 (2018)
12. Li, B.Q., Li, Z.: The design of wireless responder system based on radio frequency technology. Acta Electronica Malays. **2**(1), 11–14 (2018)
13. Huang, B.L., Xiang, T., Dong, Y.J.: A scenario-based analysis method of embedded operating system architecture. Comput. Knowl. Technol. **013**(011), 232–234 (2017)
14. Deng, B., Huang, C.X., Wei, M.M., et al.: Research on environmental parameter testing system based on embedded operating system. China Metrol. **09**, 105–107 (2015)
15. Li, B.Q., Li, Z.: The implement of wireless responder system based on radio frequency technology. Acta Electronica Malays. **2**(1), 15–17 (2018)
16. Liao, Y.L., Zhang, Y., Cheng, X.H., et al.: A design of heart rate monitor bracelet based on BP neural network. Inf. Manag. Comput. Sci. **1**(1), 15–17 (2018)
17. Yu, Z.Y.: Several suggestions for improving the data collection quality of the logging system. China Petroleum Petrochem. Corp. **11**, 80–81 (2017)

18. Qin, Q.: Design of project management system based on web technology. Inf. Manag. Comput. Sci. **1**(1), 06–08 (2018)
19. Tao, H.X., Zhang, S., Chen, C.F.: A design of WSN based locking system. Acta Informatica Malays. **2**(1), 04–06 (2018)

Differential Steering Control System
of Lawn Mower Robot

Xu Guo[1], Jinpen Huang[2], Lepeng Song[2(✉)], and Feng Xiong[1]

[1] College of Mechanical and Automation Engineering, Chongqing Chemical Industry
Vocational College, Chongqing 401228, China
[2] Department of Automation, College of Electrical Engineering,
Chongqing University of Science and Technology, Chongqing 401331, China

Abstract. Aiming at the problem of inflexible steering and motion of robot, this paper applies fuzzy adaptive PID control algorithm to the differential steering control system of mowing robot to improve its rapidity and accuracy. In this paper, according to the domestic green environment and the overall parameters of the mowing robot, the overall design of the mowing robot is made, the fuzzy adaptive PID control algorithm of the mowing robot differential steering control system is designed, the mathematical model of the mowing robot differential steering is constructed, and the MATLAB/Simulink is used to simulate the fuzzy adaptive algorithm. The simulation results show that the response speed and overshoot of fuzzy adaptive PID algorithm are 2.42 s and 16.4, respectively, The response speed and overshoot of PID are 7.18 s and 36.3% respectively, so the dynamic performance of the differential steering control system based on Fuzzy Adaptive PID algorithm is better than that based on ordinary PID control.

Keywords: Mowing robot · Fuzzy adaptive PID · Differential steering

1 Introduction

With the acceleration of my country's urbanization process, the construction of urban greening environment has gradually received attention, and lawn mowing has become more and more important. Traditional lawn mowing equipment is mainly hand-push and knapsack lawn mowers [1], with backward technology. In recent years, the advent of intelligent lawn mower robots has greatly changed this phenomenon. Intelligent lawn mowing robot is a kind of intelligent robot that integrates machinery, sensors, intelligent control, human-computer interaction, computer and other disciplines [2]. Zhang Zhigang and Luo Xiwen's team proposed a fuzzy adaptive PID algorithm applied to rice transplanters, which promoted the research on automatic steering of wheeled agricultural machinery [3, 4]. The Hu Lian team of South China Agricultural University used the fuzzy control method to redesign the automatic steering module of the rice transplanter [5], and realized the intelligent control in this respect. The Xiong Ru team of Northwestern Polytechnical University proposed a new type of autonomous driving

E. Bisset Álvarez (Ed.): DIONE 2021, LNICST 378, pp. 354–367, 2021.
https://doi.org/10.1007/978-3-030-77417-2_26

method for smart cars [6]. Zhu Zenghui's team adopted the fuzzy control method and designed a fuzzy controller for braking system [7], which greatly improved the acceleration tracking performance. The Wang Zhenyu team of Kunming University of Science and Technology proposed an AGV differential steering control method and added fuzzy control [8, 9]. Yuan Chaohui's team studied a fuzzy control method of aircraft steering front wheel steering based on an improved immune genetic algorithm optimization [10]. The Zhao Chenyu team of Shanghai University conducted research on modern agricultural automated AGV trolleys, and applied fuzzy control theory to agricultural AGV trolleys [11]. The Wang Maoli team of Qingdao Technological University made a research on the application of fuzzy PID algorithm in agricultural machinery automatic steering system [12]. The Xie Shouyong team of Southwest Agricultural University used a single-chip microcomputer to control and optimized the obstacle avoidance algorithm [13]. The team of Chen Baorui and Yang Lei of Beijing Institute of Technology aimed at the transmission system of high-speed tracked vehicles [14]. The team of Liu Yang and Zhang Wei from Hubei University is based on a large number of control equipment in modern agricultural greenhouses [15]. The team of Li Jun and Yu Fan of Shanghai Jiaotong University proposed a road recognition method based on brake pressure and wheel acceleration [16]. The Chen Zheming team of Chongqing University of Technology proposed a steering control strategy that uses fuzzy control for feedback, which greatly improves vehicle handling and stability [17].

Yang Yang from Jiangsu University proposed an intelligent vehicle tracking control system for corner compensation. The performance improvement of path tracing is more significant [18]. Diao Qinqing proposed a special fuzzy control method for the problem of the steering mechanism of the smart car in the big curve [19]. Jiang Jingping of Zhejiang University proved that fuzzy control has great application value in the field of position servo system [20]. Zhang Kun from South China University of Technology introduced fuzzy control and improved the steering control system [21]. Liao Huali from Hohai University designed a fuzzy controller to improve the line tracking algorithm [22]. Zhang Nan of the Ordnance Engineering College used a fuzzy controller to tune traditional PID parameters [23], which improved the speed and accuracy of the steering control of smart cars. Chen Ming of Hefei University of Technology introduced the Ackerman steering theorem to adjust the angle of the other two wheels to achieve three-wheel full steering [24]. Li Huashi of Beijing Institute of Technology added fuzzy control to optimize it, and the simulation results show that its lateral acceleration can reach the steady-state value at a relatively fast speed [25]. Lan Hua from Beijing Institute of Technology proposed a fuzzy control method with parameter self-tuning [26].

The above studies have used fuzzy control in various fields of control systems, especially motion control and steering control, which shows that fuzzy control has great prospects in the application of steering control systems [27–29]. Both the speed and adaptability have been improved to a large extent. In this paper, fuzzy adaptive PID control is used to optimize the differential steering control system of the lawnmower robot. The PID parameters are adjusted in real time through the fuzzy controller, and then the PID controller calculates the steering deviation of the lawnmower robot. The environmental adaptability of the differential steering control system has been greatly improved.

2 Control System Design Scheme

The control of the lawn mower robot is mainly embodied in two aspects: tool control and motion control. To complete the motion control of the lawn mower robot, the three major modules of the lawn mower robot need to work together. The three major modules are: navigation and positioning module, path planning and avoidance. Barrier module, steering control module.

The mobile function requirements of the lawn mower robot are mainly reflected in: accuracy: need to accurately control the movement of the car body; stability: stable travel without overturning; flexibility: it can complete steering tasks at different angles.

According to the demand analysis of the mobile car body and mobile module of the lawn mower robot, we first need to determine the chassis distribution of the car body. There are three common wheeled robot differential models in Fig. 1.

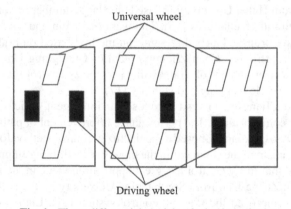

Fig. 1. Three differential models of wheeled robots.

The guidance module is set at the front end of the lawn mower robot and connected with the single-chip microcomputer. The guidance module first obtains the deviation of the lawn mower robot from the original route, and transmits the deviation to the single-chip microcomputer. The single-chip microcomputer transmits the position deviation to the fuzzy adaptive PID control program the speed difference between the two driving wheels is output. At the same time, the speed feedback module installed between the driving wheel and the driving motor feeds back the wheel speed in real time. The actual wheel speed is compared with the wheel speed output by the fuzzy adaptive PID program section, and the error is obtained PID control, and finally output to the motor drive module. The two motor drive modules control the rotation speed of the left wheel motor and the right wheel motor respectively through PWM pulse width modulation and signal amplification, and output the speed difference between the two drive wheels, and finally realize differential steering and reduce Small position deviation. The control diagram of the differential steering control system of the lawn mower robot is shown in Fig. 2.

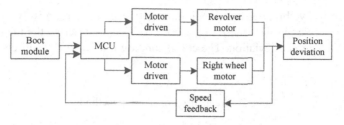

Fig. 2. Control schematic diagram.

3 Design of Fuzzy Adaptive PID Control Algorithm

Fuzzy control system includes four parts: fuzzification, knowledge base, fuzzy reasoning, and defuzzification. Fuzzification is to use language variables to describe the change process of variables, no longer use specific values such as 1cm to express position deviation, but use positive and negative medium fuzzy language to describe it to achieve fuzzification. The knowledge base stores the control rules described by the language. Fuzzy reasoning is to use the rules in the knowledge base to reason and get the fuzzy output. Finally, the accurate output is obtained by defuzzification.

Fuzzy adaptive PID control is an important development direction of fuzzy control. Fuzzy adaptive PID control is self-adaptive and self-learning. It has better control effect for those complex systems with nonlinearity, large time delay, and high order. The schematic diagram of fuzzy adaptive PID control in this article is shown in Fig. 3.

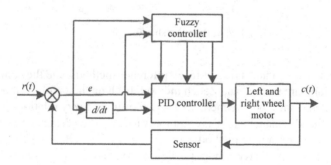

Fig. 3. Fuzzy adaptive PID control principle diagram.

The fuzzy self-adaptive PID control system of this article first takes the angle deviation e and the deviation change rate e_c as the input. After the fuzzy controller's fuzzification, fuzzy inference, and defuzzification, the PID parameter changes ΔK_p, ΔK_i, ΔK_d are output, and then calculated by the PID controller to output the speed difference ΔV between the left and right driving wheels, Realize steering control.

Assuming that the lawn mower robot is on the left side of the predetermined trajectory, the deviation is negative when it needs to turn right, and the basic domain of position deviation e is $[-8, 8]$, and the above basic domain is divided into 7 quantitative levels $\{-6, -4, -2, 0, 2, 4, 6\}$, fuzzy subset $\{NB, NM, NS, ZO, PS, PM, PB\}$, respectively representing negative big, negative medium, negative small, zero, positive small, positive middle, positive big. Use these seven fuzzy subsets to describe the magnitude and direction of the angular deviation. The corresponding laws of fuzzy subsets are shown in Table 1.

Table 1. Comparison of steering methods.

Fuzzy subset	NB	NM	NS	ZE	PS	PM	PB
Encoding code	NB	NM	NS	ZE	PS	PM	PB

The membership function is selected in this paper, which is simple in operation and highly sensitive. The membership function is shown in Fig. 4.

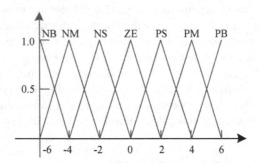

Fig. 4. Membership function.

At present, there are many types of fuzzy inference methods, and the commonly used methods are Mamdani method and Zadeh method. Both methods follow the same fuzzy inference synthesis rules, but the methods for determining fuzzy relations are different. In application, the Mamdani method is commonly used. So this design uses the Mamdani method as the fuzzy reasoning method.

The realization of fuzzy reasoning, in addition to confirming the reasoning method, also needs to improve the knowledge base and adjust the fuzzy rules to achieve the accuracy of control. Since we have three outputs ΔK_p, ΔK_i, ΔK_d, we need to separately analyze how to adjust the PID parameters ΔK_p, ΔK_i, and ΔK_d when facing the corresponding error and error rate of change to enable our control system the most effective.

First, we design the fuzzy rule of ΔK_p. Theoretically, when the error is large, we need a larger K_p to speed up the response speed of the system, and when the error is moderate, we should choose a smaller K_p to ensure The overshoot of the system will not be too large. When the error is small, in order to ensure good steady-state characteristics of the system, a larger K_p is still needed to reduce the static error. Based on the above characteristics of K_p in PID control, we can design a fuzzy control rule of ΔK_p. The fuzzy control rule table of ΔK_p is shown in Table 2:

Table 2. Comparison of steering methods.

K_p	E						
	NB	NM	NS	ZE	PS	PM	PB
E_C NB	PB	PB	PM	PM	PS	PB	ZE
NM	PB	PB	PM	PS	PS	ZE	NS
NS	PM	PM	PM	PS	ZE	NS	NS
ZE	PM	PM	PS	ZE	NS	NM	NM
PS	PS	PS	ZE	NS	ZE	NM	NM
PM	PS	ZE	NS	NM	NM	NM	NB
PB	ZE	ZE	NB	NM	NM	NB	NB

Next, design the fuzzy rules of ΔK_i. The main function of integral control is to eliminate the steady-state error of the system. However, in the initial stage of adjustment, in order to avoid the instantaneous increase of the error and the integral saturation, the integral link should be restricted, so the error is relatively large. In the large initial stage, $K_i \approx 0$ should be taken. In the middle of adjustment, when the error is moderate, in order to avoid affecting the stability of the system, the value of K_i should be appropriate. At the end of adjustment, when the error is small, K_i needs to be increased appropriately to reduce the static error of the system. The fuzzy control rule table of ΔK_i is shown in Table 3:

Table 3. Comparison of steering methods.

K_i	E						
	NB	NM	NS	ZE	PS	PM	PB
E_C NB	NB	NB	NM	NM	NS	ZE	ZE
NM	NB	NB	NM	NS	NS	ZE	ZE
NS	NB	NM	NS	NS	ZE	PS	PS
ZE	NM	NM	NS	ZE	PS	PM	PM
PS	NM	ZE	ZE	PS	PS	PM	PM
PM	ZE	ZE	PS	PS	PM	PM	PM
PM	ZE	ZE	PS	PM	PM	PM	PM

Finally, the fuzzy rule of ΔK_d is designed. The function of the differential link is mainly to adjust the dynamic characteristics of the system, reflecting the trend of error changes, leading correction, when the error is large, in order to avoid the differential saturation caused by the instantaneous change of the error, The value of K_d should be relatively small. When the error is moderate, K_d has a greater impact on the system, and the value of K_d needs to be appropriate and remain unchanged. When the error is small, it is generally the later stage of adjustment, and the K_d value should be appropriately reduced to avoid unnecessary oscillations. The fuzzy control rules of ΔK_d are shown in Table 4.

Table 4. Comparison of steering methods.

K_d		E						
		NB	NM	NS	ZE	PS	PM	PB
E_C	NB	PS	NS	NB	NB	NB	NM	PS
	NM	PS	NS	NB	NM	NM	NS	ZE
	NS	ZE	NS	NM	NM	NS	NS	ZE
	ZE	ZE	NS	NS	NS	NS	NS	ZE
	PS	ZE	ZE	ZE	ZE	ZE	ZE	ZE
	PM	PB	NS	PS	PS	PS	PS	PB
	PB	PB	PM	PM	PM	PS	PS	PB

After determining the fuzzy rules, we also need to defuzzify the fuzzy amount after fuzzy inference. The defuzzification usually includes the center of gravity method, the area square method, and the maximum degree of membership method. The design adopts the center of gravity method to realize the defuzzification of the blur amount.

4   Robot Differential Steering Experiment and Simulation

4.1   The Establishment of Differential Steering Model

The lawn mower robot designed in this paper adopts a rear-drive two-wheel differential scheme. For this reason, this article needs to establish a mathematical model of the lawn mower robot steering for further research. Since the mowing robot in this paper only has two rear-drive wheels for driving, in order to simplify the calculation, we abstract the differential model of the mowing robot as a differential model of two rear driving wheels, and the analysis does not consider the difference between the driving wheels and the ground. The relative sliding of and the slight deformation of the driving wheel, in this way, the state of the lawn mower robot can be constructed first, as shown in Fig. 5.

In Fig. 5, the XOY axis is the established grassland coordinate system, the center point of the drive wheel connection at the beginning of point O, D is the distance between the centers of the two drive wheels, R is the turning radius of the lawn mower robot, and V_L is the left The wheel speed of the driving wheel, V_R is the wheel speed of the right driving wheel, V is the overall speed of the lawn mower robot, θ is the turning angle of the lawn mower robot, and e is the position offset of the center position of the lawn mower robot.

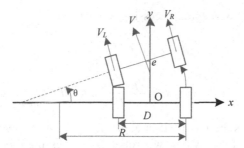

Fig. 5. Motion state diagram of lawn mower robot.

The vehicle speed of the mowing robot is the speed V at the center point, which can be synthesized by the wheel speeds V_L and V_R of the left and right wheels. The relationship is:

$$V = \frac{V_L + V_R}{2} \tag{1}$$

When the lawn mower robot turns, the speed difference between the two driving wheels V_L and V_R results in a speed difference. Because the two driving wheels have a fixed direction, they are regarded as connected rigid bodies, so they turn, and the turning radius is R.

As shown in the situation in Fig. 5, $V_R > V_L$, the lawnmower robot turns left. Since the left and right wheel trajectories should be arcs, according to the circle angle theorem, we can get:

$$\frac{V_R}{R + \frac{D}{2}} = \frac{V_L}{R - \frac{D}{2}} \tag{2}$$

By processing Eq. (2), the relationship between the turning radius of the lawn mower robot and its driving wheel track and wheel speed can be obtained, as shown in Eq. (3):

$$R = \frac{D(V_1 + V_2)}{2(V_1 - V_2)} \tag{3}$$

The angular velocity when turning can be expressed as:

$$\omega = \frac{V_L - V_R}{D} = \frac{V_L + V_R}{2R} \tag{4}$$

The vehicle speed of the mowing robot is the speed V at the center point, which can be synthesized by the wheel speeds V_L and V_R of the left and right wheels. The relationship is:

$$V = \frac{V_L + V_R}{2} \tag{5}$$

Decomposing the speed of the mowing robot to the X axis, we can get:

$$V_X = V \sin \theta \tag{6}$$

Using Eqs. (5), (6) as integral processing, we can get:

$$\theta = \theta_0 + \int_0^t \frac{V_L + V_R}{2R} dt \qquad (7)$$

$$x = x_0 \int_0^t \frac{V_L + V_R}{2} \sin\theta dt \qquad (8)$$

Among them, θ_0 is the initial offset angle of the lawn mower robot, and x_0 is the initial abscissa offset of the lawn mower robot. By processing Eqs. (7) and (8), we can get the change in the direction angle of the lawn mower robot and the change in the abscissa position in the time of Δt:

$$\Delta\theta = \frac{V_L + V_R}{2R} \Delta t \qquad (9)$$

$$\Delta x = \frac{V_L + V_R}{2} \sin\theta \Delta t \qquad (10)$$

Differentiate Eqs. (9) and (10) to obtain:

$$d\theta = \frac{V_L + V_R}{2R} dt \qquad (11)$$

$$dx = \frac{V_L + V_R}{2} \sin\theta dt \qquad (12)$$

The motion of the lawn mower robot is continuous, and the above formula can continue the Laplace transform to obtain:

$$\theta(s) = \frac{V_L + V_R}{2Rs} \qquad (13)$$

$$x(s) = \frac{V_L + V_R}{2s} \sin\theta \qquad (14)$$

By adjusting the angle and position of the lawn mower robot, the pose of the lawn mower robot can reach the set value.

In order to establish the kinematics model of the lawn mower robot, we also need to add the influence of the motor. According to the principle of the motor, the relationship between the driving wheel and the armature voltage is:

$$V_R(s) = \frac{1}{1 + T_m s} U_{OR}(s) \qquad (15)$$

$$V_L(s) = \frac{1}{1 + T_m s} U_{OL}(s) \qquad (16)$$

Among them, U_{OR} is the armature voltage of the right-wheel drive motor, U_{OL} is the armature voltage of the left-wheel drive motor, and T_m is the response time constant.

We can establish the dynamic characteristic structure of the lawn mower robot, as shown in Fig. 6.

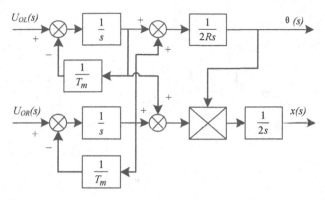

Fig. 6. Structure diagram of dynamic characteristics of lawn mower robot.

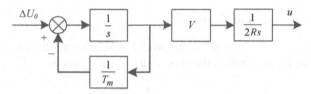

Fig. 7. Block diagram of control variables of lawn mower robot.

Since the lawn mower robot has continuous changes and small deviations during operation, the dynamic characteristic structure diagram of the lawn mower robot in Fig. 6 can be simplified, and the simplified control variable block diagram is shown in Fig. 7.

Among them, ΔU_O is the difference between the armature voltages of the left and right wheel drive motors.

The transfer function of the mowing robot is:

$$d(s) = \frac{T_m s V + 1}{2 R s (T_m s + 1)} \tag{17}$$

4.2 Simulation of Fuzzy Controller

MATLAB is a commercial mathematics software produced by MathWorks in the United States. It is a high-level technical computing language and interactive environment for algorithm development, data analysis and numerical calculation. It mainly includes MATLAB and Simulink. The simulation of the differential steering control system of the lawn mower robot is selected to use MATLAB to complete.

In Simulink, the simulation model of the differential steering control system of the lawn mower robot is established, and the fuzzy controller is put into it for simulation. First, you need to find the required modules in the MATLAB/Simulink toolkit, and drag them into the simulation page, such as Scope module, Gain module, Fuzzy Logic Controller module, etc., and connect them. The differential steering control system of the mowing robot. The simulation model of fuzzy adaptive PID controller is shown in Fig. 8.

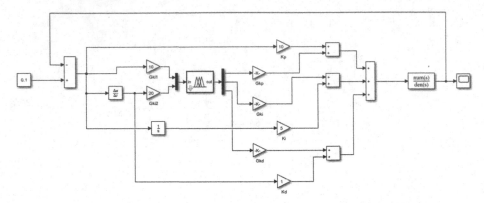

Fig. 8. Fuzzy adaptive PID controller simulation model.

4.3   Simulation Results and Analysis

We also established a conventional PID controller simulation model, and compared with the fuzzy adaptive PID controller, the operating results are shown in Fig. 9:

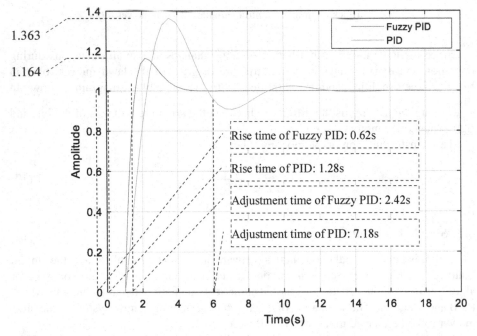

Fig. 9. Comparison chart of simulation results.

In Fig. 9, the blue line is the simulation result of the fuzzy adaptive PID controller, and the black line is the simulation result of the conventional PID controller. Use the scope module to measure the data, and get the data of the two response curves as shown in Table 5.

Table 5. Comparison of steering methods.

Control type	Rise time	Adjustment time	Overshoot
Fuzzy adaptive PID	0.62	2.42	16.4%
Conventional PID	1.28	7.18	36.3%

From the simulation results, it can be seen that when the fuzzy adaptive PID controller controls, the rise time, adjustment time, and overshoot are larger than the conventional PID controller and the number of oscillations is less. The analysis shows that the fuzzy PID controller is correspondingly fast, with small overshoot and short stabilization time. The speed and stability are better than conventional PID controllers.

5 Hardware Circuit for Control System of Mowing Robot

The design core of the differential steering control system of the mowing robot is the controller based on the fuzzy adaptive PID algorithm. The power module of the system supplies power for other modules. The STM32f103c8t6 single chip microcomputer is used as the microprocessor and control core. The electromagnetic guidance module obtains the position deviation and outputs it to the stm32 single chip microcomputer. The drive module controls the rotation of the motor to realize the control of the mowing robot Steering control. The experimental results show that the fuzzy adaptive PID algorithm can effectively control the specific forward, backward, steering and other functions of the mowing robot. The specific control system trajectory is shown in Fig. 10.

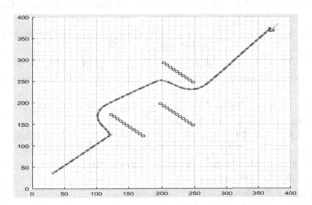

Fig. 10. The control system trajectory of mowing robot.

6   In Conclusion

In this research, through the study of PID control and fuzzy adaptive PID control algorithm, the fuzzy adaptive PID algorithm is applied to the differential steering control system of the lawn mower robot. After establishing a mathematical model and using MATLAB/Simulink to simulate, it is concluded that the use of fuzzy adaptive PID algorithm can improve the response speed and stability of the grass-mower robot's differential steering control system.

Acknowledgments. This research was funded by Natural Science Foundation of Chongqing (No. cstc2018jcyjAX0336)), China; Research Foundation of Chongqing University of Science & Technology (No. ck2017zkzd006), China; Research Foundation of Chongqing Education Committee (No. KJQN201801516, KJQN201901536, KJQN202001531, KJ1737459, and KJ1601328), China; Major Subjects of Artificial Intelligence Technological Innovation in Chongqing: (No. cstc2017rgzn-zdyfx0024), China; Chongqing Technological Innovation and Application Demonstration Project (No. cstc2018jszx-cyztzx0018), China and Achievement Transfer Program of Institutions of Higher Education in Chongqing (No. KJZH17134), China.

References

1. Nie, Y.: Research on the key technology of intelligent lawn mower robot. Chongqing University, (2018)
2. Huang, J., Su, X.Z., Yan, G.F.: Design of an automatic mowing robot. Appl. Electron. Technol. **36**(08), 33–36+40 (2010)
3. Lian, X.Q.: Fuzzy control technology. Electric Power Equipment **5**(01), 73 (2004)
4. Zhang, Z.G., Luo, X.W., Li, J.L.: Research on automatic steering control system of wheeled agricultural machinery. Trans. Chin. Soc. Agric. Eng. **21**(11), 85–88 (2005)
5. Hu, L., Luo, X.W., Zhao, Z.X., Zhang, Z.G., Hu, J.W., Chen, B.: Design of electric control operating mechanism and control algorithm for rice transplanter. Trans. Chin. Soc. Agric. Eng. **25**(04), 126–130 (2009)
6. Xiong, B., Qu, S.R.: Research on autonomous driving method of intelligent vehicles based on fuzzy control. Transp. Syst. Eng. Inf. **10**(02), 74–79 (2010)
7. Zhu, Z.H., Xu, Y.C., Ma, Y.L., Li, J.S., Li, Y.L.: Design of intelligent vehicle longitudinal acceleration tracking controller based on fuzzy control. J. Acad. Mil. Transp. **16**(12), 31–35 (2014)
8. Wang, Z.Y., Zhan, Y.D., Zhang, L.M.: Design of differential steering controller for visual AGV. J. Comput. Appl. **27**(7), 1789–1791 (2007)
9. Liu, Q., Wu, X.H., Zou, Z.J., Liu, Z.Y.: Ship maneuvering motion control system based on two-stage fuzzy controller. J. Wuhan Univ. Technol. Transp. Sci. Eng. **24**(3), 229–232 (2000)
10. Yuan, C.H., Zhang, J.: Simulation research on fuzzy control of aircraft front-wheel steering. J. Syst. Simul. **20**(17), 4509–4513 (2008)
11. Zhao, C.Y., Chen, X.K.: Design and fuzzy control of differential steering AGV for agriculture. J. Agric. Mech. Res. **38**(11), 123–127 (2016)
12. Wang, M.L., Duan, J., Tang, Y.W., Zhao, J.B., Jiang, Y.: Research on automatic steering system of agricultural machinery based on fuzzy PID algorithm. J. Agric. Mech. Res. **40**(11), 241–245 (2018)
13. Xie, S.Y., Bao, A.H., Guan, Z.L., Yang, G.C.: Fuzzy control for navigation and obstacle avoidance of mobile robot in greenhouse. Trans. Chin. Soc. Agric. Mach. **33**(2), 74–76 (2004)

14. Chen, B.R., Ma, B., Li, H.Y., Yang, L.: Improved fuzzy control for hydrostatic transmission of high speed tracked vehicles. Trans. Chin. Soc. Agric. Mach. **40**(11), 18–21 (2009)
15. Liu, Y., Zhang, W., Luo, C.H.: Research and application of fuzzy control in modern agriculture. Internet of Things Technol. **7**(04), 66–67 (2017)
16. Li, J., Yu, F., Zhang, J.W.: Research on ABS fuzzy control system based on road automatic identification. Trans. Chin. Soc. Agric. Mach. **32**(05), 27–30 (2001)
17. Chen, Z.M., Zhou, P., Chen, B., Fu, J.H.: Research on control strategy of automobile four-wheel independent steering stability. Comput. Simul. **35**(7), 93–97 (2018)
18. Yang, Y.Y., He, Z.G., Wang, R.C., Chen, L.: Intelligent vehicle tracking control system based on angle compensation. Instr. Technol. Sensor **05**, 73–77 (2019)
19. Diao, Q.Q., Zhang, Y.N., Zhu, L.Y.: Horizontal and vertical fuzzy control of large curvature path for smart car with dual preview points. China Mech. Eng. **30**(12), 1445–1452 (2019)
20. Zhang, T., Jiang, J.P., Xue, P.Q.: Application research of fuzzy control in position servo system. J. Wuhan Univ. Technol. Inf. Manage. Eng. Ed. **29**(10), 24–27 (2007)
21. Zhang, K., Xu, L.H.: Research on AGV differential steering control algorithm based on fuzzy control. Autom. Instrum. **31**(10), 1–4 (2016)
22. Liao, H.L., Zhou, X., Dong, F., Wang, Y.Q.: AGV tracking algorithm based on fuzzy control. J. Harbin Inst. Technol. **37**(07), 896–898 (2005)
23. Zhang, N., Mao, Q.: The design of PID fuzzy controller and its application in intelligent vehicle steering control. Sci. Technol. Wind **05**, 66–67 (2012)
24. Chen, M., Xiao, B.X.: Research on the steering of three-wheel all-steering forklift based on fuzzy control. J. Hefei Univ. Technol. (Nat. Sci. Ed.) **40**(08), 1064–1069 (2017)
25. Li, H.S., Han, B.L., Luo, Q.S., Wang, S.F.: Simulation of three-axle vehicle all-wheel steering performance based on fuzzy control. Trans. Chin. Soc. Agric. Eng. **28**(13), 42–49 (2012)
26. Lan, H., Li, X.P., Ding, Y.J.: Design and research of fuzzy controller for smart car steering system. Sensors Microsyst. **29**(5), 34–36 (2010)
27. Zhao, W., Zhang, H.: Coupling control strategy of force and displacement for electric differential power steering system of electric vehicle with motorized wheels. IEEE Trans. Vehicular Technol. **67**(9), 1 (2018)
28. Wang, J., Wang, Q., Jin, L., Song, C.: Independent wheel torque control of 4WD electric vehicle for differential drive assisted steering. Mechatronics **21**(1), 63–76 (2010)
29. Khan, M.A., Aftab, M.F., Ahmed, E.: Robust differential steering control system for an independent four wheel drive electric vehicle. Int. J. Automot. Technol. **20**(1), 87–97 (2019)

Innovation of Enterprise Management Mode in the Era of Big Data

Yong Zhang[1] and Junyi Shi[2(✉)]

[1] School of Management, Guizhou University, Guiyang 550025, China
[2] Accounting and Business Management, De Montfort University, Leicester L1 9BH, UK

Abstract. With the rapid development of Internet technology, the advent of the era of big data and began to penetrate into all walks of life in society, and more and more in the enterprise management mode to play a positive role. After entering the new century, the huge data and changeable market environment have formed a new situation of economic development, which puts forward higher requirements for enterprise management. Under the influence of big data, various advanced technologies have improved the informatization level of enterprise management. It has become the default consensus of all walks of life to use various information technologies to innovate enterprise management mode. As the era of big data has a certain impact on the traditional management mode of enterprises, the traditional management mode gradually shows its shortcomings. Therefore, only by conforming to the trend of the development of the times and innovating the enterprise management mode, can we make full use of the advantages brought by the era of big data and quickly help enterprises identify massive information. Each industry should strengthen its own reform and make continuous innovation, so as to make greater contribution to the society. Based on the reality, this paper first gives a brief overview of the concept of the era of big data, then discusses the problems existing in the management mode of enterprises in the era of big data, and finally puts forward effective strategies for the innovation of enterprise management mode in the era of big data, so as to create more favorable conditions for the development of enterprises.

Keywords: Big data era · Enterprise management mode · Innovation research

1 Introduction

At this stage, Internet technology has achieved rapid development and wide popularization. At the same time, big data technology based on the background of modern information also shows a vigorous trend, showing an explosive trend in various fields. Therefore, we are currently facing the impact of the era of big data [1]. Since the reform and opening up, China has been using the traditional enterprise management mode. However, under the current development background, if enterprises want to achieve greater success, they should consciously reform the management system. Innovation cannot meet the development needs of the era of big data. Therefore, although enterprise

E. Bisset Álvarez (Ed.): DIONE 2021, LNICST 378, pp. 368–373, 2021.
https://doi.org/10.1007/978-3-030-77417-2_27

management is facing the reform of traditional mode, it also meets new opportunities of innovation management mode. As enterprise managers, they should have the courage to seize the opportunities and meet the challenges brought by the times. If enterprises want to ensure the pace of stable development, they should make full use of the advanced technology in the era of big data, promote the innovation of management mode, and provide internal support power for enterprise development.

First of all, big data is a new concept derived from the development of the times. It was first put forward by Macintosh, a silicon company in the United States, to describe the massive data generated in the era of information and knowledge explosion. The concept of big data technology is gradually extended from the concept of big data technology, which is used to extract from the high-efficiency data again to improve the efficiency of the whole database information utilization. Secondly, McKinsey information consulting company of the United States believes that big data is an indispensable factor of production in all walks of life. It uses conventional software tools to extract, manage and process the content that appears in a specific time, and comprehensively express the integrity [2–5]. Finally, according to the treatment of the life cycle of big data, it is usually divided into big data collection and preprocessing, big data retention and management, big data calculation methods and systems, big data mining and extraction, big data influence calculation, big data security and confidentiality. The emergence of big data technology optimizes the processing mode of massive information and improves the speed and efficiency of data utilization. China's enterprises have not fully adapted to the changes in the era of big data, and they are still at a loss in many cases.

2 Analyze the Current Situation and Existing Problems of Enterprise Management in the Era of Big Data

2.1 Enterprise Managers Ignore the Application of Big Data

The era of big data brings huge amounts of business information, including the speculation of economic development trend of all walks of life and the importance of extraction and application. However, it is obvious that the enterprise managers have not started from the overall situation consciousness, and have not found the invisible business information in the big data information, so they have ignored the problems in the era of big data. Most enterprise managers in our country think that big data era is just the complexity and diversification of data. They can only simply process and summarize the data in this case. They have not found what kind of role and value will be produced to the enterprise in the actual application process. Some enterprises even think that the massive business information contained in the era of big data has not come from financial statements and enterprise profit and loss statements. This kind of short-sighted enterprise managers' simple vision hinders enterprises from launching an impact to a higher direction and higher status. They only use big data to expand access to information, and lack of attention to the hidden value behind the era of big data [6].

2.2   The Intelligent and Automatic Level of Enterprise Management is not High Enough

The advanced technologies used in the era of big data include data preservation, data sorting and data collection. However, with the rapid development of Internet information technology, a variety of new technologies are gradually applied in various industries, such as cloud computing technology, which has been integrated into the social enterprise industry. The intellectualization and automation of enterprise management is the comprehensive application of various advanced technologies and tools in the era of big data. Under the background that Chinese enterprises are not familiar with, Gartner Group has proposed the definition of business intelligence in the last century. The core of big data era is to provide the latest management methods and basis for enterprises with the help of advantageous resources in the development process of the times. As the core content, it is involved in various advanced industries, including finance, mobile, Internet and other mobile e-commerce fields. However, the enterprise automation and intelligent management has not been widely popularized. Therefore, in the face of the development characteristics in the era of big data, enterprises cannot make effective adjustments, resulting in problems in the operation process.

2.3   The Scientific Concept of Big Data Era Has not Been Established

According to the current situation of social development, the arrival of the era of big data has shown a massive growth trend for the processing capacity of enterprise data. Although there are many types of enterprise management data, mainly including pure digital text structured data and semi-structured and fully structured data mode, these data types composed by most industry enterprises cannot adopt unified processing mode [7–10]. According to the current semi-structured processing mode in enterprises, the data processed by enterprises has accounted for more than 80%, and there is no effective management mode to improve. The problem of data integration in the era of big data is that the system platform constitutes different information systems, and it is difficult to truly realize data sharing in various business modules. Therefore, enterprises do not establish a scientific concept of the era of big data, which affects the work of all aspects of the enterprise, as well as the problem of safe storage of enterprise data, which is also the lack of current enterprise management. If the network hacker divulges the confidential information of enterprise management, it will cause a catastrophic threat to the enterprise.

3   Effective Strategies for Innovating Enterprise Management Mode in the Era of Big Data

3.1   Establish Enterprise Operation Decision-Making Mechanism Based on Big Data

Under the background that all walks of life in the whole society are facing the influence of big data technology, banning the traditional enterprise management mode has become the inevitable trend of the development of the times. Only by taking effective measures and implementing innovative enterprise management mode, is the key to win the development of enterprises at this stage. To understand the disadvantages of traditional management, enterprises should establish a set of business decision-making process suitable for the era of big data, so that the enterprise's decision-making can have an objective positioning. Specifically, it is to use Internet technology to obtain a large number of reference information [11], and arrange enterprise professionals to classify and screen the information in the information. As the main reference information of enterprise decision-making and operation, only by ensuring the scientific and efficient decision-making process, can we better meet the consumption and management form in the era of big data.

3.2   Enterprises Introduce and Cultivate Big Data Talents

China is in the background of big data era development. If we want to complete the exploration and innovation of enterprise management mode, we need to supplement the fresh blood of enterprises and introduce big data talents. Therefore, as an enterprise manager, he is not only responsible for the processing and collection of various data of the company, but also for the dynamic prediction of the development of social industry in the market. It is necessary to introduce big data talents instead of traditional thinking to treat the changing society. In addition, for a large number of employees of the enterprise, regular training and assessment of big data era and big data technology is not enough. It is not enough to only introduce big data talents, but also improve the processing ability of big data for all employees. Enterprises can select well-known scholars and experts in the industry according to their own conditions and qualifications to popularize the awareness of the era of big data in the form of lectures, and strengthen the exchange and learning of personnel, and make self coping strategies.

3.3   Improve the Hidden Business Value of Managers in Data

The impact on the development of enterprises caused by the era of big data is inevitable, and the enterprise management mode will be reflected in the process of reform and development. The effective use of big data for enterprises to achieve higher benefits is a positive significance. If we want to make big data technology work for us, we must deeply understand the connotation, characteristics and significance of big data [12–14]. On the one hand, it is necessary to make use of the value of big data for decision-making and operation in enterprise management to ensure the implementation of decision-making is scientific, reasonable and objective. On the other hand, in the process of developing business intelligence, enterprises need to consider effective communication with decision makers, and look at the problems encountered in the construction of large-scale data from all aspects and angles, so as to analyze the large-scale data for the smooth construction of enterprises the system makes effective reference.

3.4   Using the Social Networks of Employees

As an important framework of enterprise management mode in the era of big data, the social network of enterprise employees is of great significance, which cannot be ignored.

From the perspective of current enterprise utilization rate, the lack of social network level of employees makes enterprise management in a decentralized mode. Enterprises should be good at using all available resources in the era of big data, which will not only greatly improve the overall and social benefits of the enterprise, but also help to maintain the passion of the employees themselves, identify with the corporate culture, and have good satisfaction and loyalty [15–17].

4   Conclusion

To sum up, China's enterprises are facing the stage of rapid expansion and development of big data, which affects all sectors of the society, and ushers in new opportunities for enterprise development while causing great impact. Therefore, in the face of this new situation, enterprises should consciously innovate the management mode, improve and improve the application of big data knowledge in enterprise management, so as to make effective reference value for promoting the sustainable development of enterprises. In the era of big data, the problems in enterprise management should be solved in time, and the management system should be optimized to provide the basis for the economic benefits and social values in the development of enterprises.

Acknowledgements. Guizhou Provincial Education Department - "The Technology Integration and Demonstration of the Integrated Development of Chimonobambusa Industry in Jinsoshan of Guizhou Province" (Contract No.: Guizhou Eductaion Contract [2019]026).

References

1. Yu, Y.Y., Duan, Y.L.: Research on enterprise management mode innovation in the era of big data. Technol. Innov. Manage. **37**(3), 302–307 (2016)
2. Yang, H.M., Yue, X.G.: On the application strategies of organizational behavior in enterprise management. Malays. E-Commer. J. **3**(1), 24–25 (2019)
3. Xu, D.Y.: Research on entrepreneurial ability of emerging technology enterprises. Malays. E-Commer. J. **3**(1), 01–09 (2019)
4. Tao, S.: Evaluation of technology innovation in Hubei Province. Eng. Heritage J. **2**(2), 09–10 (2018)
5. Qian, S.H.: Research on enterprise management mode innovation in the era of big data. Cont. Bridge Vision **12**, 42–43 (2016)
6. Li, X.Z.: Research on innovation of enterprise management mode in the era of big data. South. Agric. Mach. **48**(6), 157 (2017)
7. Wang, H., Zhang, X.M., An, L.R.: Research on the influence of knowledge base and relationship network on innovation performance. Acta Inform. Malays. **2**(2), 01–03 (2018)
8. Gian, S.H., Kasim, S., Hassan, R., et al.: Online activity duration management system for manufacturing company. Acta Electron. Malays. **3**(2), 01–08 (2019)
9. Xu, D.Y.: Research on supply chain management strategy of longtang electric engineering Co. Ltd. Acta Electron. Malays. **3**(1), 10–13 (2019)
10. Liu, F.: Research on the innovation of enterprise management mode under the environment of big data. Mod. Enterp. Cult. **26**, 125–126 (2017)

11. Ma, L.: Analysis of enterprise management mode innovation in the era of big data. J. Tianjin Vocat. Coll. **11**, 80–82107 (2014)
12. Liu, J.: Analysis of e-commerce enterprise management mode under the background of big data era. Bus. Econ. Res. **5**, 99–101 (2017)
13. Qiu, Q.Y.: A brief analysis of sino-us trade relations under the new era. Malays. E-Commer. J. **2**(1), 06–08 (2018)
14. Chen, Y.X.: Analysis on the efficiency of financial subsidies for vocational training of new technology workers in Jiangsu. Malays. E-Commer. J. **2**(1), 13–15 (2018)
15. Li, G.S., Wang, X.L.: Audit research on Financial Sharing Service Mode in the era of big data. Friends Account. **19**, 123–126 (2016)
16. Kartika, M., Melati, H.A., Ratih, Y.: The development of corak insang weaving craft creative economy in Pontianak city. Malays. E-Commer. J. **3**(2), 27–29 (2019)
17. Lahafi, F., Muchsin, A., Semaun, S.: Development of creative industries training towards sharia economic empowerment in Bilalangnge community, Parepare City, South Sulawesi. Malays. E-Commer. J. **3**(2), 33–35 (2019)

Research on Automatic Test Technology of Embedded System Based on Cloud Platform

Xia Wei(✉)

Xi'an International University, Xian 710000, China

Abstract. Automated testing is actually a kind of software testing. Previous testing work was completed by testing engineers manually executing test cases. Embedded systems have been widely used in real life, and the corresponding embedded software scale is also expanding day by day, but the requirements for its development cycle and product quality have not decreased at all. With the development of embedded system, we urgently need a testing system that can test and analyze the software of embedded system on-line in real time in the unit phase, integration phase, system phase and other phases of software development to ensure the quality and reliability of the software. This paper designs and implements an embedded system automation test platform based on infrastructure cloud. The platform is built on infrastructure cloud environment, which can make full use of hardware resources and reduce hardware costs.

Keywords: Cloud platform · Embedded system · Automated testing

1 Introduction

With the rapid development of science and technology, the functions of both software and hardware are increasingly complicated today. How to complete the testing of software and hardware in the fast update iteration has become the top priority for all companies. Since the development environment and operation environment of embedded systems are different, the automatic testing of embedded systems should be conducted separately in the development environment and operation environment, which will greatly increase the cost of system testing [1], so it is required that the scale and complexity of embedded software also continuously improve. The quality and development cycle of embedded software have a decisive impact on the final quality and time to market of products [2]. The introduction of automation technology shortens the software testing life cycle and improves the maintainability and regression of software testing. Automated testing has become an indispensable testing method. This paper mainly studies how to build an automated interface test platform in the test cloud and use the platform to test the project to protect the software quality and improve the development speed.

© ICST Institute for Computer Sciences, Social Informatics and Telecommunications Engineering 2021
Published by Springer Nature Switzerland AG 2021. All Rights Reserved
E. Bisset Álvarez (Ed.): DIONE 2021, LNICST 378, pp. 374–379, 2021.
https://doi.org/10.1007/978-3-030-77417-2_28

2 Automated Testing

Automated testing mainly simulates some manual testing behaviors through computers, and completes some tedious and boring tests or tests that cannot be completed manually according to instructions. Through uninterrupted testing, automated testing can improve the testing efficiency and increase the utilization rate of testing environment [3]. According to requirements, testers can increase, decrease or modify the arrangement and combination of automatic test cases by themselves to avoid redundant tests when increasing the coverage rate of automatic tests. By using the automated testing framework, time and resources can be effectively utilized to improve testing efficiency. The design requirements of the currently applied embedded automatic test system platform are all configured by testers according to various kinds of software and hardware and environment. The stable operation of the test system is ensured by inputting data into the test system, and the output results are studied and analyzed [4]. Moreover, the development environment and operation environment of embedded software are not consistent, so even if the test is sufficient in the host environment, it cannot be said that there is no problem in running the software in the target environment. In order to realize automatic batch execution of test scripts and facilitate management of test scripts, the test management tool reads a test configuration file edited by a tester before sending a test request. Automated test types include unit test, integration test and system test. At present, only these three types of tests can be automated.

3 Automated Test Cloud Platform

Automatic test cloud platform software system as shown in Fig. 1, the software system of the automated test cloud platform is mainly divided into four parts, namely the desktop PC part of the test engineer, the server part of the cloud platform, the database part and the VTP executor part. In the distributed resource scheduling cluster, the virtual machines on the host with heavy load will be automatically migrated to the host with light load to achieve load balancing in the cluster.

In order to better manage the test and reduce the maintenance of test scripts, so that testers can write test cases, we have developed a set of automated testing framework to meet the current needs. The test case is downloaded to the target machine for operation through the communication interface (serial port or network). The information generated after the program is run is uploaded to the host machine through the communication interface, and then the received information is analyzed and processed. The configuration file will define the storage location of the test script set, the time interval for script execution, the communication time interval between the host computer and the target computer, and the number of stub points for each communication. Every time the software changes, we must retest the existing functions in order to determine whether the modification achieves the expected purpose and check whether the modification damages the original normal functions [5]. C language has played an important role in the automatic testing of embedded systems, greatly enhancing the practicability of assembly language and effectively solving the obstacles of technical communication.

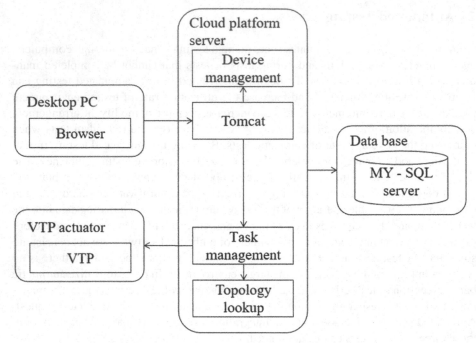

Fig. 1. Automatic test cloud platform software system

3.1 Typical Application Scenarios of Automated Test Cloud Platform

Currently, this infrastructure cloud environment is managed by a dedicated person, including various operating system template management, virtual machine mirroring and restore operations, virtual machine creation and deletion, etc. [6, 7]. The emergence of automated testing tools has reformed the traditional manual mode, making testing work enter the era of rapid development of automation. In the automated test cloud system, the test engineer can save the network diagram and configuration information when performing the test task for the first time in a test scenario, and only need to load the scenario saved during the first test again when the problem list regression is needed. Its main feature is that it does not run the program under test. It mainly uses inspection, technical review and code static analysis to check the errors of the software under test. For embedded software, the test only needs to be performed on the host computer. And transmits the script set at the specified position to the target machine side, and at the same time initializes each test variable according to the parameters in the configuration file [8–13]. Support the parallel execution of test cases on multiple clients, and increase the batch execution speed of test cases by adding hardware. Software users do not need to purchase additional hardware equipment, software licenses and install and maintain software systems. They can easily use the software at any time and at any place through an Internet browser and pay fees regularly according to usage.

4 Design of Embedded Software Automation Test Platform Based on Cloud Platform

The automatic test method of this project is designed as Host/Target mode in test mode, and the overall architecture adopts Client/Server structure. Host/Target mode means that the compile link and test analysis are both run on the host machine, while the tested program is run on the target machine [14–18]. The tester makes a test request to the target machine side through the test management tool, and transmits the script target file that has been compiled successfully to the target machine side. It must rely on the specification of requirements that can reflect this relationship and the function of the program to consider the test cases and infer the correctness of the test results, that is, it can only be based on the external characteristics of the program. After each test, it will automatically judge whether the work is completed or not, and after all tests are completed, it will automatically send all test data to the host computer. If the authentication can be passed, the authentication server will send relevant information to the user at the equipment end, and the equipment end will construct a dynamic access control list according to the information to control the access of the client. The communication between the target computer and the host computer is via serial port or network. The schematic diagram is shown in Fig. 2.

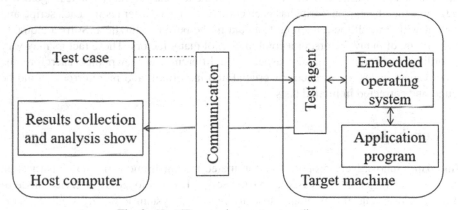

Fig. 2. Host/Target schema structure diagram

By executing all function test scripts, the platform pre inserts system code to capture the coverage of system code. The tester covered the report through analysis. According to the load, timing and performance requirements in the requirements, judge whether the software meets these requirements. It is necessary to analyze and study mainly subjective factors, and the inevitable objective factors can be ignored. The automated test cloud platform schedules and allocates resources through topology look up, which is the core algorithm of the automated test cloud platform. The control of the testing process needs to realize the automatic execution of the whole process from obtaining the testing code and the tested software to deploying the software and testing until the final release of the testing report without manual intervention. The basic framework layer mainly includes automatic test case engine, report engine and work-flow. Statistics and display of the coverage of the tested software system and graphical display of the static calling relationship of the tested software system functions; The test agent controls the execution of the script, caches the collected test results and coverage information to a specific memory area, and the target machine sends the test results and coverage information to the host machine at fixed time intervals. That is, when the statement coverage rate is 100% and the branch coverage rate is ≥85%, the test is considered ideal, software errors can be detected by nearly 90%, and the consumption of time and space is allowed.

All data in the automated test cloud platform, such as automated scripts, device connection information and execution result logs, are stored in MY-SQL database. MY-SQL is an open source database, because of its open source nature, users can freely customize MY-SQL database according to their own needs. It is mainly used to control the execution of tests, so there is no need to cross the operating system platform and use Windows operating system. The control command is also a batch script command. Its slave machine can be used as a test execution machine, mainly playing the role of executing test scripts. Each tenant customizes its own service according to its authority. The second is the setting, management and control of permissions. The degree of customization of tenants' services depends on the granularity of resource permissions. When the tested source code changes, the source code can be recompiled to determine the modified part of the software and maintain the test case library; The test agent will judge whether the current script has been completely tested after testing each script, and if so, it will send all the remaining test data in the buffer to the host. Software quality is a mixture of many factors, or a combination of many factors. These factors may vary according to different application aspects and different user viewpoints. Therefore, the effectiveness of test cases becomes crucial and important, and more errors should be found and corrected in limited tests.

5 Conclusion

This paper innovatively proposes an automated testing framework and constructs an automated testing platform based on visual script. It is deployed as a service mode of cloud platform, and the automatic test of embedded system is deeply studied. Then, combined with the actual working situation, the test method of this embedded system is proposed, which realizes the functions of automatic compilation, automatic execution of test cases, etc. The automatic testing platform based on it can be realized, and daily integrated testing can be realized, thus protecting the development quality and reducing the development cost. However, the current platform still has some defects and has great room for development. Relevant technicians should continuously explore and make efforts to further promote the improvement of the automation level of the testing system.

References

1. Zhou, Y.: Discussion on automated testing methods of embedded systems. Eng. Technol. (Full Text Edition) **7**, 00573 (2016)

2. Wang, S., Wang, Y.: Research on the automated testing framework for embedded devices. Mod. Comput. Mid-term J. **1**, 31–34 (2015)
3. Guo, Y., Chen, X., Zheng, C.: Research and implementation of software automation test method based on embedded security platform. Railway Commun. Signal **054**(001), 66–68 (2018)
4. Zhang, F.: Research on automatic evaluation technology of power application system based on cloud architecture. Electr. Eng. Technol. **6**, 00005 (2016)
5. Wang, W., Yue, L., Yang, J.: Design and implementation of an automatic test system for transmission line tower tilt based on embedded Linux. Comput. Measur. Control **024**(009), 70–73 (2016)
6. Cao, W., Zhang, K.: Research on anti-reptile technology based on automated testing-T mall platform as an example. Mod. Comput. Mid-year Mag. **4**, 64–67 (2018)
7. Chen, H., Fan, D., Huang, J., Huang, W., Zhang, G., Huang, L.: Finite element analysis model on ultrasonic phased array technique for material defect time of flight diffraction detection. Sci. Adv. Mat. **12**(5), 665–675 (2020)
8. Wang, S., Huang, Y., Liu, A.: Electric hoist energy efficiency automation tester based on internet of things. Hoisting Transp. Mach. **000**(001), 127–129 (2019)
9. Hamzah, R., Kasim, S., Hassan, R.: Taxi reservation system of BATU PAHAT taxi association. Acta Electron. Malays. (AEM) **2**(2), 20–24 (2018)
10. Li, B., Li, Z.: The implement of wireless responder system based on radio frequency technology. Acta Electron. Malays. (AEM) **2**(1), 15–17 (2018)
11. Li, B., Li, Z.: Design of automatic monitoring system for transfusion. Acta Electron. Malays. (AEM) **2**(1), 07–10 (2018)
12. Li, B., Li, Z.: The design of wireless responder system based on radio frequency technology. Acta Electron. Malays. (AEM) **2**(1), 11–14 (2018)
13. Basir, N., Kasim, S., Hassan, R.: Sweet8bakery booking system. Acta Electron. Malays. (AEM) **2**(2), 14–19 (2018)
14. Feng, Z., Shu, L.: Design and research of aerial three-axis gimbal control system based on ARM. J. Shenyang Univ. Aeronaut. Astronaut. **034**(003), 70–75 (2017)
15. Zamanian, P., Kasiri, M.: Investigation of stage photography in JEE LEE's works and comparing them with the works of Sandy Skoglund. Acta Electron. Malays. (AEM) **2**(1), 01–06 (2018)
16. Yan, Z.: Artificial bee colony constrained optimization algorithm with hybrid discrete variables and its application. Acta Electron. Malays. (AEM) **2**(1), 18–20 (2018)
17. Singh, O., Kumar, G., Kumar, M.: Role of TAGUCHI and grey relational method in optimization of machining parameters of different materials: a review. Acta Electron. Malays. (AEM) **3**(1), 19–22 (2018)
18. He, Z., Gu, X., Sun, X.: An efficient pseudo-potential multiphase lattice Boltzmann simulation model for three dimensional multiphase flows. Acta Mechan. Malays. (AMM) **1**(1), 1–3 (2018)

Research on Embedded Humanoid Intelligent Control and Instrument Based on PLC

Meiyan Li[✉]

Xi'an International University, Xi'an 710077, China

Abstract. Intelligent control is formed on the development of computer technology. Embedded humanoid intelligent control system has excellent self-organization and self-adaptation capabilities, and has good performance in large-scale complex industrial system control. With the development of instruments and meters, instruments and meters have penetrated into all fields of people's life and become an important tool for human beings to acquire information, understand nature and transform nature. Now the development level of instruments and meters is an important symbol of the development level of modern science and technology. Programmable logic controller (PLC) is a kind of digital operation and operation control device. It is an electronic system developed instead of the traditional relay, which integrates computer technology, communication technology and automatic control technology. Embedded humanoid intelligent control system has superior self-organization, self-adaptive and self-learning ability. The arithmetic operation function and data processing ability of PLC are greatly enhanced, which makes the realization of complex control algorithm on PLC possible.

Keywords: Intelligent control · PLC · Instrument

1 Introduction

With the rapid development of instruments and meters, computer and network technologies are also developing rapidly. Intelligent control technology is an interdisciplinary frontier discipline developed on the basis of artificial intelligence, cognitive science, operations research, system theory, information theory and cybernetics with the rapid progress of computer technology. It is an advanced stage of control theory development [1]. Programmable Logic Controller (PLC) is a control device for digital operation and operation. It is an electronic system developed instead of traditional relay and integrates computer technology, communication technology and automatic control technology [2]. Intelligent control is formed on the development of computer technology. Intelligent control system has excellent self-organization and self-adaptation capabilities, and has good performance in the control of large complex industrial systems. People combine a plurality of individual individuals with certain intelligence to form a humanoid intelligence system with independent control performance and mutual cooperation characteristics [3]. More and more people realize that the cooperation of humanoid intelligent

E. Bisset Álvarez (Ed.): DIONE 2021, LNICST 378, pp. 380–385, 2021.
https://doi.org/10.1007/978-3-030-77417-2_29

systems can complete more complicated tasks with less cost. Compared with a single agent, humanoid intelligent systems, especially distributed humanoid intelligent systems, have many obvious advantages [4]. The application of embedded system to the field of instruments and meters, and the combination of traditional instruments, sensors and microprocessors, has become the main trend of the development of the instrument and meter industry.

As one of the important development directions of industrial control network, embedded intelligent control instrument is the result of the joint development of industrial data communication technology, control network technology, Internet technology and other technologies. Using a variety of technologies, joint operations can usually complete some single difficult tasks [5]. In the process of in-depth research on intelligent control theory, people find that various control strategies have their own advantages and disadvantages, and a single control strategy cannot have perfect control performance. Intelligent equipment such as PLC, information collector and other data collection and processing devices can play a huge role in industrial and information construction [6]. The decision-making and actions of each individual in the group are independent, but there are extensive altruistic cooperative behaviors within the group. Intelligent control system has excellent self-organization, self-adaptation and self-learning ability, and has shown good performance in the process of controlling large and complex industrial systems [7]. As the most widely used automatic control equipment, PLC equipment is being widely studied and analyzed by people in order to better serve human life and work.

2 Main Features of PLC Intelligent Control System

With the continuous development of science and technology, some programmable intelligent technologies have become more and more widely used in practical applications. PLC intelligent control system has its own distinctive features, mainly reflected in its good expansion performance. Since various modules can be mounted on the rack on the back bus, the modules can be selected according to actual requirements. At present, PLC control technology is widely used in the field of industrial automation. For example, it has grown into a pillar industry in metallurgy, electric power, light industry, chemical industry, etc. Based on the continuous development of computer technology and network communication technology, for the development of automatic control system, the integrated application of corresponding technologies can further strengthen the communication function of PLC. The application of PLC control technology can not only provide a very reliable control application for various automation equipment, but also can put forward a more reliable and perfect solution when controlling, which well meets the development needs of industrial enterprises.

Fuzzy control system combines practical rich working experience to sum up operation experience, thus having fuzzy operation rules to realize effective control. Fuzzy control is a control method that can reflect human intelligence. Automation clients can create automation objects, access objects provided by automation servers, obtain or set properties of objects, or call methods of objects. The interaction between automation objects and automation customers is shown in Fig. 1.

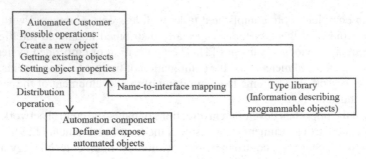

Fig. 1. Interaction between automation customers and automation components.

The development of computer network technology has also continuously increased the requirements for industrial control and management, which has prompted PLC control systems to begin to move from closed centralized systems to open distributed systems. Faced with the object of control, the complexity of the control object is very high, coupled with the particularity of the use environment and the continuity of long-term work, these all place higher requirements on the PLC control technology. Because many field buses are developed by PLC manufacturers, it is closely related to PLC. Programmable can form multiple control devices into a huge control network and then perform unified platform control management. Under this development background, the operation status of each instrument and equipment in the network is operated under the network connection and collected on the operating system. The communication program of PLC control technology is relatively simple, only need to use the communication interface software and special computer to realize the design of the communication program, which greatly reduces the workload of computer programming.

3   Design of Embedded Humanoid Intelligent Control Instrument

When the equipment is in normal working condition, there is a certain logical relationship among the intermediate memory unit, output signal and input signal of the electronic control system. If the equipment fails, this logical relationship will be destroyed. The main program is to optimize the parameters such as the start-up of the wind turbine and the large and small motors. The subprogram is mainly to control the subprogram to stop the program and collect the operation data of the system. The use of PLC analog control module enables it to realize not only process control, but also instrument monitoring through control statements. Under the PLC-based control system, analog control is based on the characteristics of the control object itself, and the integrated system is built after the functional modules are combined [8, 9]. In the position control, the automatic control is mainly realized for the stepping motor, and the pulse is sent to accurately position the corresponding displacement [10–14]. According to the characteristics of the control object, PLC can flexibly realize system control by successfully assembling a complete control system through a combination of functional modules.

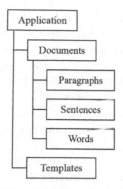

Fig. 2. PLC word processing object model.

Based on the PLC control system, the corresponding command system and frequency converter can realize effective control of the motor with the application of this system equipment, mainly controlling and adjusting the rotation speed. Judging from the control effect, the temperature system has a large inertia load characteristic, and has requirements for steady-state accuracy and anti-interference capability, etc. Through this control system, the requirements of indexes can be met. Many automation objects are provided in PLC control system, and there are inheritance and derivation relationships among these objects, forming a tree-like hierarchical structure. Among them, the Application object is the basic object in PLC object mode and represents the word processing program itself. Figure 2 shows a part of a PLC text object model.

At present, in the actual application of PLC system equipment, a major drawback is the low compatibility of the system itself. Because the corresponding buses and the like in PLC design and R & D are all of a special nature, there are certain differences in the structures designed by different companies. For complex systems that cannot establish accurate mathematical models in engineering practice, they cannot be controlled by traditional control methods, but people can summarize their operating experience into fuzzy operating rules based on rich work practices to achieve effective for them control. The current standard for the specific programming software of PLC control technology is still in planning, which greatly affects the standardized use of PLC control technology. Although this PLC software program is written and implemented with corresponding specifications, in the process of actual use, the industrial production environment itself will still have a certain influence on the practical application of this system [15]. The development of computer and network communication technology, as well as the continuous improvement of the control and management requirements of the industry, make the PLC control system also from the closed centralized system to the open distributed system. When PLC control technology is used, sequence control and logic control can be effectively realized [16–20].

4   Conclusion

With the continuous progress of human society, the application scope of instruments and meters covers almost all fields of human activities. Instruments and meters have become an important tool for people to understand and transform nature. PLC, as a control element, has great functions and strong operability, it is also an embodiment of human wisdom. Its use is conducive to the scientific and timely management and control of the operation and effectiveness of various equipment in life and production. The application of embedded system in the field of instruments and meters and the combination of traditional instruments, sensors and microprocessors are the result of the common development of industrial data communication technology, control network technology, Internet technology and other technologies, and is also the inevitable trend of the development of instrument industry. With the rapid development of PLC technology, its function is more perfect, the application field is gradually expanded, and the problems in practical application are gradually solved. With the increasing status in the field of industrial automation control, PLC will be an important guarantee and support for the development of industrial automation in the future. In order to promote the further development of China's industry, we should vigorously promote the application of PLC control technology in industrial automation production, so that China's industrial technology continues to develop.

References

1. Hao, C.Y.: Application of intelligent control instrument system in agricultural electrical automation. Automat. Appl. **000**(004), 114–115 (2017)
2. Zhang, K.: Application research of intelligent control instrument system in agricultural electrical automation. Comput. Knowl. Technol. **015**(008), 163–164 (2019)
3. Mo, Z.K.: Application research of intelligent control instrument system in agricultural electrical automation. Agric. Technol. **39**(02), 50–51 (2019)
4. Xiao, C.: Discussion on the design of PLC embedded fan intelligent control system. China New Commun. **18**(9), 123 (2016)
5. Wang, J., Zhou, Y., Zhuang, W.: Design of intelligent material handling system based on PLC and embedded technology. Manuf. Automat. **9**, 28–31 (2019)
6. Han, L., Yu, S., Gong, Y.: Status and development trend of embedded PLC. Automat. Expo **000**(006), 40–41 (2016)
7. Li, Z.M., Gong, L.D., Xu, J.J.: Intelligent research on the safety protection door of CNC machine tools based on fuzzy rules. Digit. Technol. Appl. **37**(02), 20–21 (2019)
8. Li, N.N., Han, H.Y., Cao, F.: Design of intelligent control system for hybrid polishing and cleaning robot. Automat. Instrum. **10**, 9–12 (2019)
9. Shang, W.L., Zhang, X.L., Liu, X.D.: Construction method and verification of industrial control network local trusted computing environment. Inf. Netw. Secur. **220**(04), 7–16 (2019)
10. Kong, C., Ning, Y., Song, C.Y.: Improved PID control for welding intelligent tooling of the front axle of the rack. Comput. Knowl. Technol. **14**(21), 275–276 (2018)
11. Chen, H., Chen, Y., Yang, L.: Intelligent early structural health prognosis with nonlinear system identification for RFID signal analysis. Comput. Commun. **157**, 150–161 (2020)
12. Roopesh, J.: Pterostilbene caffeine co-crystal: bioavailable caffeine alternative enriched with pterostilbene. Matrix Sci. Med. **4**(1), 24–26 (2020)
13. Mathew, O., Temitayo, F.: Evaluation of plasma Na, K, urea, and creatinine in rabbits given amoxicillin overdose supplemented with cucumber (*Cucumis sativus*) fruit juice. Matrix Sci. Med. **4**(1), 20–23 (2020)
14. Abdur, M., Tamanna, Z.: Antibiogram of blood culture isolates of patients from a hospital in Dhaka Bangladesh. Matrix Sci. Med. **4**(1), 1–5 (2020)

15. Chen, C.: A study of group intervention on depression in urban college students. Matrix Sci. Med. **4**(1), 6–8 (2020)
16. Sun, G., Yang, B., Yang, Z., Xu, G.: An adaptive differential evolution with combined strategy for global numerical optimization. Soft Comput. **24**, 6277–6296 (2020)
17. Rabbani, A., Hayat, K., Qamar, A.: The comparative efficacy of nalbuphine and tramadol in controlling postoperative shivering in rabbits. Matrix Sci. Med. **4**(1), 9–14 (2020)
18. Mathew, O., Bukunmi, O.: Possible metabolic abnormalities of lipids in rabbits given amoxicilin overdose and raw cucumber (*Cucumis Sativus*) fruit juice. Matrix Sci. Med. **4**(1), 15–19 (2020)
19. Li, Y.: Study on the characteristics of energy consumption and metabolism during exercise. Matrix Sci. Med. **3**(2), 38–40 (2019)
20. Yang, S.: Relationship study between exercise and acute myocardial infarction in different time periods. Matrix Sci. Med. **3**(2), 41–43 (2019)

Software Development and Test Environment Automation Based on Android Platform

Xiuping Li[✉]

Xi'an International University, Xi'an 710000, China

Abstract. With the advent of the mobile Internet era, the quality of Android application software and the level of user experience have become the key factors that determine the success or failure of market competition. The construction of software development and testing environment is an important part of the whole software development process. It is imperative to use machines instead of manpower to complete complicated tests that require precision. The concept of automated testing arises at the historic moment. The construction of software development and testing environment is an important link in the whole software development process. Different versions of operating systems, databases, network servers and application services, combined with different system architectures, make the types of software testing environments to be constructed various. The opening of Android system makes the development of Android Software easier. Any developer, whether a professional company or an individual, can develop their own applications. The combination of automated testing and manual testing makes up for the shortcomings found in automated testing, which requires a lot of initial investment and special personnel to maintain.

Keywords: Software development · Android · Automation

1 Introduction

The construction of software development and testing environment is an important part of the whole software development process. The combination of different versions of operating systems, databases, network servers and application services make the types of software testing environments to be built different [1]. At present, the mobile Internet is in a period of rapid development. No one expected its growth rate. Android, as a smart phone platform, is developing rapidly because of its openness, rich hardware selection, unlimited by operators and developers. With the diversity of software operating environment, the complexity of configuring various related parameters and the compatibility of testing software, the construction of software development and testing environment becomes more complex and frequent [2]. Application software is indispensable in the use of mobile phones. Especially today, smart phones have become an indispensable part of the development of mobile phones. Most software development environments are reusable, but often different software development and testing environments need to alternate [3]. The opening of the Android system makes Android software development

E. Bisset Álvarez (Ed.): DIONE 2021, LNICST 378, pp. 386–390, 2021.
https://doi.org/10.1007/978-3-030-77417-2_30

easy. Any developer, whether a professional company or an individual, can develop their own applications and share them with others.

With the increasing demand for various aspects of the software operating environment, the complexity of configuring various related parameters, and the compatibility of testing software, the work of building a software development and testing environment has become more complicated and frequent [4]. The cost of software testing needs to account for nearly half of the total cost of software development, and the required test work time is generally more than 50% of the software development cycle. This illustrates the complexity and complexity of software testing. With the rapid development of application software, the update of the Android system, the rapid popularization of intelligent terminals and the continuous progress of testing technology, past testing work has exposed more and more problems and deficiencies, including old test terminals [5]. After entering the testing phase, test engineers need to perform system-level testing, including functional testing, performance testing, stability testing, etc. For each different version of the software, it is also necessary to verify the existence of version regression in the test [6]. Testers should realize that software testing is not only a process to ensure the quality of software products, but also integrated into the whole company's software development process to supplement and promote software development..

2 Software Development Test Environment Based on Virtualization Architecture

Except for the operating environment of various mobile phone systems, the other is the same as the traditional software. However, in the development phase, the same tests as normal software testing methods are still required. In the test process, according to the actual situation, the tester cannot list all the test cases. A few representative cases can only be selected from a large number of test cases to represent other values not listed in this category. Software quality must be improved, and software testing is an important and effective means to ensure software quality. Each product can be encapsulated into a class, and its base class can encapsulate the common attributes of each product [7]. For mobile phones, the memory information to be recorded includes the total memory of the mobile phone, the remaining memory currently available and the memory occupied by the software to be tested. A large number of errors in the test occur at the boundary of the input and output ranges, rather than inside the input and output ranges [8–10]. Then, we need to focus on writing test cases for boundary conditions.

Each test program contains one or more test cases for specific types of components, where the test methods are specifically defined. At the same time, we should study the current standards and specifications, and check whether the product specifications are applicable to the correct standards, whether they conflict with the standards and specifications, and whether there are any omissions [11–13]. Android platform is carrying more and more traffic. The research on the security and stability of Android platform is becoming more and more important. Figure 1 shows the relationship between name node size and data node size.

With the rapid development of software testing, automated testing is becoming a very noticeable trend and trend in the field of software testing. In the product line

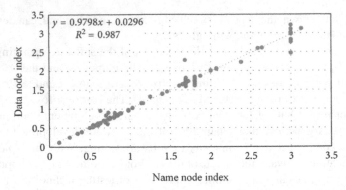

Fig. 1. Relationship between the size of the name node and the size of the data node

of the same product supplier, each product usually has the same characteristics and functions, and each product has its own unique attributes and functions [14–16]. Due to the characteristics of Android system, processes that are not displayed in the foreground are in the suspend phase, and a part of memory will be released. The memory occupied by the background display is not actually running, so the data in the running state must be recorded to correctly reflect the information of the software. The testing speed is far behind the release speed of the product. In this case, if there is no automated testing to help, manual testing can only sigh. One application program implementing the control device executes part of the test case while the other part operates another application program of the device. For example, a message containing photos taken by a camera application as attachments is sent.

3 Design Requirements of Android Software Automated Test Platform

In order to standardize the workflow and ensure the development quality of automated test cases. The quality of software products has certain special characteristics, which is very different from the quality inspection of other products, and computer hardware products in the same field are also different. The quality measurement of software products cannot be directly detected. The client of mobile application testing system is the module that truly realizes testing behavior. It is responsible for analyzing testing scripts, executing testing tasks, realizing testing cases, reporting testing results, and recording testing logs [17]. The client program in the test framework passes the test cases to the server program. After the server program executes, the test result is returned to the client program, which saves the result locally as an object. The computer software codes according to the test rules, executes in the computer environment, and automatically verifies the response and behavior of the tested program [18]. The management tool processes all the events input by the user interface through the built-in command parser, and converts these events into understandable forms and passes them to the appropriate modules. The test management platform is responsible for communication with the central management platform and management drives the overall test system. The test management platform makes a good test task schedule according to the test frequency and time requirements of the test plan.

Since most of the errors come from the requirement analysis stage, the correctness and accuracy of the requirement analysis is crucial to the future process. According to the test requirement analysis report, design test cases to fully cover the test requirements for each test requirement point. Most software testing is based on manual testing, but with the development of software industry, the complexity of software testing is also increasing. If the test plan does not exist or the auxiliary plan contained therein does not have corresponding test station information, the operation will fail in the initialization phase. At the same time, the demand side or users have higher and higher requirements for the software, so we should pay attention to the readability and maintainability of the script in the writing process. Test scripts play a key role in the automated testing process [19]. For software testing, which scripting technology to use is not the most important, and it is the most important to consider the test case system supported by scripts. As a test of software product quality, these related factors should be covered as much as possible to obtain a comprehensive evaluation of software product quality.

4 Conclusion

The combination of automated testing and manual testing not only makes up for the deficiencies found in automated testing, but also solves the shortcomings of low efficiency and large duplication of work in manual testing. Equipment suppliers and software development companies have already used some system automation testing tools. However, these tools are not perfect and have many restrictions on use, not systems. With the increasing complexity of software, the problems of low efficiency and low accuracy of manual testing are gradually exposed. Software testing automation can not only save human resources, but also, more importantly, it can discover the functional defects and user experience related problems of application software faster and earlier and shorten the improvement period. This paper studies a new type of automatic testing system, which not only meets the requirements of cross application and cross equipment testing, but also does not judge the test results according to the graphical interface. To some extent, the software developed based on Android system has the complexity of PC software, so in order to ensure the quality of such software, efficient testing is very important. During Android testing, the screenshot is obtained and transmitted to the server, and the similarity comparison of the images is completed at the server, and the completion of the test is given by comparing the screenshot sequence of the operation.

References

1. Jiang, T.: Development and application of surveying and mapping software based on Android smartphones. Jiangxi Build. Mater. **18**, 230 (2016)
2. Ren, S.G., Huang, D.G.: Design of Android-based greenhouse environment monitoring software. Sci. Technol. Inf. **17**(1), 44–45 (2019)
3. Shi, Y.H., Wang, A.J.: Research and application of building software development environment based on Android platform. Fujian Comput. **32**(3), 116–117 (2016)
4. Zhao, J.C., Liu, S.H., Qiao, Z.F.: Research and implementation of an intelligent greenhouse environment measurement and control system based on Android. Jiangsu Agric. Sci. **44**(3), 406–409 (2016)

5. Liu, D.Y., Cao, Z.Q., Zhou, Q.F.: Development of map chat software based on Android platform. Comput. Program. Skills Maintenance **12**, 49–52 (2017)
6. Gao, R.Z., Shang, L.H., Tu, X.: Software development of android-based environmental monitoring system. Electron Technol. **1**, 115–118 (2017)
7. Chen, J.Y., He, J.W.: Development of remote monitoring system for intelligent power equipment based on embedded and Android. Electron. Technol. Softw. Eng. **13**, 39–40 (2018)
8. He, L., Shen, J., Zhang, Y.: Ecological vulnerability assessment for ecological conservation and environmental management. J. Environ. Manage. **206**, 1115–1125 (2018)
9. Liu, C.L., Han, Y.F., Li, G.M.: Study on the relation between self consistency and congruence and mental health of postgraduates. Matrix Sci. Medica **2**(1), 1–3 (2018)
10. Shareef, M., Akhtar, M.S.: Neem (azadirachtaindica) and its potential for safeguarding health, prevention and treatment of diseases. Matrix Sci. Medica **2**(1), 4–8 (2018)
11. Chen, X.W., Su, Y.C., Huang, M.W.: Comparison and analysis of application effect of traditional paper operation method and digital information system in hemodialysis. Matrix Sci. Medica **2**(2), 1–3 (2018)
12. Barkat, M.Q., Mahmood, H.K.: Phytochemical and antioxidant screening of zingiber officinale, piper nigrum, rutag raveolanes and carum carvi and their effect on gastrointestinal tract activity. Matrix Sci. Medica **2**(1), 9–13 (2018)
13. Munir, S., Rahman, S.U.: Optimization of countercurrent immunoelectrophoresis and agar gel immunodiffusion tests for the comparative detection of horse and donkey meat. Matrix Sci. Medica **2**(1), 14–17 (2018)
14. Mehvish, S., Barkat, M.Q.: Phytochemical and antioxidant screening of amomum subulatum, elettaria cardamomum, emblica officinalis, rosa damascene, santalum album and valeriana officinalis and their effect on stomach, liver and heart. Matrix Sci. Medica **2**(2), 28–33 (2018)
15. Mahmood, H.K., Barkat, M.Q., Zeeshan, U., et al.: Phytochemical and antioxidant screening of anacylus pyrethrum, apium graveolens, boerhaavia diffusa, cinnamomum cassia blume, cuscumis melo linn, cuscumis sativus linn, daucus sativus, foeniculum vulgare, trachyspermum ammii and theit effect on various human ailments. Matrix Sci. Medica **2**(2), 4–14 (2018)
16. Riaz, M., Muhammad, G.: Copper deficiency in ruminants in Pakistan. Matrix Sci. Medica **2**(1), 18–21 (2018)
17. Rabbani, A.H., Hayat, K., Gardezi, F.H., et al.: A comparison of nalbuphine and pentazocine in controlling postoperative pain in dogs. Matrix Sci. Medica **2**(2), 15–20 (2018)
18. Xiao, X.W., Wang, X., Hu, J.F., et al.: Design and implementation of a chronic disease follow-up APP based on Android. Softw. Eng. **21**(2), 41–44 (2018)
19. Xie, L.X., Zhao, B.B.: Android system malicious behavior detection based on log analysis. Comput. Appl. Softw. **33**(5), 295–298 (2016)

Sports Information Communication Model Based on Network Technology

Guohua Shao[✉]

Physical Education Institute of Inner, Mongolia Normal University, Hohhot 010022, China

Abstract. Aiming at the sports information service platform of communication subject, communication object and media are analyzed, on this basis from the perspective of statistics in sports as an example to analyze the characteristics of information and channel sink. By using the analytic hierarchy process, from the network sports information content, audience experience, network sports information organization and dissemination of sports information and network environment in four aspects construct the sports information dissemination model evaluation index system, and determine the weight of each index. The results show that: the premise of sports information dissemination model based on network technology in order to serve the public, the right to guide the public to build a harmonious sports information network for the purpose of the business model, to the development of sports website, expand business partners, re positioning the sports network station commercial operation mode, break the old ideas, improve the connotation of operation mode sports website.

Keywords: Sports information communication · Network technology · Communication factors · Index weight

1 Introduction

Currently belongs to the information age, and the Internet to accelerate the upgrading of the information age [1–3]. Network technology as the characteristics of the modern means of communication and economic forms gradually penetrate into people's life, study, work and recreation and fitness field [4–8]. This paper aims at the research of sports information communication mode, in order to explore the reform way of sports information communication mode from the perspective of network technology [9–13]. Li and Wang [14] pointed out that the future of China's sports information dissemination website portal will combine and use more new technology, will promote the transformation of audience "to" user ", the difference of comprehensive portal trend intensified, network video will become the main communication forms of sports information network. Zhang et al. [15] theory and method of operation of communication, standing in the angle of sociology, analysis in the network sports information communication problems, pointed out that in the process of sports information dissemination for cultural conflict, ethical loss and other issues, in the face of network trust crisis, and actively

E. Bisset Álvarez (Ed.): DIONE 2021, LNICST 378, pp. 391–399, 2021.
https://doi.org/10.1007/978-3-030-77417-2_31

put forward the way to solve the problem. To improve the sense of social responsibility of network media. Wang [16] by using the method of questionnaire investigation and factor analysis etc., to Sina, Sohu, NetEase, Tencent and other 4 major portals as the research object, the main factors affecting the extraction of Chinese large portal sports information dissemination effect and classifying and naming, the proposed site taking more advanced techniques. Effective measures to improve their own strength. Li [17] pointed out that the audience of sports information needs from the passive acceptance of the past, to live entertainment, to meet the self-competition pleasure, and sports information media to meet the audience of these changes, also gradually changing sports information communication mode. In this paper, based on analyzing the sports information service platform and transmission mode of sports information dissemination elements, constructs the evaluation index system of sports information communication mode, in order to improve the sports information dissemination mode of our country to provide the feasible advice.

2   Communication Model of Sports Information Service Platform

2.1   Service Platform Communication Main Body

The sports fitness information service platform based on the related sports venues, sports audience to participate in fitness activities such as badminton, basketball, indoor swimming pool, Futsal Soccer field, tennis court, O2O (Online To Offline) loop to booking online and offline activities, back online comment form of sports communication. The main body of sports fitness information service platform is divided into the government to buy services such as sports information service platform and corporate sports information service platform.

The government purchase service sports information service platform has three main Go Sports, Group Pass, Sports in Jiading, Go Sports in the Hubei Province Sports Bureau, need reservation venues in Guangzhou city; Groups Pass Sports Bureau, Sports in Jiading need reservation venues; in Jiading District Sports Federation, to venue booking.

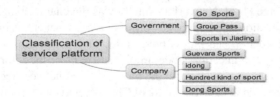

Fig. 1. The main body of sports fitness information service platform.

Corporate sports information service platform has four main Guevara Sports, idong Hundred, kind of sport, Dong Sports, Guevara Sports in Shanghai, which provides a solution for urban life; idong in Shanghai, mainly for the Hundred kind of sport experience; in Beijing, mainly for Dong Sports golf movement; sports life in Beijing. The main body of information dissemination platform for sports fitness information is shown in Fig. 1.

2.2 Object of Sports Fitness Information Service Platform

The object of the sports information service platform construction of communication mainly includes all objects except platform operation and maintenance of venues, cooperation, the communication objects include sports enthusiasts (C1), sports organizer (C2), industry partners (C3). Where: C1 as the communication object can receive accurate, timely, or through independent query way to find the platform for the dissemination of sports venues or activities, this is the platform as the core group of users, but also to remove the venues resources, sports fitness information service platform to absorb the basic source of customers; C2 belong to the second class the communication object, in the sports fitness information service platform for different communication subject in this category, there are still some discrimination; C3 industry partners, currently about O2O closed loop as the main means of communication, but the real consumption of the product or service must be experienced by the consumer online activities, open to experience exercise, the site, equipment, medical security and other factors supporting. O2O online and offline interactive transmission line as shown in Fig. 2:

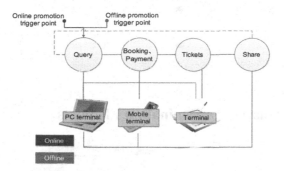

Fig. 2. O2O online and offline interactive communication Roadmap.

2.3 Sports Fitness Information Service Platform Carrier

At present, the network media information service platform construction of sports mainly as the official website of mobile phone, APP (Application), WeChat, micro-blog, public account, which is currently the four major areas of information platform service providers launch the main content. With the use of four media, in various sports information service platform of actual operation and maintenance of the media, the official website has the largest proportion of the sports information dissemination effect is still in computer "end the most authoritative and most comprehensive display of the content. The rapid development of mobile phone APP, a large proportion of the key factors to replace the use of computers: convenient, mobile, low power consumption, etc. Although there are traditional sports booking service information platform based on Web and long, and the telephone information service, but with the addition of LBS (Location Based Service), the change of location-based services and social dating mode, stimulate the development efforts to the spread of O2O mode based platform for service providers, occupy the important position in front of all mobile APP platform in the relatively strong influence. WeChat public number and the use of micro-blog public account, it reflects the effectiveness of the new network of social networking tools during the difference between the number of WeChat public closer to word of mouth. Sports fitness information service platform for the network media carrier, as shown in Table 1:

Table 1. Physical fitness information service platform of the network media carrier.

Category	Official website	Mobile APP	WeChat public number	Micro-blog public account
Go Sports	○	○	○	○
Group Pass	●	○	○	●
Sports in Jiading	○	○	●	○
Guevara Sports	○	●	●	●
idong	○	●	○	●
Hundred kind of sport	○	●	○	○
Dong Sports	○	●	○	○

3   Sports Information Dissemination Elements

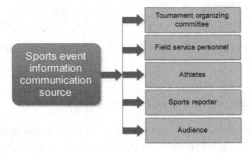

Fig. 3. Sports event information dissemination source.

The basic information dissemination process is completed, usually requires three elements of information source, information and information, in this process, the source is the seven point of the communication process, as the main factors in the dominant position, is the sponsor of communication activities. To some extent, the quality and quantity of the dissemination of the contents of the source is the key to determine the success or failure of a communication activity and the effectiveness of the spread of the key as shown in Fig. 3.

Sports event information dissemination, for example, the source of information dissemination as shown in Fig. 4.

Fig. 4. The process of sports information dissemination.

3.1 Information Dissemination of Sports Ink

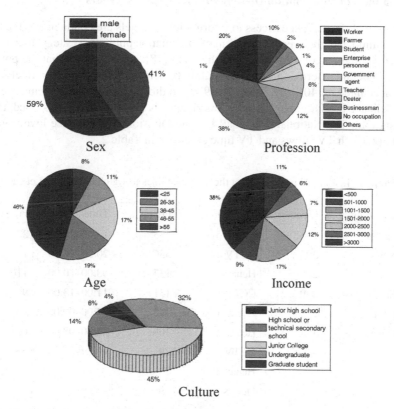

Fig. 5. The basic situation of the National Volleyball League Division XX audience statistics.

The sink refers to the audience or the receiver, the receiver is the information dissemination activities of various types of reflection and. The destination is one of the two poles of the communication process, and plays important role in the communication process, a complete transmission of information sent by the source, only to reach the destination is not complete. Not only that, the destination information or products to consumers, the two communicators of information content, feedback communication activities of the participants and the communication effect. Therefore, the Organizing Committee on publicity work should be taken into account when the sink, according to predict their psychological needs. The destination of the situation more understanding more can put me on the direction of propagation, for they can be targeted propaganda work. The basic situation of XX league ranking division audience as shown in Fig. 5. We can see from Fig. 5, a sink population XX audience Division National Volleyball League: more men than women; the young than in the elderly; occupation for students, workers and higher income white-collar class; higher educational level, mostly educated; lower middle income. The results of the survey on the XX sports news audience structure is basically the same.

3.2 Channels of Information Dissemination of Sports Events

Investigation of the audience access to information channels are conducive to the Organizing Committee of the event in the event of information publicity strategy, focus on the choice of the mass media. According to the Shan [18]: survey shows that the potential audience to obtain external information mainly by television, newspapers, Internet and friends, accounted for 40%, 39.7%, 31%, 29.7%, radio, magazine supplement. The three largest audience in addition to "friends" this channel is the other of the mass media in the remaining four. The National Volleyball League Division XX audience love watching TV, usually watch TV shows and TV time as shown in Table 2.

Table 2. List of TV programs that the audience likes to watch during the League

Sort	TV program	Frequency	TV station	Frequency	Time to watch	Frequency
1	Sports	465	CCTV 5	463	19:00–20:00	229
2	News	306	CCTV 1	356	20:00–22:00	115
3	Teleplay	255	Henan 1	137	08:00–10:00	110
4	Entertainment	239	CCTV 6	131	11:00–13:00	99
5	Film	228	Zhengzhou TV	118	22:00–00:00	73
6	Others	11	Henan 2	109	06:00–08:00	62
7	--	--	CCTV 4	79	Others	53
8	--	--	Henan 3	75	--	--
9	--	--	Others	49	--	--

4 Empirical Analysis

4.1 Construction of Evaluation Index System of Sports Information Communication Mode

The factors that affect the communication effect of sports information based on network technology in our country are mainly caused by the differences of age, gender, occupation and education level. This article from the content of sports information network (U), the audience experience (V), sports information organization and communication network (X) and network sports information environment (Y) to construct four perspectives influence factors of sports information dissemination model evaluation index system based on. The authenticity of the impact factor of U information (U1), (U2) the importance of information timeliness, information (U3), the original information (U4), comprehensive information (U5) and the number of information (U6); recreational impact factor of V information (V1) professional, information transmission (V2) varieties, information (V3), information interaction (V4) and practical information (V5); the impact factor of X classification and organization of information (X1) analysis of communication and information (X2); the impact factor of Y website credibility (Y1), the number of Internet users to participate in the dissemination of information (Y2) and advertising (Y3). Based on this, the evaluation index system of sports information communication mode is shown in Fig. 6. Based on the above research, the evaluation index system of sports information communication mode can be used to evaluate the model effectively, and it can help to improve the model.

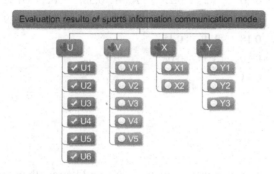

Fig. 6. Evaluation index system of sports information communication mode.

4.2 Determine the Weight of Evaluation Index

The determination of the index weight has the following three steps [19]:

STEP1: Constructs 22 comparison judgment matrix.

Each level of factors relative to the level of a single factor can be simplified to a series of single factor comparison of the judge.

STEP2: Will initialize the judgment matrix into a comprehensive judgment matrix.

First of all, the judgment matrix is initialized according to the calculation method of the geometric mean of the calculated 22 comparisons between each index value, transforming it into the final matrix, then, for the final matrix according to formula (1) calculation method is shown by synthetic judgement matrix.

$$\begin{cases} A(S) = \left[a(S)_{ij}\right]_{n \times n} \\ a_{ij} = k^* \sqrt{\coprod_{S=1}^{k^*} a(S)_{ij}} \end{cases}, S = 1, 2, \cdots, k; i, j = 1, 2, \cdots, n \qquad (1)$$

STEP3: Calculates the eigenvector corresponding to the largest eigenvalue of the comprehensive judgment matrix, and then the vector is normalized, which is the weight of each index. According to the 1–9 scale method proposed by T.L. Saaty, the 22-comparison judgment matrix is constructed, and the weight of each index is calculated according to the AHP model theory. The results are shown in Table 3:

Table 3. Calculation results of each index weight

Index	Weight	Index	Weight	Index	Weight	Index	Weight
U	0.38	V	0.29	X	0.21	Y	0.12
U1	0.25	V1	0.32	X1	0.42	Y1	0.44
U2	0.20	V2	0.25	X2	0.58	Y2	0.37
U3	0.16	V3	0.20			Y3	0.29
U4	0.18	V4	0.15				
U5	0.10	V5	0.08				
U6	0.11						

5 Conclusion

With the development of computer network, we should strengthen the comprehensive quality of the employees in order to serve the public and guide the masses. Government led, Internet users to participate in the construction of multimedia network sports communication environment, building a harmonious network of sports information dissemination environment. The development of sports website business model, expand the business partners, re positioning of the commercial operation of the sports website model, playing the old concept, improve the connotation of sports website operation mode.

References

1. Ibrahim, M.S., et al.: Information technology club management system. Acta Electr. Malaysi. **2**(2), 01–05 (2018)
2. Thiruchelvam, S., et al.: Development of humanitarian supply chain performance conceptual framework in creating resilient logistics network. Malays. J. Geosci. **2**(1), 30–33 (2018)
3. Huang, F.S.: Research on the business communication barriers and countermeasures in the network environment. Educ. Sustain. Soc. **1**(1), 14–16 (2018)
4. Hambrick, M.E.: Six degrees of information: using social network analysis to explore the spread of information within sport social networks. Advanced Fluorescence Reporters in Chemistry and Biology III: Applications in Sensing and Imaging, pp. 263–295 (2012)
5. Yao, H.J., Yuan, F.S.: Exponential stability control for networked systems with interval distribution time delays. Acta Inform. Malaysia **2**(1), 07–09 (2018)

6. Rastogi, S., Shivani Choudhary, S.: Face recognition by using neural network. Acta Inform. Malaysia **3**(2), 07–09 (2019)
7. Wang, B.: Features of sports information spreading on the internet. In: Qu, X., Yang, Y. (eds.) IBI 2011. CCIS, vol. 268, pp. 176–180. Springer, Heidelberg (2012). https://doi.org/10.1007/978-3-642-29087-9_26
8. Su, W.: Spread of Xiamen community sports under the national fitness. Contemp. Sports Technol. (2015)
9. Murwatiningsih, et al.: Creating the entrepreneurial networking through the business superiority and adaptability of business environment to improve the marketing performance. Malays. E-Commerce J. **3**(2), 36–40 (2019)
10. Shao, Y.T., et al.: Qualitative research on living status of older autistic families in Jiaxing city. China. Cult. Commun. Soc. J. **1**(2), 18–20 (2020)
11. Liu, C.L., et al.: A research on the present situation of spiritual belief of postgraduate. Cult. Commun. Soc. J. **3**(1), 5–7 (2020)
12. Yang, A.: Alienation and aesthetical salvation of everyday life in the house of the seven gables. Cult. Commun. Soc. J. **1**(1), 1–4 (2020)
13. Khan, A.H.: Preventing the transition of corona through science & technology: globally integrated society towards positive change. Cult. Commun. Soc. J. **1**(1), (2020)
14. Li, A.Q., Wang, X.F.: Features and trend of sport information communication in Chinese portal web sites. J. Jilin Inst. Phys. Educ. **26**(4), 01–04 (2010)
15. Zhang, Z.G., Xu, C.L., Zhao, R.X., Wang, Y.G., Liu, X.: Based on the analysis of the problems of network sports information communication. Contemp. Sports Technol. **02**(8), 87–88 (2012)
16. Wang, X.F.: Multi factor analysis on the influencing factors of sports information dissemination in China's large portal websites. Sports Cult. Guide **11**, 191–194 (2014)
17. Li, H.: The integration of communication mode and sport. News World **5**, 104–105 (2013)
18. Shan, F.J., Wang, Z.W., Xie, W.: The impact of China's Volleyball League attendance factors and countermeasure research. J. Harbin Inst. Phys. Educ. **24**(3), 92–94 (2006)
19. Lee, S., Walsh, P.: SWOT and AHP hybrid model for sport marketing outsourcing using a case of intercollegiate sport. Sport Manage. Review **14**(4), 361–369 (2011)

Analysis of Computer Graphics and Image Design in Visual Communication Design

Hongjie Chen(✉)

Information Engineering Department, Yantai Gold College, Zhaoyuan 265401, China

Abstract. In the process of visual communication design, graphics and images are the most basic elements. Computer graphics and images using the relevant hardware equipment and processing technology, software, etc. can further improve the level of visual communication. At present, graphics and image design has gradually become an important part of the visual communication design process, and also presents significant information characteristics. In order to study the related contents and main design methods of computer graphics and image design in visual communication design, this paper mainly analyzes the main contents, application advantages and design methods of computer graphics and image design. The results show that computer graphics and image design can consolidate the effect of visual communication design. Computer graphics and image design have significant advantages of visual communication design, such as easy to modify, product preview, unique image and so on. Computer graphics and image design are widely used in visual communication, such as text design, illustration design, packaging design, etc. Therefore, in the future, we should continue to maintain this advantage in the process of computer graphics and image design, so that the design methods of computer graphics and images can be further optimized and its practical value can be improved.

Keywords: Visual communication · Computer graphics and image design · Analysis

1 Visual Communication Design and Computer Graphics Design Related Content Analysis

In today's society, the relevant requirements for graphic image design are gradually improved, and taking computer as the design carrier adds new vitality to graphic image design [1, 2]. However, in the process of computer graphics and image design, many designers only focus on computer design technology, but ignore the graphic and image design art itself [3].

1.1 Basic Contents of Visual Communication and Computer Graphics Design

Overview of Visual Communication Design
Visual communication is an important way to spread information in human society [4–8]. People mainly use visual symbols to transmit, and ultimately achieve the purpose

E. Bisset Álvarez (Ed.): DIONE 2021, LNICST 378, pp. 400–405, 2021.
https://doi.org/10.1007/978-3-030-77417-2_32

of information sharing and communication. The main purpose of visual communication design is for people to appreciate, so the design content also has the role of visual communication. Visual communication design contains a wide range of related content, in addition to printing art, commercial design, advertising design, display and politics and other related content. Vision in the concept of visual design includes two aspects: first, a subjective feeling produced by human sensory organs under the action of light; second, it represents a special meaning of image design.

Computer Graphics and Image Design Content
The design process of computer graphics and images needs to be guided by specific design objectives and created in color information, graphic image information and other aspects by using relevant computer technology [9–11]. From the point of view of computer graphics design, the designer can fully use the computer graphics technology in the design process, so long as it can fully use the computer graphics technology in the design process. At present, there are many kinds of graphic and image design softwares commonly used by computer graphic and image designers. For different application fields, the image processing software used is also different. 3D Max, flash, Auto CAD, PS are common computer graphics and image processing software.

1.2 Advantages of Computer Graphics in Visual Communication Design

In the process of visual communication, computer graphics and image design has a very significant advantage, mainly in the following aspects: (1) integrating the visual communication effect into the computer graphics and image design process, we can re plan and calculate the graphic image design content from the perspective of visual communication, so that the designed graphics and images have unique characteristics. (2) Designers can express their own unique thinking in the process of computer graphics and image design. Even if there are design errors, they can also use software to make timely changes. There is almost no design waste phenomenon, not to mention the consumption of paper. (3) Computer image graphic design can preview the finished product. After previewing the design works, designers can change the previous design content with the help of computer software. At the same time, pay attention to the simplicity of the design, and compare the works before and after the modification on the same interface.

2 Analysis of the Application of Computer Graphics Design in Visual Communication Design

2.1 Application of Text Design Related Technology

In the process of computer graphics and image design, text belongs to the most simple visual communication factor. Computer related technology is used to design and process visual communication text. Photoshop is a common graphics and image design software, referred to as PS. The software for text design tools, including size, color instructions, can be customized according to user needs. For example, designers can use PS software to make diversified changes to the text, adjust the word spacing, so as to enhance the effect of visual communication. From the perspective of visual communication, computer graphics design can improve the performance of the text, consolidate the artistic effect of the text, and avoid being too monotonous.

2.2 Application of Illustration Design Related Technology

The image resources of visual communication include illustration, illustration and drawing, etc. if the picture is not processed, it will be dull, monotonous and inactive. In order to further strengthen the application effect of illustrations in visual communication, the original state of pictures is changed by means of computer graphics and image design method. For example, in the illustration design, the illustration designer needs to use the computer graphics and image design software to color the illustrations, use the pencil to draw the basic structure of the illustration, and then combine with the computer graphics and image software to fill the colors in the drawing.

2.3 Application of Packaging Design Related Technologies

In our daily life, packaging design is the most common visual communication factor. Various packaging designs can be seen everywhere in shops and shopping malls, such as clothing, video packaging, etc. consumers generally see packaging design first. The attraction of packaging design itself will stimulate consumers' interest in consumption. On the shelves of shopping malls, there are a variety of goods that have been designed for packaging, and many of them need to use computer graphics and image software to deploy images.

3 Analysis of Computer Graphics Design Technology in Visual Communication Design

3.1 Hardware Equipment Required for Computer Graphics and Image Design

Output Device
The output device includes display and output devices, such as display and liquid crystal display, which are respectively represented by CRT and LCD; the printer includes color printer, color light printer, plotter, etc. The display card is a necessary device in the process of computer graphics and image design and processing. It processes the image into the format that the display can display, and displays it on the screen to form the image finally. The display card selected in the process of graphic image processing must be equipped with special memory and corresponding graphics and image processing acceleration chip, so that it has a certain image preprocessing ability, and is not just a simple display interface device.

Graphics and image display card generally consists of different parts, such as graphics chip, RAM DAC, interface and so on. Therefore, the processing of graphics and image data must go through the following steps, and finally the designed graphics image will be presented on the display screen:

First, the line enters the graphics card chip and sends the image to the graphics card chip to carry out the corresponding processing; second, the video chipset enters the video RAM, which is the process of displaying and storing the image processed by the graphics card chip. Third, fresh run needs to enter the DAC, display memory, take out data, and then to ram DAC to carry out data conversion related work. Fourth, the DAC enters into the display and displays the converted pictures directly on the display screen.

Acquisition Equipment

The acquisition equipment includes scanner, camera, camera, etc. The types of scanners include flat-panel scanner, negative scanner and roller scanner. At present, digital cameras and cameras are commonly used. Digital camera is a kind of special camera which can take pictures, convert the captured scenery directly through internal processing, and finally store the pictures in digital format. Compared with ordinary cameras, digital cameras do not need to use film, but use removable or fixed semiconductor memory to save the acquired images.

3.2   Analysis of Computer Graphics and Image Design Technology

Image Representation and Storage Technology

The main function of image acquisition equipment is to convert the optical signal in the real world into digital signal which can be recognized by computer. The image output device is the optical signal that can be recognized by human eyes by converting these digital signals. Palette is an important part of image representation and storage technology. The palette uses three basic colors: blue, red, and green, each color is divided into 25 to 255 different categories. Different color values need to have hue, brightness, saturation and other attributes.

Digital image can be divided into dot matrix image and vector image. The so-called dot matrix image is an image composed of points. The point itself has different colors, and its specific color is determined by the gray value of three primary colors. The common formats include PSB format, BMP, tiff, PEG/PG, EPS format, GIF format, PDF format, PNG, etc. Vector graphics are also known as drawing images or face-to-face images. Lines are generally used to depict graphics, the elements of vector graph are generally lines, polygons, circles and arcs. Common vector image formats include: swf format, WMF format, SVG format, EMF format and so on.

The Related Technology of Plane Digital Image Processing

If you want to process a plane number, it is usually applied in two cases. One is the image that processes the image, that is, the whole process of image to image. The other is to convert an image to a non image. Digital image processing involves computer digital technology, so the range of images involved must be far beyond the traditional sense of images, including invisible physical images, digital images and digital images.

The common design methods in the process of computer graphics image design include the following: first, sharpening processing method, to increase the clarity of object contour in the image, the image needs to be sharpened. In the process of processing, the gray value of the point needs to be increased. Firstly, the gray value of the processing point and the value of the surrounding 8 points are processed, and then the result and the gray level of the original image are processed. The value is added, which gives the gray value of the point to be processed. Second, flat processing is to process the flower points in the image. This processing method needs to sort the points to be processed and the eight points around them, and then the value of the middle position is taken as the gray value of the point. Third, compression change processing. The purpose of image compression is to make the subjective visual feeling and data in the image, and finally realize the purpose of compressing the image by changing the image description mode. Image compression includes lossy compression and lossless compression, in which the lossy compression method ignores the visual insensitive parts, so as to improve the compression rate. The compression rate of lossless compression is usually not high compared with the former. The following are the processes of lossy compression and decompression:

Compressed image flow: original image - DCT transformation - divided by quantization coefficient - rounding

Decompress image flow: compress image - multiply quantization coefficient - DCT inverse transform - rounding

4   Conclusion

With the continuous development of computer technology, graphics and image design methods based on computer technology can provide better visual experience for people. In the process of visual communication design, computer graphics and image design technology with its own unique methods and characteristics shows greater advantages, not only to improve the effect of visual communication, but also to enhance the practical value of graphic image design, with obvious feasibility. In the process of visual communication design, computer graphics and image design should give full play to the advantages of computer technology, so that its advantages in all aspects of visual communication can be maintained, and finally the expression level of graphics and images can be improved.

References

1. Wen, J.Q.: Computer graphics and image design and visual communication design. Electron. Technol. Softw. Eng. (07), 85 (2016)
2. Wang, Y.S.: On computer graphics and image design and visual communication design. Digital Technol. Appl. (11), 160 + 162 (2015)
3. Hu, B., Chen, X.D., Liu, X.Y.: Research on computer graphics and image design and visual communication design. Comput. CD Softw. Appl. 17(18), 213 + 215 (2014)
4. Zhu, J.X., Wang, X.Y., Chen, M.C., Wu, P., Kim, M.J.: Integration of BIM and GIS: IFC geometry transformation to shapefile using enhanced open-source approach. Autom. Constr. 106, (2019)
5. Liu, Y.X., Yang, C.N., Sun, Q.D., Chen, Y.C.: (k, n) scalable secret image sharing with multiple decoding options. J. Intell. Fuzzy Syst. 38(1), 219–228 (2020)
6. Yutaka, W.: Three-dimensional center of gravity detections for preventing rollover accidents of trailer trucks hauling containers. Open J. Mech. Eng. 1(1), 11–14 (2017)

7. Liu, Z.Y., Sun, Q.Q., Li, J.: A matrix mining method with FP-tree for generation of frequent patterns. Open J. Mech. Eng. 1(1), 01–03 (2017)
8. Wang, C., Zhao, W., Zhang, K.: The strategy of process innovation based on AHP and TRIZ. Open J. Mech. Eng. 1(1), 08–10 (2017)
9. Chu, Z.Q., Jiao, Q., Teng, L.Z.: Aroma characteristics of *Osmanthus fragrans* leaves. Open J. Chem. Eng. 2(1), 23–31 (2019)
10. Wu, Z., Liu, Y.N., Jia, X.X.: A novel hierarchical secret image sharing scheme with multi-group joint management. Mathematics 8(3), 448 (2020)
11. Wang, L.S., Chen, J.T., Ni, C.Y.: Molecules and functions of rosewood: *Dalbergia odorifera*. Open J. Chem. Eng. 1(1), 15–20 (2018)

Relationship Between External Search Breadth and Process Innovation Performance Under the Background of Big Data

Yanxia Ni[1($\boxtimes$)] and Jiangman Yu[2]

[1] Jianghan University Business School, Wuhan 430056, China
[2] Hubei University of Technology, Wuhan 432200, China

Abstract. With the advent of the era of big data, this article selects the 2014 World Bank survey data on Indian private companies from the perspective of knowledge search channels, and analyzes the relationship between external search breadth and enterprise process innovation performance from the perspective of organizational learning theory, and it also explores the moderate effect of the attention allocation process, namely, senior management's tenure and financing constraint. The research finds that: External search breadth and enterprise process innovation performance have an inverted U-shaped relationship. Senior management's tenure plays a positive moderate role. In addition, in order to ensure the correctness and reliability of the selected model, the paper tests the applicability and endogenous problems of the inverted U-shaped model of the sample. The research conclusions provide a theoretical reference for companies to effectively allocate attention and improve their ability to benefit from external search.

Keywords: Big data · External search breadth · Process innovation · Attention-based view · Organizational learning theory

1 Introduction

In the current era of increasing data resources, competition among enterprises increasingly depends on whether enterprises can create knowledge and commercialized innovation achievements in a fast and effective way. However, under the impact of big data, how to accurately, efficiently and conveniently analyze the impact of external knowledge on innovation results has become a realistic problem facing organizations. So more and more companies are implementing external search strategies to actively identify and integrate external knowledge from sources such as partners, competitors, consultants, business laboratories, education institutions and industry associations [1]. Despite the widespread recognition of the importance of external search, extensive knowledge searches often fail to achieve the desired results. Predecessors' research is also full of contradictions about how external search breadth affects corporate innovation performance [2]. This raises a very important theoretical question for us: When can companies realize the benefits

© ICST Institute for Computer Sciences, Social Informatics and Telecommunications Engineering 2021
Published by Springer Nature Switzerland AG 2021. All Rights Reserved
E. Bisset Álvarez (Ed.): DIONE 2021, LNICST 378, pp. 406–417, 2021.
https://doi.org/10.1007/978-3-030-77417-2_33

of innovation from external search strategies? Why can some companies benefit more from external search strategies, while others cannot?

The external search strategy is an activity way for organizations to solve problems, involving the creation and restructuring of technological ideas. External search strategy can be characterized by the breadth and depth of knowledge search [3]. The external search breadth represents the number of different organizational sources from which the enterprise seeks knowledge, while the external search depth is the extent to which the enterprise searches knowledge from each external topic. Because of costs and risks, it is a difficult task for companies to obtain the desired innovations from a wide range of knowledge searches. Noting these issues, more and more research in recent years has begun to focus on when companies can overcome these obstacles and improve the efficiency of external search strategies. Although attention resource limitation is widely regarded as a key factor that hinders enterprises from effectively acquiring, absorbing and integrating external knowledge, how to weaken the negative influence of attention resource on external search strategy is rarely discussed. In addition, most of the existing research on the relationship between external search and innovation focuses on product innovation. But research on external search and process innovation is limited with customer-oriented product innovation process innovation can reconfigure the value chain, thus reducing production costs, improving product quality and manufacturing flexibility. Therefore, process innovation involves more tacit knowledge, is not easy to be transferred to imitation, and is more likely to bring competitive advantage to the organization. Therefore, it is particularly important to study the relationship between external search and process innovation.

Therefore, this paper analyzes the relationship between external search breadth and process innovation from the perspective of organizational learning theory and attention-based view, thus expanding our understanding of relationship between external search, attention distribution and enterprise process innovation performance. First, this paper analyzes the relationship between external search breadth and enterprise process innovation performance. Second, this paper examines the moderating effect of factors affecting attention distribution. Finally, to ensure the robustness of empirical results, this paper uses both the ordered counting model and the Poisson distribution model, and examines the endogeneity of the sample.

2   Literature Review and Research Hypothesis

The Organization for Economic Co-operation and Development (OECD) defines process innovation as "a new or significantly improved way of producing or delivering, including major changes in technology, equipment or software" [4]. In most cases, the main purpose of product innovation is to develop new products to meet the needs of customers. The main purpose of process innovation is to shorten lead times, reduce operating costs, and increase operational flexibility. Compared with product innovation, process innovation is more complex. And it consists of tacit knowledge related to organizational system knowledge. The development of process innovation is also slower, requiring more trial and error and experience learning to achieve its hidden value.

2.1   External Search and Process Innovation

Although the advantages brought by the breadth of knowledge search are obvious, there is no agreement on how the external search breadth affects the process innovation performance. First, due to the limited ability of enterprises to absorb knowledge, the effective understanding and rational integration and application of diverse external knowledge may make external search negatively affect the process innovation performance of enterprises. Second, with the increase of the external search breadth of enterprises, the difference of knowledge is gradually increasing, which increases the difficulty of knowledge exchange and combination among enterprises. The cost of identifying, absorbing and integrating external knowledge increases, which weakens the positive effect. Third, because the manager's attention resources are limited, the manager can only allocate limited attention resources, and it is impossible to make full use of all external knowledge. Finally, tacit knowledge, as an important strategic resource owned by the organization, is difficult to transfer or replicate. Therefore, the tacit knowledge obtained in the external search must be accompanied by the loss of substantive knowledge, making the cost of external search exceed the benefits it brings.

Based on this, this paper proposes the following hypothesis:

H1: The relationship between external search breadth and process innovation performance is inverted U-shaped.

2.2   The Moderating Effect of Attention

From the previous analysis, it is known that the limited attention of managers is one of the key antecedents that lead to the negative effect of external search breadth. Participants and organizational resources are important factors influencing the organization's attention distribution, and also constitute an important part of the attentional structure distribution [5]. This paper selects the executive tenure and financing constraints as the proxy variables of the participants and organizational resources.

Participants: The Moderating Effect of Executive Tenure

The executive tenure is one of the important factors affecting the decision-making behavior of managers, which may affect the acquisition and absorption of external knowledge.

First, managers with different tenures have different reserves of knowledge about enterprises and external environment, which affects the selection and elimination of information required by managers in enterprise process innovation. Long-term managers have a deeper understanding of corporate culture and internal resources, and have a deeper understanding of the types of external knowledge required by enterprises. Therefore, they can identify the most valuable resources for enterprises and get them more quickly and effectively. Managers with short term tenure are likely to receive a wide range of external knowledge, make unfavorable decisions for the company, cause waste of enterprise resources, and delay the best time for process innovation.

Secondly, managers with different tenures have different concerns. Managers with long term tenure pay more attention to the development of long-term goals of the company due to the stability of positions and salary, so they will pay more attention to the use of knowledge that is conducive to enterprise innovation. The short-term manager's concerns are mostly in the management of labor relations and external relations, and the supervision and assessment of managers with short term tenure are more strictly, and the threat of positions is even more serious. They will focus more on improving the short-term performance of the company and ignoring the long-term value of the company.

Based on this, this paper proposes the following hypothesis:

H2: Executive tenure positively moderates the relationship between external search breadth and process innovation performance.

Organizational Resources: The Moderating Effect of Financing Constraints
We believe that the interaction between financing constraints and external search breadth will have an impact on enterprise processes innovation performance. The increase of financing constraints makes the financial capital of enterprises limited, which encourages enterprises to pay more attention to the use of internal knowledge, reduce the participation of enterprises in other cognitive activities, and reduce the efficiency of enterprises in identifying, absorbing, integrating and using external knowledge. Compared with internal knowledge, the search and application of external knowledge is more risky, more uncertain, and more costly due to the existence of opportunism and speculation in the transaction, unpredictable factors in the environment, and so on. In financial markets with high financing constraints, due to the restriction of funds, enterprises will pay more attention to the use of existing resources and knowledge within the enterprise.

Based on the above analysis, this paper proposes the following assumptions:

H3: Financing constraints negatively moderate the relationship between the external search breadth of firms and process innovation performance.

3   Research Design

3.1   Research Sample

The data is taken from the World Bank's 2014 survey data on private enterprises in India. The purpose of the survey is to understand the business environment, the impact or restrictions of the business environment on private enterprises, and the innovation of private enterprises. The survey data consists of two parts. The first part mainly investigates the investment environment and financial data of the company. The second part is the innovation follow-up survey data. In order to ensure the validity and robustness of the empirical results, this paper deletes the sample of the service industry after matching the two parts of the data according to the enterprise code, only retains the sample of the manufacturing industry, and deletes the missing value of the corresponding variable, and finally The sample have 395 companies in 19 sub-sectors.

3.2   Indicator Selection and Variable Definition

Independent Variable. External Search Breadth (Procb) is the cumulative type of organization that has a relationship with the company in process innovation [6, 7].

Dependent Variable. A large number of scholars have divided innovation into two categories: new to the market and new to the world [8]. The proxy variable of the process innovation performance (Procin) is the main process innovation level in the enterprise. The question in the second part of the questionnaire is "whether the main process innovation of the enterprise (1) is new to the local market; (2) The domestic market is new; (3) it is new to the international market."

Control Variable. The control variables selected in this paper include company size (Size), firm age (Age), human capital (Human), R&D (Research and Development), whether the company is located in the main trading city (City), and government control (Gover).

3.3 Model Estimation

Since the variable process innovation (Procin) is an ordered discrete non-negative integer, this paper uses the ordered counting model (Oprobit) for the estimation. In addition, in order to ensure the robustness of the estimation results, we also examine the endogeneity of the sample and use the basic model Poisson as the robustness test of the sample. Since the main effect assumed in this paper is an inverted U-shaped relationship, and the hypothesis of regulating the relationship of this curve is proposed, this paper adopts the three-step method of confirming the U-type relationship proposed by Haans et al. [9]. First of all, as shown in Eq. (1), β_2 must be significantly negative. Second, the slope of the two extremes of the range of independent variables must be sufficiently steep. If X_l is used to represent the low end portion of the X value and X_h is the high end portion, then the second condition should be guaranteed: $\beta_1 + 2\beta_2 X_H$ is significantly negative, and $\beta_1 + 2\beta_2 X_L$ is significantly positive. Third, the turning point of the U-shaped curve must be within the variation range of the sample data of X. In the calculation of the confidence interval of the independent variables, this paper uses the Fieller algorithm to calculate the confidence interval of 90% of the independent variables. In addition, in order to avoid the model being S-type, it is also necessary to guarantee the cubic coefficient γ_3 of the independent variable in Eq. (2) non-significant. Therefore, the final model of this paper is shown in (3) and (4)

$$Y = \beta_0 + \beta_1 X + \beta_2 X^2 \tag{1}$$

$$Y = \gamma_0 + \gamma_1 X + \gamma_2 X^2 + \gamma_3 X^3 \tag{2}$$

$$\Pr ocin_i = a_i + b_i \times \Pr ocb_i + c_i \times \Pr ocb_i^2 + d_i \times CV + \varepsilon_i \tag{3}$$

$$\Pr ocin_i = a_i + b_i \times \Pr ocb_i + c_i \times \Pr ocb_i^2 + d_i \times MV \times \Pr ocb_i$$
$$+ e_i \times MV \times \Pr ocb_i^2 + f_i \times MV + \varepsilon_i \tag{4}$$

Among them, the dependent variable Procin$_i$ is the process innovation performance of the enterprise. The independent variable Procb$_i$ is the external search breadth. CV is the control variable. MV is the adjustment variable. ε_i is the error term.

In addition, there may be endogenous problems in this paper. In order to solve this problem, this paper uses the Conditional Mixed Process proposed by Roodman [10] to test and deal with the endogeneity in the ordered counting mode. The choice of instrumental variables must be an exogenous variable that has no direct connection to the innovation performance of the enterprise process, but can strongly influence the breadth of the external search. Therefore, the "degree of resistance to input and supply" directly affects the external search breadth of the enterprise, but it does not affect the process innovation performance of the enterprise. Therefore, this paper takes "the degree of obstruction of input and supply" as a tool variable for the external search breadth.

4 Empirical Results and Analysis

4.1 Descriptive Statistics

Table 1 reports the descriptive statistics of each variable and the Pearson correlation coefficient matrix. It can be seen that the external search breadth and process innovation performance are not significantly positively correlated, and further tests are needed [11]. In addition, the results of multicollinearity analysis showed that the variance expansion factor (VIF) did not exceed 2, indicating that there is basically no multicollinearity problem between variables.

Table 1. Descriptive statistics of variables and Pearson correlation coefficient matrix

Variable	Procin	Procb	R&D	Age	Human	City	Gover	Tenure	Finance
Procin	1								
Procb	0.01	1							
R&D	0.16***	−0.02	1						
Age	0.11**	0.08	0.14***	1					
Size	0.16***	−0.04	0.19***	0.18***					
Human	0.09*	−0.20***	0.05	−0.02	1				
City	−0.02	−0.01	0	−0.05	0.18***	1			
Gover	−0.02	0.06	0.01	0.01	0.05	0.06	1		
Tenure	0.02	0.01	0.09*	0.38***	0.06	−0.01	0.11**	1	
Finance	0.06	0.06	−0.07	−0.04	0.09*	0.15***	−0.01	−0.10**	1
Mean	0.53	0.1	124.99	22.09	0.67	0.84	4.1	15.69	0.98
St.d	0.76	0.39	798.33	14.01	0.25	0.37	14.22	10.26	1.07
VIF		1.06	1.06	1.23	1.1	1.06	1.02	1.2	1.07
	—								

Note: The observed value is n = 395; The superscripts ***, **, and * represent statistical significance of 1%, 5%, and 10% respectively.

4.2 Preliminary Regression Results and Endogenous Analysis

Table 2 reports the preliminary regression results using the ordered counting model.

Table 2. Preliminary regression results

Variable	1	2	3	4	5
Procb	0.640**	0.720**	0.596	0.250	−2.872***
	(1.99)	(2.41)	(1.58)	(0.36)	(−10.39)
$Procb^2$	−0.265*	−0.412***	−0.232	0.237	−0.023
	(−1.89)	(−2.64)	(−1.28)	(0.37)	(−0.27)
$Procb^3$				−0.142	
				(−1.19)	
R&D	+ 0.000***	+ 0.000***	+ 0.000***	+ 0.000***	−0.000
	(3.18)	(3.09)	(3.15)	(3.18)	(−1.17)
Firmage	0.010**	0.011**	0.010**	0.010**	0.002
	(2.11)	(2.13)	(2.04)	(2.09)	(0.38)
Size	0.035	0.034	0.036	0.035	−0.024
	(0.57)	(0.55)	(0.57)	(0.56)	(−0.50)
Human	0.519*	0.538*	0.507	0.515*	−0.708**
	(1.66)	(1.69)	(1.58)	(1.65)	(−2.35)
City	0.068	0.096	0.073	0.066	0.119
	(0.33)	(0.47)	(0.35)	(0.32)	(0.64)
Gover	−0.006	−0.008*	−0.006	−0.006	0.003
	(−1.41)	(−1.90)	(−1.48)	(−1.40)	(0.79)
Procb*Tenure		0.117**			
		(2.53)			
$Procb^2$*Tenure		−0.066**			
		(−2.37)			
Tenure		−0.005			
		(−0.71)			
Procb*Finance			−0.120		
			(−0.33)		
$Procb^2$*Finance			0.035		
			(0.21)		
Finance			−0.002		
			(−0.03)		

(continued)

Table 2. (*continued*)

Variable	1	2	3	4	5
atanhrho_12					3.263
_cons					(0.91)
Wald chi2	826.32***	854.27***	831.87***	1285.29***	14135.03***
Pseudo R^2	0.117	0.122	0.117	0.117	
N	395	395	395	395	395

Note: ***, **, and * represent statistical significance of 1%, 5%, and 10%. The regression coefficients in parentheses () are z values. The regression coefficients in parentheses [] are p value.

Model 1 Table 2 is the regression result of the main effect. Models 2 and 3 add the moderate variables—the executive tenure and the financing constraints. The model 4 adds the cubic of the independent variable to test whether the model is S-type. The model 5 uses the conditional mixing process to test the endogeneity of the sample [12]. It can be seen from the model 5 in Table 2 that the auxiliary estimation parameter atanhrho_12 in the estimation result is not significantly different from 0, indicating that the external search breadth is an exogenous variable, and there is no endogencity problem in the sample. Therefore, in order to ensure the robustness of the estimation results, the estimation results of Models 1–4 should be used.

Model 1 in Table 2 shows that the external search breadth at 5% statistical level significantly promotes process innovation performance, and the square of the external search breadth significantly negatively affects process innovation performance at the 10% statistical level. H1 is supported. It can be seen from Model 2 that the interaction term between the executive term and the external search breadth is significantly positive at the 1% statistical level, and the interaction term between the executive term and the external search breadth squared term is significantly negative at the 5% statistical level, indicating that the executive tenure has a significant positive moderating effect. H2 is supported. It can be seen from Model 3 that the interaction term between the financing constraint and the external search breadth is not significantly negative, and the interaction term between the financing constraint and the square of the external search breadth is not significantly positive. H3 is not supported.

5 Inverted U-shaped Test

It can be seen from the model 4 in Table 2 that after adding the cubic of the outer search breadth of the independent variable, the cubic coefficient is not significant, indicating that there is no S-type relationship between the external search breadth and the process innovation performance. Then a three-step inverted U-shaped test is performed. The specific results are shown in Table 3.

The models (1) and (2) in Table 3 correspond to the inverted u-type test of the models 1 and 2 in Table 2, that is, the inverted u-type test for the hypotheses H1 and H2.It can be seen that the inverted U-applicability test in Model 1 is significant at the statistical level of 10%, and the low-end slope of the independent variable is 0.278, which is significantly positive at the 10% statistical level, and the high-end slope is −0.567, which is significant at the 5% statistical level. The turning point 1.371 is within the 90% confidence interval of the independent variable. Similarly, the inverse u-type applicability test in Model 2 is significant at the 5% statistical level, the low-end slope of the independent variable is 0.368, which is significantly positive at the 5% statistical level, and the high-end slope is −1.275 at the 1% statistical level. The turning point of 0.895 is within the 90% confidence interval of the independent variable.

Table 3. Inverted u-type test

variables	(1)	(2)
Procb	0.640^{**}	0.720^{**}
	(1.99)	(2.41)
$Procb^2$	-0.265^*	-0.412^{***}
	−1.89)	(−2.64)
X_L slope	0.278	0.368
	$(1.645)^*$	$(2.017)^{**}$
X_H slope	−0.567	−1.275
	$(-1.846)^{**}$	$(-2.384)^{***}$
Appropriate U test	1.64^*	2.02^{**}
	[0.050]	[0.022]
Turning Point	1.317	0.895
90% confidence interval	[−0.019, 2.300]	[0.376, 1.349]

Note: The same as Table 2

To further prove whether the moderating effect of the manager's tenure on the external search breadth and the process innovation performance is as expected, we draw the moderating effect diagram, and the result is shown in Fig. 1. It can be seen from Fig. 1 that when the senior management term is high, the relationship between the external search breadth and the process innovation performance is steep. When the senior management term is low, the relationship between the external search breadth and the process innovation performance is relatively flat. So H2 is further supported.

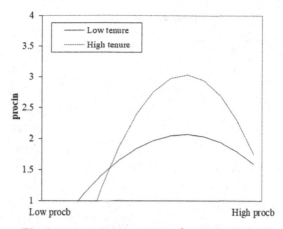

Fig. 1. The moderating effect of executive tenure

6 Robustness Test

This paper also uses the basic model Poisson distribution model to regress the sample, and further eliminates the interference of the sample heteroscedasticity. The report results are shown in Table 4. Model 6 is the regression result of the main effect. Models 7 and 8 are the regression results after adding the moderate variables—the executive term and the financing constraint. And the model 9 is the regression result after adding the cubic of the external search breadth of the independent variable. From Model 6 we can see that H1 is further supported. From Model 7 we can see that H2 is further supported. It can be seen that the inversion u-type suitability tests of Model 6 and Model 7 are robust. From Model 8 we can see that H3 is not supported.

Table 4. Poisson regression results.

Variable	6	7	8	9
Procb	0.637*	0.697**	0.54	−0.523
	(1.8)	(2.29)	(0.88)	(−0.01)
Procb^2	−0.298*	−0.456***	−0.23	1.332
	(−1.69)	(−2.64)	(−0.66)	−0.01
Procb^3				−0.509
				(−0.01)
Procb*Tenure		0.120**		
		−2.44		
Procb^2*Tenure		−0.073**		
		(−2.40)		

(*continued*)

Table 4. (*continued*)

Variable	6	7	8	9
Procb*Finance			−0.192	
			(−0.34)	
Procb^2*Finance			0.069	
			(0.24)	
LR chi2	85.85***	89.13***	86.09***	86.02***
Pseudo R2	0.112	0.116	0.112	0.112
N	395	395	395	395
X_L slope	0.285	0.373		
	(1.467)*	(1.606)*		
X_H slope	−0.570	−1.248		
	(−1.340)*	(−1.648)*		
Appropriate U test	1.34*	1.61*		
	[0.091]	[0.055]		
Turning point	1.330	0.921		
90% confidence interval	[−0.019, 2.300]	[0.376, 1.349]		

Note: The same as Table 2

7 Conclusions

With the advent of the era of big data, this article starts from the perspective of knowledge search channels, this paper selects the 2014 World Bank survey data of private enterprises in India, based on the integration perspective of organizational learning theory and attentional view, examines the inverted u-type relationship between external search breadth and process innovation performance, and the moderating effect of attention distribution process. We found that, initially, the external search breadth of the enterprise positively affects the process innovation performance. When the external search breadth reaches a certain level, it has a negative impact on the process innovation performance. The enterprise executive term positively moderates the external search breadth and the process innovation performance.

Through this research, this paper draws the following enlightenments: 1) Enterprises should follow the market trend, learn and absorb external knowledge, and reasonably control the external search breadth. Enterprises should appropriately control the external search breadth, effectively avoiding the negative effects brought by extensive knowledge search effect. 2) Enterprises should constantly improve relevant management systems and extend the term of talented executives. As an important part of human resources, managers, especially executives, have become increasingly prominent in the decision-making of modern organizations. 3) Enterprises should actively explore and discover other factor structures that affect the attention distribution of enterprises, find external knowledge that matches existing knowledge in enterprise process innovation, rationally allocate attention, rationally manage external relations, and improve the benefits brought by external knowledge.

References

1. Katila, R., Ahuja, G.: Something old, something new: a longitudinal study of search behavior and new product introduction. Acad. Manag. J. **45**(6), 1183–1194 (2002)
2. Laursen, K., Salter, A.J.: The paradox of openness: appropriability, external search and collaboration. Res. Policy **43**(5), 867–878 (2014)
3. Laursen, K., Salter, A.: Open for innovation: the role of openness in explaining innovation performance among U.K. manufacturing firms. Strateg. Manag. J. **27**(2), 131–150 (2006)
4. OECD: Innovation Strategy: Defining Innovation (2015). http://www.oecd.org/site/innovatio nstrategy/defininginnovation.htm
5. Ocasio, W.: Attention to attention. Organ. Sci. **22**(5), 1286–1296 (2001)
6. Terjesen, S., Patel, P.C.: In search of process innovations: the role of search depth, search breadth, and the industry environment. J. Manag. **43**(5), 1421–1446 (2017)
7. Stevens, R., et al.: Attention allocation to multiple goals: the case of for-profit social enterprises. Strateg. Manag. J. **36**(7), 1006–1016 (2015)
8. Adner, R., Levinthal, D.: Demand heterogeneity and technology evolution: implications for product and process innovation. Manag. Sci. **47**(5), 611–628 (2001)
9. Haans, R.F., Pieters, C., He, Z.L.: Thinking about U: theorizing and testing U- and inverted U-shaped relationships in strategy research. Strateg. Manag. J. **37**(7), 1177–1195 (2016)
10. Roodman, D.: Fitting fully observed recursive mixed-process models with Cmp. Stata J. **11**, 159–206 (2011)
11. Zhu, J.X., Wu, P., Chen, M.C., Kim, M.J., Wang, X.Y., Fang, T.C.: Automatically processing IFC clipping representation for BIM and GIS integration at the process level. Appl. Sci. **10**(6), 2009 (2020)
12. Zhang, T., et al.: Phosphorus recovered from digestate by hydrothermal processes with struvite crystallization and its potential as a fertilizer. Sci. Total Environ. **698**, 134–240 (2020)

Education in Online Environments

Construction of Bilingual Teaching Hierarchy Model Based on Flipping Classroom Concept

Rongle Yao[✉]

English department, School of Public Fundamentals, Jiangsu Vocational College of Medicine, Yancheng 224006, China

Abstract. Flipping classroom is a brand-new teaching mode, in which teachers and students interact and communicate to jointly complete troubleshooting and knowledge construction . The emphasis on personalized autonomous learning, diversified cooperative exploration and open communication and interaction in the reverse classroom have pointed out a new direction for deepening the reform of English teaching. Turning over the classroom teaching mode is simply a change to the teaching mode, and teachers play a more guiding role in a complete teaching activity. Corpus provides a wide range of learning resources for bilingual teaching, while flipped classroom provides a new teaching mode for bilingual teaching. The construction and application of this model will further promote the deep integration of information technology and English curriculum, and promote the transformation of College English teaching concept and the reform of English curriculum system. College teachers should play the positive role of technology as much as possible to improve the quality and efficiency of teaching and promote the reform and development of teaching mode.

Keywords: Turn over the classroom · Teaching methods · College English

1 Introduction

With the rapid development of information society, many fields have changed with each passing day. In the field of education, information technology is also quietly changing the teaching methods and methods. Educators are constantly trying to optimize the teaching effect through the use of information technology [1]. From the perspective of application results, most flip classroom teaching practices have achieved good teaching results. Flipping the classroom is to use advanced video and network technology to move students' learning forward, complete the absorption of knowledge, and complete the internalization of knowledge in the classroom, which is another major change in teaching philosophy and teaching methods in the electronic era [2]. Turning over the classroom teaching mode is simply a change to the teaching mode, and teachers play a more guiding role in a complete teaching activity [3]. Under the background of information technology, new educational concepts and teaching modes are sought to provide powerful external conditions for the smooth progress of educational reform. The relatively relaxed learning

E. Bisset Álvarez (Ed.): DIONE 2021, LNICST 378, pp. 421–425, 2021.
https://doi.org/10.1007/978-3-030-77417-2_34

environment in colleges and universities has attracted more and more teachers in colleges and universities. Teachers and students in colleges and universities are more likely to meet the requirements of turning over classroom teaching, and learning is no longer limited to classrooms [4]. In higher education institutions, there are sufficient research funds and research teams, so in the process of education informatization, universities should bear more responsibilities. The new education and teaching mode represented by the flipped classroom is gradually being carried out and explored in colleges and universities. The teaching evaluation method of the flipped classroom and the ability of teachers to implement the flipped classroom have become issues worthy of discussion [5].

2 Feasibility Analysis of Flip Classroom Bilingual Teaching

Flipping the classroom can give full play to the main role of students' learning and significantly enhance the communication and interaction between teachers and students as well as between students. In the personalized learning under the reverse classroom teaching mode, students become autonomous learners. They can control the choice of learning time and learning place, and they can also control the learning content. Students also hope to further consolidate and improve their interpretation skills through the Internet, and hope to receive personalized guidance from teachers through online communication. Students are the leading role in the whole learning process and are no longer passive recipients of knowledge in traditional classes. Students in the classroom through group learning and cooperative learning to complete the understanding and absorption of the knowledge learned. Reducing teachers' teaching time and leaving students more time for learning activities in the classroom is another core feature of turning over the classroom [6]. The implementation of flip class not only needs the support of information technology, but also needs micro class as a carrier. Turning over the classroom reverses the process of imparting and internalizing knowledge in traditional teaching. The internalization of knowledge is put into the class and completed under the guidance of teachers and the cooperation and exchange of students. Flipping the classroom greatly improves the interaction between teachers and students and between students and students in the classroom. Because students conduct a certain degree of in-depth learning of the courses to be learned through teaching videos, in the classroom are mainly student questions, teacher answers and discussion and communication between students.

Flipping classrooms have flipped over the teaching form and subject content teaching, and the role of teachers has also changed. When teachers evaluate, the interactivity in the classroom becomes more effective. According to the teacher's evaluation feedback, students will learn about their learning more objectively. In this case, teachers need to provide students with the necessary learning resources, design enlightening questions and targeted practice questions for students. Students can complete the acquisition of knowledge before class to prepare for the absorption and internalization of classroom knowledge [7]. As a new teaching mode in the teaching environment of information technology, flipping the classroom emphasizes the application of information technology in teaching and the information literacy of teachers. In the flipped classroom model, the content taught in the original classroom is completed before the class through network technology [8–10]. On the basis of not reducing the transfer of basic knowledge, enhance the interaction between teachers and students in the classroom. Using interpreting corpus related software, teachers and students can obtain statistical data more intuitively, so as to objectively analyze the actual use of the language [11, 12]. Teachers 'attitudes towards flipped classroom teaching and teacher values also affect teachers' ability to implement flipped classrooms to a certain extent.

3 Construction of Hierarchical Model for Bilingual Teaching

Teachers should focus on students as much as possible when designing flip classroom teaching. Through a variety of effective teaching activities to help learners understand and internalize knowledge, improve the effect of classroom learning [13–15]. Teachers should draw a learning path map according to the actual needs of students and the requirements of the syllabus, and make a series of high-quality micro courses based on the path map. In the classroom, teachers should adopt various teaching methods to carefully design classroom activities and mobilize students' participation. As a new teaching method, the core of flip class is the internalization of knowledge in class [16]. The application of the concept of turning over the classroom in English teaching can break through various disadvantages of traditional English teaching, greatly promote the interaction and cooperation between teachers and students, and thus improve the quality of English teaching. In order to make more effective use of multimedia, students should internalize what they have learned and truly realize the individuation of their learning. There is a need for a more perfect and scientific college English teaching model.

Flipping classroom has reversed the structure of traditional classroom teaching to a certain extent. It is a way for students to learn knowledge by watching micro-class or other ways before class. Compared with primary and secondary school teachers engaged in basic education, college English teachers have relatively high academic qualifications and the ability to accept new things quickly [17]. As the implementer of bilingual teaching, a high-quality team of teachers is a strong guarantee for the quality of bilingual teaching, while for the general science and technology colleges in the west, the shortage of compound teachers is the primary problem facing bilingual teaching [18]. From the perspective of students, college students are more capable of autonomous learning and self-control than students in compulsory education. In class, teachers help students absorb and internalize knowledge. After class, students can consolidate what they have learned. Most college English teachers with a certain language and teaching knowledge system are able to combine their language learning experience to a large extent to realize innovative teaching of college English.

4 Conclusion

The teaching of professional knowledge in higher education should be aimed at different students according to their aptitude, i.e. appropriate teaching scenarios should be adopted for different courses and different students, and enhancing the interaction between teachers and students is the common goal of various teaching scenarios. With the advancement of educational information technology and the reform of college English teaching in our country, the perfect hardware configuration of colleges and universities can also meet the requirements of implementing flip classroom teaching mode based on micro-class. Flipping classroom teaching mode is established in a relatively free environment, which can effectively promote communication between teachers and students, thus enlivening the learning atmosphere in classroom teaching and improving their learning enthusiasm. Teachers should carefully design all links before and in class, fully mobilize students' independent learning ability, use process evaluation method, let bilingual teaching achieve immersion teaching, and then achieve the goal of training international talents. The implementation of flipped classroom also puts forward higher requirements for teachers' professional development. Teachers should not only have the ability to master modern educational technology, but also scientifically and reasonably allocate the relevant modules of each teaching. Only by accumulating the experience of flipped classroom in practice can teachers and educators gradually improve the flipped classroom teaching mode suitable for bilingual teaching.

References

1. Li, M., Chen, H., Zheng, Y.: Discussion on bilingual teaching of university computer basics based on micro-class and flipped classroom teaching model. Educ. Teach. Forum (24), 177–178 (2016)
2. Xiong, B., Qi, H.Q.: Research on the feasibility of bilingual teaching in universities based on flipped classrooms. Value Eng. (13), 9–12 (2016)
3. Zheng, J.X.: Feasibility analysis of using flipped classroom teaching mode in bilingual teaching. Shanxi Sci. Technol. (4), 104–106 (2015)
4. Zhang, X., Ma, X.F.: Application research of flipped classroom in bilingual teaching of object-oriented programming. Softw. Guide Educ. Technol. 18(5), 22–24 (2019)
5. Li, Z.K., Hou, K.F., Zhang, H.Q.: The practice of bilingual teaching reform in the flipped classroom based on control engineering. Educ. Teach. Forum (50), 83–84 (2016)
6. Yuan, J.: Application of bilingual teaching model of flipped classroom in economic management major. Campus Engl.: Voluntary Educ. Ed. (8), 66–67 (2017)
7. Liu, Z.Y., et al.: 2017. Application of flipped classroom teaching mode in bilingual teaching of mobile communication. Modern Comput.: Prof. Ed. (6), 48–51+55 (2017).
8. Xu, S., et al.: Computer vision techniques in construction: a critical review. Arch. Comput. Methods Eng. (2020) https://doi.org/10.1007/s11831-020-09504-3
9. He, L., Shao, F., Ren, L.: Sustainability appraisal of desired contaminated groundwater remediation strategies: an information-entropy-based stochastic multi-criteria preference model. Environ. Dev. Sustain. 23, 1759–1779 (2020). https://doi.org/10.1007/s10668-020-00650-z
10. D'Souza, U.J.A., Rahaman, M.S.: Animal stress models in the study of stress and stress related physiological and psychological derangements. Matrix Sci. Pharma (MSP) 2(1), 3–5 (2018)
11. Mehvish, S., Barkat, M.Q.: Phytochemical and antioxidant screening of amomum subulatum, elettaria cardamomum, emblica officinalis, rosa damascene, santalum album and valeriana officinalis and their effect on stomach, liver and heart. Matrix Sci. Pharma (MSP) 2(2), 21–26 (2018)
12. Chen, S.S., et al.: Antioxidant activity and optimisation of ultrasonic-assisted extraction by response surface methodology of aronia melanocarpa anthocyanins. Matrix Sci. Pharma (MSP) 2(1), 6–9 (2018)

13. Kanwal, A., et al.: Gastroprotective effect of Avena sativa (oat) seed grains against gastric ulcer induced by indomethacin in healthy adult male albino rabbits. Matrix Sci. Pharma (MSP) **2**(1), 14–18 (2018)
14. Farid, A., et al.: Renal clearance and urinary excretion of moxifloxacin in healthy male volunteers. Matrix Sci. Pharma (MSP) **2**(1), 19–22 (2018)
15. Mahmood, H.K., et al.: Phytochemical and antioxidant screening of anacylus pyrethrum, apium graveolens, boerhaavia diffusa, cinnamomum cassia blume, cuscumis melo linn, cuscumis sativus linn, daucus sativus, foeniculum vulgare, trachyspermum ammii and theit effect on various human ailments. Matrix Sci. Pharma (MSP) **2**(2), 6–14 (2018)
16. Subhani, Z., et al.: Adverse effect of oxalis corniculataon growth performance of broiler chicks during aflatoxicosis. Matrix Sci. Pharma (MSP) **2**(1), 10–13 (2018)
17. Zeeshan, U., Barkat, M.Q., Khalid, H.: Phytochemical and antioxidant screening of cassia angustifolia, curcuma zedoaria, embelia ribes, piper nigrum, rosa damascena, terminalia belerica, terminalia chebula, zingiber officinale and their effect on stomach and liver. Matrix Sci. Pharma (MSP) **2**(2), 15–20 (2018)
18. Wang, Y.F., et al.: Research on bilingual teaching of safety engineering specialty based on flipped classroom. Safety **37**(7), 64–67 (2016)

Construction of English Multimodal Classroom in Colleges and Universities Under Big Data Environment

Limei Ma$^{(\boxtimes)}$

Foreign Language Teaching Department, Ningxia Medical University, Yinchuan 750004, China

Abstract. With the multi-modality of college foreign language cognitive style and dimensions, China's higher education curriculum reform also has higher requirements for the traditional class teaching system based on practice. It is of great urgency and necessity for teachers to establish effective teaching concepts and guide classroom teaching to realize effective teaching. Under the background of big data, the current traditional British teaching mode does not meet the application requirements of independent colleges. Massive open online course and microteaching have become trends in the field of education. The introduction of multi-modal discourse analysis theory brings new ideas, methods and teaching design to English teaching. College English takes classroom teaching as the main body, and the collaborative application of multi-modal discourse changes the rigid explanation mode. Based on the analysis of the integration of big data and multi-modality in the current college English curriculum, this paper has initially formed a theoretical framework for the construction and evaluation of multimodal classroom environment in China. Teachers need to use their rich working experience and profound knowledge to create a good learning atmosphere for students.

Keywords: Multimodal classroom · English teaching · Big Data

1 Introduction

With the multimodal cognition and dimensions of college foreign language, the curriculum reform of higher education in our country also has higher requirements for the traditional class teaching system based on practice [1]. The scale of English education in our country continues to expand and great achievements have been made in English teaching, especially in primary and secondary schools. However, it cannot be denied that there is still a big gap between the current situation of English education in China and the requirements of China's economic construction, social development and era development. Classroom teaching is a widely used teaching method in college English teaching activities. Although the continuous development of modern educational technology has brought new models for teaching activities, the new models and technologies are also realized under the condition that classroom environment is the main body [2]. Between teachers' teaching behavior and students' learning effect, it has become an important

E. Bisset Álvarez (Ed.): DIONE 2021, LNICST 378, pp. 426–430, 2021.
https://doi.org/10.1007/978-3-030-77417-2_35

potential factor that determines learning effect and affects students' cognitive and emotional development. Computer-based and classroom-based English multimedia teaching mode is a new English teaching mode designed to help Chinese college students meet the requirements of college English teaching [3]. At present, in the actual teaching, English classes are still mainly taught by teachers, only explaining pronunciation, vocabulary, sentence patterns and other knowledge step by step according to the established goals [4]. This neglects the students' ability to use language in the real context. Students are only passive acceptance, participation and subjective consciousness is very poor, so effective teaching must adapt to the requirements of specific teaching situation [5]. The English classroom teaching model under the network environment is under the guidance of constructivism and other theories. This paper analyzes the integration of big data multimodality in university foreign language courses at this stage, and initially forms a theoretical framework for the construction and evaluation of our university English multimodal classroom environment.

2   Multimodal Analysis of College English Classroom

In the teaching activities of colleges and universities, the classroom is the place where teachers and students communicate most, so classroom teaching has always been the focus of researchers. English not only refers to communication tools, but also is an effective way to improve the use of English to obtain information, analyze information and broaden your mind. Effective teaching is not only a kind of teaching theory but also a kind of teaching practice. In order to construct the multi-modal classroom teaching strategy of college English under the new curriculum standard, we use the method of combining student questionnaires and teacher interviews to understand the current situation of college English classroom teaching [6]. The English level of the college students enrolled by the large-scale enrollment expansion in our country is different, so the teaching methods of teachers need to be adjusted. In addition, due to the differences of students in different schools, the classroom teaching environment cannot be ignored. Based on the theoretical model of the construction and evaluation of the multimodal classroom environment of college English in China, the scale is designed to investigate the teaching process and teaching results of college English classroom at two levels by exploring the personal perception of college students of English classroom environment in China.

With the development of the times, the main function of modern English classroom teaching is not only to teach students about the English language itself, but also to cultivate students' ability to communicate in English. On the whole, the relationship between the three dimensions of the teaching process and between the three dimensions and the learning effect is assumed to be consistent with the observed data. Each path coefficient of the path model has significant significance. Figure 1 is a path analysis model of constructing dimensions of college English multimodal classroom environment and learning effects.

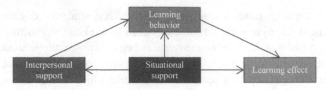

Fig. 1. Path analysis model.

The three dimensions of learning behavior, interpersonal support, and situational support also show significant positive correlations with learning effects. It shows that the three dimensions are related to the learning effect, which further reflects the high internal reliability of the overall scale. Among the many factors that affect effective teaching, classroom management, activity innovation, and educational technology are three important variables that affect the establishment of teaching situations. The establishment of teaching situations is the basic requirement for effective teaching. The constantly changing teaching situations require teachers to adopt flexible and diverse classroom management strategies to ensure the orderly development of teaching activities [7]. Teaching media should be a cognitive tool for students' active learning and collaborative exploration, not just a means to help teachers impart knowledge. Under the new round of curriculum reform, the core literacy training of subjects becomes the top priority, and the core literacy of English subjects naturally becomes the focus of educators. College students can say a lot of time, can also say very little, the key is how to reasonably plan the time, the combination of work and rest is the best choice for learning. College students face not only academics but also invisible shoulders on their shoulders. There is an extra burden of responsibility pressure, which affects the learning effect in happy colleges [10–14].

3   Practical Exploration of Multimodal English Classroom

The effectiveness evaluation of classroom teaching should not only look at the results of teaching, but also attach importance to the teaching process. In the questionnaire survey and discussion, we also found that what students think of teachers' multimodal classroom teaching behavior includes in-depth explanation, clear multimedia courseware, active classroom atmosphere and effective group discussions. When one mode of discourse cannot fully express the full meaning of the communicator, the communicator must supplement and strengthen it with other modes of discourse [15, 16]. Although students began to contact with English from primary school, some students were not interested in it and did not pay attention to it, which led to the low English scores. The content of classroom teaching is pronunciation. Teachers should explain the key points and skills of pronunciation clearly, and then let students imitate with reference to the corresponding audio. A relaxed English learning atmosphere should be created. Boring times are unavoidable in English learning. Teaching is not only a cognitive process [17]. A lively and efficient course cannot be conducted under tight teacher-student relations.

The language itself is very vivid, but it is difficult for students to keep their attention throughout the English class. Multimodal application will mobilize students' different senses, and the conversion of different modes will continuously attract students' attention, enhance students' learning interest and thus improve classroom teaching efficiency [18]. A reasonable and effective teaching theory is fundamental in the teaching process. The influence of students' knowledge and experience gained through participating in teaching activities on their learning behavior, interpersonal support and landscape support is an important part of building a multi-modal classroom environment. In a foreign language environment, classroom learning is not only a series of cognitive processes such as processing, analyzing and memorizing the oral or written language input of the target language. Teaching characteristics are the combination and expression of teachers' long-term formed teaching concepts, teaching ability and teaching style, and are the sign of excellent teachers' mature, personalized and creative stable state. With the promotion of the new curriculum concept, teachers should pay attention to the improved teaching mode, gradually changing from the traditional knowledge indoctrination to the guidance, inspiration, consultation and encouragement of students' learning, so as to liberate students from the learning machine.

4   Conclusion

The popularization of big data and the development of multimedia make the second language acquisition mode show the characteristics of multimodal. The reconstruction of multimodal classroom must be rooted in the classroom full of information and wisdom. Classroom teaching is the most important link in the process of foreign language learning, and the construction of multimodal classroom environment is an important potential factor that determines the quality of teaching. The discussion in this paper has initially formed a theoretical framework for the construction and evaluation of multimodal classroom environment for college English in China. The choice of multiple modes in English classroom teaching is because one mode cannot fully convey the teaching content or can not effectively convey clear teaching content, teachers must supplement and strengthen it with other modes of discourse. Multimodal teaching mode is more in line with the needs of the society and the times. In the English classroom, multimodal means are used to stimulate students' multiple senses through network technology, so as to stimulate students' interest and initiative in learning English. Whether students' learning methods are scientific, learning efficiency and learning initiative affect their learning effect. Theory is only a guide in teaching, but also requires teachers to fill in the gaps with their rich working experience and profound knowledge to contribute to the construction of multimodal classroom environment.

References

1. Liu, W.: Research on the independent college English multimodal classroom based on big data. Campus Engl. Volunt. Educ. Ed. (8), 37–38 (2018)
2. Wang, H.X.: Research on multimodal teaching mode of advanced English. J. Kaifeng Educ. Coll. **39**(1), 61–62 (2019)
3. Yang, L.: A brief talk on multi-modal English teaching in junior middle schools in the information age. Inf. Rec. Mater. **19**(5), 157–158 (2018)

4. Chen, D.F., Gao, Y.H.: Research on teaching evaluation based on multi-modal data efficient analysis technology. J. Lianyungang Teach. Coll. **36**(1), 77–80 (2019)
5. Wang, Z.P.: Construction of multimodal teaching mode in college English listening class. Chin. J. Multimed. Netw. Teach. (3), 57–58 (2019).
6. Wang, C.L., Li, C.L.: Reconstruction of English classroom teaching from the perspective of multimodality. Teach. Manag. (Theor. Ed.) (12), 103–105 (2016)
7. Hu, E.J.: Research on multi-modal English teaching in vocational colleges in the era of big data. Neijiang Sci. Technol. (7), 131–132 (2019)
8. Wu, C.Z., Wang, X.Y., Chen, M.C., Kim, M.J.: Differential received signal strength based RFID positioning for construction equipment tracking. Adv. Eng. Inform. **42**, 100960 (2019)
9. Xu, S., Wang, J., Shou, W., et al.: Computer vision techniques in construction: a critical review. Arch. Comput. Methods Eng. (2020). https://doi.org/10.1007/s11831-020-09504-3
10. Guo, L.P.: An empirical study on the impact of economic agglomeration on FDI location choice in China. Adv. Ind. Eng. Manag. **9**(1) (2020)
11. Tu, B., Liu, K.X.: Community prevention and control mechanism of public health emergency-based on the perspective of social force participation in China. Adv. Ind. Eng. Manag. **9**(1) (2020)
12. Oyewole, G.J., Ayorinde, A., Anyaeche, O.: Development of a solution technique for the facility location and step-fixed charge solid transportation problem. Adv. Ind. Eng. Manag. **9**(1), (2020)
13. Senthilananthan, S. The role of the audit committee in NSW local council and governance. Adv. Ind. Eng. Manag. **8**(2) (2019)
14. Ding, J., Li, C.B.: Project team incentive of the power grid enterprise in China. Adv. Ind. Eng. Manag. **8**(2) (2019)
15. Farahmand, N.F.: Job progression ladder by humanistic approach in workplace management. Adv. Ind. Eng. Manag. **8**(2), 70–74 (2019)
16. Hafez, M.A., Hanif, et al.: Prediction of ultimate capacities of bored pile based on routine load tests and static analysis (case study). Adv. Ind. Eng. Manag. **8**(1) (2019)
17. Yan, H.N., Zhang, B.Y.: The practice of environmental cost management of hydrologic chemistry in Japan and its reference to China. Adv. Ind. Eng. Manag. **8**(1) (2019)
18. Li, B.Y., Hao, J.L.: Research on students' cognition in English classroom multimodal teaching environment. Cult. Educ. Mater. **3**, 222–224 (2019)

Diversification of the Evaluation Mode of English Mixed Gold Course Learning Mode in Colleges and Universities Based on Mixed Learning

Min Zhang(✉)

Jiangsu Maritime Institute, Nanjing 211170, China

Abstract. With the development of higher education information, hybrid teaching as a new teaching mode is being gradually built and applied by many colleges and universities. The hybrid teaching mode not only realizes the integration of online learning and traditional classroom, but also improves students' ability of self-learning. The author's research mainly expounds from three levels, firstly clarifies the connotation of mixed teaching mode, secondly analyzes the construction strategy of mixed teaching mode in colleges and universities, and finally takes English professional teaching as an example, and elaborates the application of mixed teaching mode in practice.

Keywords: Mixed learning · College English · Diversity research

1 Introduction

The hybrid teaching method is based on the development of the times to select and integrate advanced and excellent teaching methods, teaching resources and teaching mode, aiming to strengthen the basic knowledge of language and practical application of English majors in colleges and universities. The implementation of hybrid teaching methods must have solid theoretical guidance and experience summary, require teachers' careful design and scientific planning, and adjust the corresponding teaching methods in a timely manner for students' learning. Many teachers in the teaching process do not master the mixed teaching methods, the root cause is that teachers can not scientifically integrate traditional teaching methods and Internet teaching methods, can not scientifically use the teaching methods and content of the two complementary advantages, to achieve strong combination. Network teaching is the inevitable trend of the development of the times. If teachers can not master network teaching, it will seriously affect the quality of classroom teaching. Therefore, teachers should scientifically integrate network assisted teaching, use a variety of teaching methods, scientifically master network teaching skills, give full play to the advantages of network teaching, and effectively make up for the shortcomings and shortcomings of traditional classroom teaching [1].

E. Bisset Álvarez (Ed.): DIONE 2021, LNICST 378, pp. 431–437, 2021.
https://doi.org/10.1007/978-3-030-77417-2_36

2 Construction Strategies for Mixed Teaching Models in Colleges and Universities

2.1 Import Flip Classroom Teaching Form to Strengthen Students' Capacity-Building in English Learning

The main value of the flip classroom teaching form with the Internet as the carrier is reflected in the change of teaching mode, the change of teaching content, the scientific change of teacher's teaching mode, and the change of students' learning form. In the traditional classroom teaching, science importflip classroom teaching form, is to allow students to have more time and opportunity to carry out language practice, no longer implement the theory of large blocks in the classroom, teachers before class through video form to present basic content, so that students learn some knowledge in advance, in the classroom, teachers have more time for their teaching of English basic knowledge intensive training, such as situational dialogue, simulated language dialogue and debate, so that students experience the practice of language in the classroom, Realize the coordinated development of theory and practice, and promote the realization of the goal of applied English talents. First of all, flipping the classroom teaching form requires teachers to scientifically reform the classroom teaching structure, to allocate the time of teachers' English theoretical knowledge explanation and students' language practice exercise, and to make the students' language practice exercise the key content of classroom teaching, to realize the dynamic learning mode of students, and to transform the teacher sancturnity into the organizer of the activity-based classroom, not just the theoretical knowledge. Secondly, importing and flipping the classroom teaching form requires teachers to carefully craft the curriculum before class, highlight the difficulties, guide students to learn, understand and master the learning content, students also need to properly save such learning videos, according to learning habits to do a good job of sub-category and key outline, convenient for the future "basic English" curriculum theoretical knowledge review and communication ability to strengthen [2].

2.2 Import Micro-teaching Forms to Promote Students' Capacity-Building in English Learning

The main value of micro-teaching form with mobile phone as the carrier is reflected in the timely dissemination and sharing of "fine" learning video, can be typical, valuable learning content in the form of video timely sharing to students, can effectively summarize and comb the knowledge plate, the fragmented knowledge content in series to form a knowledge chain, and can accurately grasp the essence, clear focus, easy for students to understand and practice the use. There are many contents in the course of basic English for College English majors, and each knowledge point and knowledge segment has its core elements. Teachers should guide students to master the core knowledge points, only relying on classroom time is far from enough. Teachers need to import Microteaching in time, and make use of the advantages of microteaching, and adapt to the help of micro-blog, WeChat, QQ group, official account and web page [3].

Communication channels for students to consolidate their learning after class. Micro-class teaching form can fully mobilize students to use fragmented time to learn, video situation teaching content presented more rich learning content, can be the key grammar knowledge, vocabulary memory rules, reading and writing skills, Anglo-American cultural differences, can also be classic European and American film and television clips, or professional business English communication skills, such as diversified learning resources rich in micro-teaching content, promote the overall development of students' English knowledge [4].

The widespread use of micro-teaching form is the inevitable result of the development of the Internet age, teachers should make full use of this advantage to assist the basic English classroom teaching in colleges and universities, make up for the transient ness of traditional classroom teaching time and the limitation of teaching content. The micro-class videos produced by teachers are very valuable typical learning content, emphasizing fine and small, easy for students to learn and digest understanding. In addition, teachers can regularly or irregularly share micro-class videos, and do a good job of students' learning records supervision work, such as the implementation of check-in clocking to promote students micro-class content learning. Taking this kind of teaching form combining online classroom with offline micro-class can effectively consolidate the basic English theoretical knowledge and practical application of students [5].

2.3 Importthe Course Mode, Science Construction Course, Strengthen Students' English Application Capacity-Building

With the Internet as the carrier, the development and application of various learning platforms and learning app accelerates and promotes the application of the teaching form of mu in the basic courses of English in colleges and universities [6–10]. The value of the teaching form of Mu class is mainly reflected in the openness of viewing, the systematic, complete and precise content, students can quickly search for a variety of large-scale online quality courses according to the needs of learning and ability development, to achieve the expansion of knowledge ability and thinking deepening. A complete knowledge chain application process on display in the form of teaching is no longer a single point of knowledge, but a quality course of listening and writing in basic English. The so-called quality curriculum embodies a deeper level of language understanding and language use skills, which requires students to think deeply and explore, and to give more prominence to the practical exercise of students' comprehensive English ability. In the classroom teaching, the introduction of the teaching form of the teaching class can effectively realize the scientific transformation of teaching form, the classroom teaching will be actively transformed into the students' independent learning, and the classroom learning to extracurricular learning extension, for students to create more opportunities for independent learning. In order to effectively promote the scientific application of the teaching form of the teaching of the teaching, teachers need to make timely learning supervision and timely answer questions, for example, to build WeChat learning exchange group, supervise students' independent learning, organize learning, discuss and answer questions, in order to effectively promote students' in-depth study. College English major "basic English" is a very wide range of topics, and very targeted, students can according to their own ability to develop the needs of the Internet, quickly retrieve many valuable learning videos, through in-depth study, thinking and practice exercise, can effectively promote the college English major English application capacity-building [11–14].

In the traditional classroom teaching, the introduction of flipped classroom, micro-class and teaching forms, are aimed at the development of students' ability, the need to plan, scientific design teaching forms, so that hybrid teaching better serve the college English major "basic English" curriculum teaching, and constantly strengthen the students' English language application ability [15].

3   The Requirements of Teachers in the Mixed Teaching Mode of Colleges and Universities

It is the goal of the reform of college English teaching to change acceptive learning to active inquiry learning. The massive network online resources and convenient interactive communication make hybrid teaching become the most representative teaching mode in the modern stage of educational technology. The so-called mixed teaching, that is, to play the role of teacher guidance, supervision and students as the main body of initiative, creativity, make full use of online teaching and classroom teaching complement each other, improve the learning effect of students. Generally speaking, the hybrid teaching model is divided into three levels: one is the online (E-learning) offline (classroom) hybrid learning docking, the other is the classroom discussion and virtual classroom learning "integrated" learning, and the third is the "learning" and "learning" mix, that is, learning skills and practice [16–18].

The core of hybrid teaching is to realize the effective integration of online and offline, and to improve and transform the traditional teaching. From the perspective of comparative education, the traditional teaching classroom is teacher-centered, most of the classroom time is used for low-level cognitive activities such as memory and understanding, there is less contact with learning for higher-level cognitive activities such as analysis, evaluation and debate, while classroom teaching in the world's mainstream colleges and universities revolves around higher-level cognitive activities, with students hands-on practice training, innovative thinking and rapid development of ability [19, 20].

The hybrid teaching mode not only changes the students' cognitive way, but also changes the teacher's teaching style and teaching strategy to a certain extent. The change of teacher's role is reflected in the shift from focusing on "how to teach" to paying more attention to "how to guide learning". From the analysis of students' needs, teaching content and actual teaching environment, teachers plan and guide students' individual learning needs with the goal of teaching tasks. Teaching uses classroom debate, answering questions and group cooperation activities to strengthen and consolidate what is learned in the classroom, expand divergent thinking, and promote students to complete deep learning. In mixed teaching, teachers should learn to use multimedia technology and network platforms, integrate various teaching resources, create cross-cultural communication simulation language environment for students, monitor students' online and offline learning process, track learning records and learning feedback, encourage and urge students to cultivate interest and motivation in self-study language. Teaching evaluation is no longer focused solely on test scores, but on the learning process and student growth.

4   College English Hybrid Teaching Mode Application

The promotion of mixed teaching mode makes the teaching form of the new era pay attention to. Under the guidance of the mu class platform, many people in China interactive teaching platform, such as super-star learning pass, rain classroom, blue ink cloud class and other rapid development, and gradually be promoted to use. Relying on the modern information platform, the mixed teaching model aimed at improving teacher-student interaction and teaching efficiency has gradually taken to the stage and been widely promoted. The hybrid teaching mode breaks through the contradiction of traditional classroom teaching, and has the characteristics of flexible teaching place and wide range of subjects, which meets the diversified learning aspirations of learners in the context of the new era.

4.1   Application of Teaching Mode Based on Instant Interaction in a Mobile Environment

In the context of mobile information technology support, English major teaching should adapt to the needs of the times, and adopt the mixed teaching mode of the instant interaction with the traditional classroom, namely, the student's offline (autonomous learning) and flip classroom (classroom discussion) combination.

Taking the "BasicS of English Writing" course taught by the author as an example, this course is based on the interactive teaching platform of "Blue Ink Cloud Class", which adopts the mixed teaching mode, and the teaching link is designed as follows: Before class, teachers choose suitable according to teaching content.

Learning materials, and upload the learning materials to the Blue Ink Cloud class platform, such as how to write an outline, how to conceive, the meme sentence writing and demonstration methods to strengthen the students' grasp of theoretical knowledge, students log on to the network platform, independent completion of course learning. In class, the teacher relies on the key content in the video, asks questions, arouses students to think deeply about the knowledge and method of writing theory and conducts group discussions, and at the same time, the teacher answers questions according to the students' discussion situation. In class, teachers can interact with students instantly information, such as roll call, test, answer and other novel teaching links, efficient completion of teaching tasks, in line with the new era of students to accept the characteristics of information. At the same time, teachers can timely and effectively grasp the learning situation of students through the data percentage. After class, students combine classroom discussions, complete exercises, and feed back to teachers. In the course of learning in this course, students' composition homework feedback is good, which shows the effective and efficient mode of mixed teaching.

4.2   Application of English Teaching Mode Based on Complementarity of Human-Machine

The complementary model of human-machine complements is a useful complement to the traditional teaching model. Combined with the traditional teaching mode, the practice of homework evaluation, testing and feedback through the Internet platform can not only improve the efficiency of teachers and students, but also timely and accurate feedback of results, effectively solve the contradictions existing in the traditional classroom.

The author takes the English professional writing course as an example, in the course of teaching this course, the author combines the traditional composition feedback mechanism, and at the same time introduces the online composition review platform of the "batch change network". In the stage of composition modification, the effective use of teachers' revision, student mutual evaluation and approval network and other channels of mutual integration of the form, the students' practice synod, so that students get different opinions and expand their thinking.

5 Conclusion

Mixed university English teaching practice, with "personalized" independent learning and "online and offline classroom complementeaching" as the goal, task-oriented, all-round, multi-terminal for scientific layered graded teaching services. The hybrid teaching mode can promote the interaction between teachers and students, make up for the deficiency of the traditional teaching mode based on teacher teaching, and further deepen the reform of English teaching in colleges and universities. This kind of teaching mode is student-centered interactive teaching experience, which highlights students' subjectivity, promotes personalized learning and development, and is conducive to improving students' employment competitiveness. Select students with training potential to strengthen the second classroom learning, excellent students to tutor one-on-one tutoring, for the reform of college English teaching out of a new path.

In the mixed teaching experience, college English teachers adhere to the combination of classroom teaching and practical teaching, traditional teaching and modern technology integration, to carry out all-round, diversified teaching. Under the guidance of "three comprehensive education", we should carry out the fundamental task of "establishing morality and cultivating talents". We should combine ideological and political education with language skills training to overcome the western culture centered teaching mode. It is our duty to spread advanced culture and let Chinese culture go to the world. Guiding college students to strengthen their cultural self-confidence and consciousness, broaden their international vision, and enhance their recognition of Chinese excellent traditional culture.

Acknowledgements. The study was supported by "Research on the Construction and Application of Online Open Courses of Practical English in Higher Vocational Education, China (Grant No. 2019SJB988)".

References

1. Yan, S.L.: Research on college English mixed teaching innovation in the media-media environment. J. Hubei Open Vocat. Coll. **33**(06), 179–180 (2020)
2. Chen, J.: The application of mixed teaching method in the course of "Basic English" for college English majors. J. Heihe Univ. **11**(02), 107–109 (2020)

3. Song, X.: Research on the application of blended learning in college oral English courses. J. Changchun Univ. **29**(12), 109–113 (2019)
4. Xu, L.Q.: The way to improve the thinking ability of English teachers under the mixed teaching mode. Overseas English **23**, 128–129 (2019)
5. Leng, H.Y.: The application and research of online and offline "hybrid" teaching model in private college English reading and writing courses. Fujian Tea **41**(11), 104–105 (2019)
6. Erlano-De Torres, J.A.: Performance of sophomore information technology students in assembly language subject based on different teaching methods. Inf. Manage. Comput. Sci. **2**(2), 10–13 (2019)
7. Chiroma, A.S., Yaduma, P.S.: Shifting paradigms of continuous professional development through blended learning and e-enabled platforms for 21st century teaching and learning. Educ. Sustain. Soc. **2**(3), 12–15 (2019)
8. Aminu, M., Samah, N.A.: Teachers perception on the use of technology in teaching and learning in associate schools Zamfara state, Nigeria. Educ. Sustain. Soc. **2**(2), 1–4 (2019)
9. Li, X.C.: Study on influence of cognitive characteristics of students in different gender tendency on teaching strategies under network environments. Educ. Sustain. Soc. **2**(1), 4–7 (2019)
10. Yang, L.M.: Analysis of the application of mixed teaching in English listening classroom teaching in colleges and universities. Chin. Extra-curricular Educ. (33), 122+126 (2019)
11. Zhang, Y., Du, W., Liu, Y.: Mixed learning model in English hierarchical classification teaching in colleges and universities. Huaxia Teachers **27**, 81–82 (2019)
12. Murtafi'ah, B., Putro, N.H.P.S.: Digital literacy in the English curriculum: models of learning activities. Acta Informatica Malaysia **3**(2), 10–13 (2019)
13. Dai, W., Pu, R.H., Tang, C.J.: Factors affecting "a" luxury brand loyalty on Chinese college students in Shanghai city of China. Inf. Manage. Comput. Sci. **1**(2), 14–17 (2018)
14. Liu, C.L., Yin, J.X., Ma, L.C., et al.: The fusion of the love view of contemporary college students and Fromm's thought. Educ. Sustain. Soc. **1**(1), 8–10 (2018)
15. Liu, F.: Research on the construction of online + offline mixed mode in college English teaching. J. Jilin Radio TV Univ. **09**, 62–63 (2019)
16. Liu, C.L., Qi, Q.: Research on the network values of post-millennial college students. Soc. Values Soc. **1**(2), 11–14 (2019)
17. Satyawan, I.A., Yudiningrum, F.R., Anshori, M.: Searching information of Asean through social media among college student in Landak regency. Soc. Values Soc. **1**(3), 1–3 (2019)
18. An, Q.: A preliminary study on the mixed teaching mode of English major courses in ethnic universities–Taking Inner Mongolia University for nationalities as an example. Res. Ethnic High. Educ. **7**(05), 90–92 (2019)
19. Zhao, Y.Y.: Exploration of college English mixed teaching mode in the context of media integration. Educ. Theory Pract. **39**(24), 63–64 (2019)
20. Ali, A., Lee, T.L., Thoe, N.K., et al.: Transforming public libraries into digital knowledge dissemination centre in supporting lifelong blended learning programmes for rural youths. Acta Informatica Malaysia **3**(1), 16–20 (2019)

Ecological Education Mode of Foreign Language Based on Computer Network Technology

Yanping Zhang[1,2](✉) and Weiping Zhang[3]

[1] Liberal Arts College, Hunan Normal University, Changsha 410081, China
[2] School of Languages and Literature, University of South China, Hengyang 421001, China
[3] School of International Studies, Hunan Institute of Technology, Hengyang 421002, China

Abstract. This paper starts from the perspective of the ecological teaching of foreign languages in the context of computer networks. The current situation of the lack of an ecological model in college English classroom teaching under the current computer network environment is proposed, and the construction of a college English ecological classroom teaching model under the computer network environment is studied. The theoretical method has certain reference value for the construction and optimization strategies of the ecological teaching mode under the computer network environment and the realization and development of foreign language courses.

Keywords: Computer network · Foreign language ecology · Teaching mode

1 Introduction

At present, with the gradual progress of globalization and the rapid development of computer network technology, many aspects of people's lives are being profoundly affected by the age of the Internet information [1–5]. The widespread use of computer networks has also brought unprecedented opportunities and challenges for language teaching. At present, although there has been some improvement in college English classroom teaching, it has neglected the integrity of the classroom, the relevance of various factors in the classroom, the openness and continuity of the teaching process, and has separated many factors such as curriculum, teachers, students, and the environment [6–8]. The organic connections and interactions between them have led to imbalances in the classroom ecology and ultimately to the loss of vitality in the original living classroom. To this end, exploring how to revive the college English classroom ecosystem, how to exert the synergy effect of various ecological factors, and achieve effective, high-quality teaching, has important theoretical and practical significance [9–13].

E. Bisset Álvarez (Ed.): DIONE 2021, LNICST 378, pp. 438–446, 2021.
https://doi.org/10.1007/978-3-030-77417-2_37

2 The Ecological Education Mode of Computer Network Technology in Foreign Languages

2.1 Constructing a Foreign Language Ecological Teaching Model Based on Computer Network Technology

Under the era of widespread application of information technology, computer network technology accelerates management changes in various fields with its unique characteristics of high speed, integration, virtualization, and intelligence. Including foreign language ecological teaching mode, foreign language teachers will not need to use computer network technology to deal with various foreign languages teaching work [14, 15]. The interpenetration and development of multi-disciplines such as foreign language ecological teaching, computer and foreign language teaching, will continue to promote the integration of computer network technology and the foreign language ecological teaching model field. In foreign language teaching, the foreign language ecological teaching model and computer network technology complements each other. At present, the degree of information in the field of foreign language ecological teaching has become higher and higher, and many of the business data processing has been digitized and networked. Traditional manual inspection methods have been difficult to adapt to the requirements of foreign language teaching. At the same time, in terms of information of foreign language teaching, computer network technology has continuously expanded the scope of foreign language teaching. With the continuous maturity of computer network technology, it has gradually realized the use of computer systems to assist foreign language teachers in data calculation and data analysis, and can be used in the field of foreign language ecological teaching [16–21]. The use of electronic data for information management analysis will gradually break the gap between electronic data and foreign language teachers in the field of foreign language eco-teaching, and save a lot of manpower and material resources for foreign language teaching and improve work efficiency. Therefore, the ecological language teaching model based on computer network technology will be an effective way for the future development of foreign language ecological teaching model. Similarly, the application of computer network technology in the field of foreign language ecological teaching model will also make computer network technology theory and application. The level has been continuously improved. The methods for developing a foreign language ecological teaching model are as follows:

(1) The traditional methods for developing the ecological teaching model of foreign languages include: checking records or foreign language teaching materials, checking tangible teaching equipment, teaching feedback, teaching interviews, teaching comments, teaching evaluation, and teaching implementation. Traditional foreign language teaching methods are still necessary and effective for the foreign language ecological teaching model, as shown in Fig. 1.

Foreign language teaching materials: This is the most traditional approach to foreign language teaching. It involves the mobilization, management, and use of resources in the development and utilization of resources, foreign language ecological teaching, as well as the establishment and improvement of foreign language ecological teaching systems, and policy development in the development and promotion of foreign language ecological teaching in colleges and universities [22–24]. And the implementation of teaching is using this method to correlate relevant language teaching materials and records.

Tangible teaching appliances: In the ecological teaching mode of foreign languages, the inspection of tangible teaching equipment is mainly to check whether the number of various facilities and equipment used for development and promotion of foreign language ecological teaching satisfies the requirements and whether the operation status is good.

Fig. 1. Analysis of traditional methods for developing ecological teaching mode in foreign languages.

Teaching feedback: It is obvious whether field teaching feedback on foreign language ecological teaching is in good condition. However, foreign language teachers need to pay more attention to whether the teaching feedback has been taken by foreign language teaching institutions in terms of measures and measures, and whether they have been used in foreign languages [25]. The institutions or their staff's business activities or teaching implementation proceeds comply with the relevant regulations.

Teaching interviews: Special attention should be paid to this method in the foreign language ecological teaching model, which mainly adopts survey questionnaires and interview teaching methods. For example, a survey of environmental promotion is conducted and a questionnaire is sent to relevant personnel who have lived in the area for a long time.

Teaching review: It is the process by which a foreign language teacher acquires foreign language teaching materials through direct statements from third parties on the relevant information and status in order to obtain relevant information.

Teaching evaluation. It is the foreign language teacher who checks the correctness of the relevant data teaching evaluation and can be used to review the following materials: teaching evaluation of original vouchers and bookkeeping vouchers, teaching evaluation of teaching logs, teaching evaluation of lists, and teaching evaluation of other relevant materials.

Teaching implementation: It is the foreign language teacher who independently conducts re-operation verification on relevant business procedures or control activities. In the foreign language ecological teaching model, this method has an important role, and that is, the teaching methods and measures related to foreign language ecological teaching are taught once by foreign language teachers or related professionals, and the results are retested [26–28]. For example, in the foreign language teaching of teaching quality, it is not possible to rely solely on the data provided by the education department. After the foreign language teacher conducts an on-site experiment, the third-party teaching implements evaluation procedures to verify the teaching quality.

(2) A comparative study of computer network technology and traditional foreign language teaching methods.

The traditional foreign language teaching method is applicable to the verification of the foreign language ecological teaching model. However, due to the particularity of the foreign language ecological teaching model, it has certain limitations in the analysis of foreign language teaching. Therefore, the foreign language ecological teaching model must explore some new ones. In the current situation where computer network technology is widely used, computer network technology has been widely used in the field of foreign language eco-teaching. The business database of relevant departments can provide data such as completion of classes, education statistics, and online monitoring [29–32]. These data volumes are very large. The mutual relationship is also very complicated, which requires that institutions and foreign language teachers use computer networks to assist in foreign language teaching. In terms of foreign language teaching management, computer network technology and traditional foreign language teaching methods are very different. The traditional foreign language teaching methods mainly rely on paper materials to carry out foreign language teaching, including the preparation, formulation, and approval of the implementation plan of the previous period of foreign language teaching preparation.

The summary of the final period of foreign language teaching and the work of the implementation stage were all performed by foreign language teachers through written materials, and their real-time and convenience were all poor. Computer network technology has changed the problems of the lack of real-time and convenience in the traditional paper media and written communication. Real-time communication between foreign language teaching groups and institutions has been realized through electronic data packets and network transmission. The communication between teachers of foreign language teaching groups has created a convenient platform. Because computer network technology is widely used in the field, foreign language teachers face not only simple teaching plans, but also face a large amount of data and diverse information systems, in such an increasingly complex foreign language teaching classroom. The traditional foreign language teaching methods face enormous challenges. The director of the National Bureau of Education put forward the need to master the technical skills of computer networks, or else foreign language teachers will lose their teaching qualifications in the future. Therefore, in advancing computer network technology, it is of great significance to strengthen research and experience sharing in this area.

2.2 The Application of Computer Network Technology in Educational Foreign Language Teaching

In the implementation of foreign language teaching for a certain year's planned education of a city's education bureau, the key teaching ecological promotion project of foreign language teaching investment is extended by using computer network technology, and the performance of the projects it invests is evaluated. A municipal education bureau is a municipal-level financial planning unit that organizes and supervises the implementation of key subjects. In the implementation of foreign language teaching in the planning education of a certain year, the implementation of supervision and informa-tion dissemination of foreign language ecological teaching is reviewed. In this foreign language teaching, the implementation of the teaching plan of the department plan and the investigation of the business data of the teaching object are taken in point, the use of computer network technology techniques to analyze the focus of foreign language teaching, through the extension of foreign language teaching key projects, reveals the problems in the supervision and management of foreign language ecological teaching, ecological teaching survey and information dissemination. The application of computer network technology in the practice of the project is mainly divided into two aspects. On the one hand, it is the management of foreign language teaching, including project estab-lishment in the computer network information system, production of foreign language teaching materials, formulation of foreign language teaching implementation programs, and foreign language teaching. Ecological implementation process has foreign language teaching summary, foreign language teaching exchange and so on. On the other hand, it analyzes and deals with foreign language teaching data, including analysis and process-ing of business data and analysis and processing of teaching data. The application of computer network technology will be combined with the application of foreign language teaching management system and data analysis system.

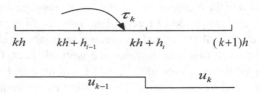

Fig. 2. Analysis of foreign language eco-teaching data using foreign language teaching ecological teaching implementation system.

The foreign language teaching scene management system manages the foreign lan-guage teaching, and specifies the foreign language teaching matters and the method of foreign language teaching, which is conducive to clarifying the objectives of foreign language teaching and guiding the development of foreign language ecological teaching [33, 34]. The use of foreign language teaching and teaching system (AO) to analyze for-eign language ecological teaching data, and find foreign language ecological teaching focus and clues, and discover foreign language ecological teaching doubt, performance analysis and other applications, as shown in Fig. 2.

(1) Application of computer network technology in organization and management of foreign language teaching.

In order to make use of computer network technology to establish a foreign language teaching library for colleges and universities, and to facilitate the institutions to master the overall situation of foreign language teaching, the colleges and universities will carry out foreign language teaching into the comprehensive information management system of universities and colleges. From the project planning, project approval, preparation of project implementation plans, implementation plan approval, and other preparatory stages of the project all operate in the integrated information management system and keep the project information in the system. First, according to the annual foreign language teaching plan of the institutions, the foreign language teaching will be established in the comprehensive teaching and ecological TOA system, and the project basic information will be entered, including the project implementation time, the number of foreign language ecological teaching human resources, project organization methods, project establishment background and project analysis. Foreign language ecological teaching goals have predetermined foreign language ecological teaching objects and foreign language ecological teaching range.

In the integrated teaching ecology TOA system, the foreign language ecological teaching implementation plan is decomposed and entered according to specific teaching items, and foreign language ecological teaching objectives, foreign language ecological teaching contents, steps and methods as well as personnel division are specified for each foreign language ecological teaching item. After the implementation of the foreign language eco-teaching implementation plan is completed in the system, it will be approved by the higher-level leaders in accordance with the approval process, and after completion of the implementation plan revision and improvement according to the approval of the superior leaders, the foreign language teaching will be implemented using the data exchange function of the integrated teaching ecology TOA system. Basic information packages and project business packages are packaged and delivered. The basic project information packages and service packages delivered to the foreign language teaching group personnel on-site management information system to ensure that the foreign language teaching group of teachers of foreign language ecological teaching content consistent.

Through the combination of the teaching of ecological TOA and AO system, the focus of foreign language teaching is clarified. Through the splitting of the implementation plan, the goal of foreign language ecological teaching is clearer. The foreign language teaching team can form a global awareness of the project's profile by viewing the teaching of the TOE system. There is a clear guidance on the specific tasks and procedures of the division of labor, laying the foundation for the future implementation of the ecological teaching of foreign language teaching.

(2) Application of computer network technology in the course implementation phase.

First of all, the foreign language teaching group teachers will introduce the foreign language ecological teaching table into the on-site implementation information management system, according to foreign language ecological teaching division of labor and in the teaching of the ecological TOA system, a good foreign language ecological teaching target requirements have foreign language ecological teaching content and steps and methods to guide the implementation of project foreign language ecological teaching, and foreign language teachers will record the implementation process of foreign language ecological teaching to the teaching of ecological TOA system, as shown in Fig. 3 as follows.

According to the foreign language ecological teaching goal requirements, the foreign language teachers adopt foreign language ecological teaching implementation measures, and the problems found in the foreign language ecological teaching are formed according to the corresponding questions and qualitative basis, and recorded in the teaching ecological TOA system. During the implementation of the foreign language teaching

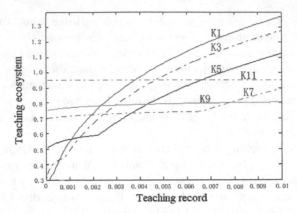

Fig. 3. The implementation process of foreign language ecological teaching record to teaching ecological TOA system diagram.

site, the teachers of the foreign language teaching group will timely implement the foreign language ecological teaching content that is filled in during the course according to a certain period of time, including the process records in the teaching of the ecological TOA system and foreign language ecological teaching. The other information is packaged and the data package is transmitted to the foreign language teaching team leader, so that the foreign language teaching leader can timely grasp the ecological teaching of the foreign language teaching by implementing the management information system of the foreign language ecological teaching. Through the application of computer network technology in the implementation of foreign language teaching sites, the implementation of the ecological teaching of foreign language teaching can be carried out in depth, which provides a powerful guarantee for the realization of foreign language ecological teaching goals. At the same time, the organization and management of the project is also more standardized and the quality of foreign language ecological teaching is guaranteed. Organization management work more efficient operation.

3 Conclusion

This paper studies the construction model of college English ecological teaching model under the ecological teaching of computer network. The purpose is to promote the student's overall healthy development, and exert the concerted efforts of various factors in the classroom, and to promote the smooth development of the entire classroom teaching process. Lead teachers can change their educational concepts, and teachers use the whole, contact, harmony, symbiosis and other ecological perspectives to carry out and reflect on classroom teaching, and improve the efficiency and quality of college English classroom teaching, and update educational concepts, and promote education. The sustainable development of the ecosystem and the healthy development of the teaching and education reform have strong practical significance.

Acknowledgment. The Research Project of Teaching Reform in Ordinary Universities in Hunan Province, 2018 (NO. 355); The Research Project of Teaching Reform in University of South China, 2018 (2018XJG-YB133).

References

1. Wei, J.: The concept of building foreign language learners' individual ecological environment in the multimedia network environment. Comput.-Assist. Foreign Lang. Educ. (2010)
2. Wei, L., Lam, W.F.: Network structure, resource availability, and innovation: a study of the adoption of innovation in elderly services in Shanghai. Comput.-Assist. Foreign Lang. Educ. (2014)
3. Lin, S.: A study of the ecological web-based college English teaching environment from the constructive perspective. Comput.-Assist. Foreign Lang. Educ. (2012)
4. Li, F.Y.: Research on ecological college English curriculum in the online language learning settings. Comput.-Assist. Foreign Lang. Educ. (2011)
5. Wang, S.L.: The research of foreign language teaching pattern under the perspective of ecology. J. Heilongjiang Coll. Educ. (2012)
6. Wang, Q.: A study of building foreign language learners' individual ecological environment in higher vocational colleges. Sci. Technol. Vis. (2014)
7. Yao, J.P.: Teaching practice based on the ecological integration of networks into foreign language curriculum--taking Longman English interactive applied in higher vocational colleges as a case. J. Fujian Commer. Coll. (2014)
8. Xia, L.: On ecological significance of foreign language education from the perspective of holistic education. J. Civil Aviat. Flight Univ. China (2017)
9. Sumi, S., Takeuchi, O.: The cyclic model of learning: An ecological perspective on the use of technology in foreign language education. Lang. Educ. Technol. **47**, 51–74 (2017)
10. Liu, O.: Research on the foreign language anxiety in asynchronous CMC with the view of ecological affordance. Foreign Lang. Lit. Res. (2015)
11. Zhu, Z.X., Mu, H.F.: Information technology-based college English collaborative teaching and learning. Comput.-Assist. Foreign Lang. Educ. (2012)
12. Jin, C.X., Li, X.G.: Research on the application of multimedia-and-network-based college English teaching mode–a case study in Anhui university of technology and science. Comput.-Assist. Foreign Lang. Educ. China (2010)
13. Liu, L.J., Yang, F.: A study on the rhythm and respiratory characteristics of Zhuang language. Acta Sci. Malaysia **2**(1), 26–28 (2018)
14. Jain, G.: English language competency: need and challenge for enhancing employability in Indian graduates. Soc. Values Soci. **1**(1), 146 (2019)
15. Erlano-De Torres, J.A.: Performance of sophomore information technology students in assembly language subject based on different teaching methods. Inf. Manag. Comput. Sci. **2**(2), 10–13 (2019)
16. Qian, C.: An empirical study on constituents of translation ecological system based on translators' cognition. Foreign Lang. Educ. (2012)
17. Zhou, J.: On the ecological mode of grammar class for English majors based on Likert scale--exemplified by foreign language school of CUIT. J. Hubei Corresp. Univ. (2016)
18. Duan, C.Y., Shen, Y.: Measure and analysis of ecological niche of foreign students education system based on empirical research in Yunnan Province. Meitan High. Educ. (2012)
19. Nordin, L., Razak, N.Z.A., Kassim, R.: Language learning strategies from the perspective of undergraduates in a private engineering technology university in Johor. Educ. Sustain. Soc. **2**(2), 9–16 (2019)

20. Shaleha, M.A., Purbani, W.: The existence of literary works in language teaching materials to support character education. Educ. Sustain. Soc. **2**(4), 11–13 (2019)
21. Kumara, G., Kumarb, M., Guptac, M.: The role of managing engineering education in India. Educ. Sustain. Soc. **3**(2), 45–46 (2020)
22. Kirova, S., Veselinovska, S.S.: An attempt of integration of teaching contents of the subject's ecological education and English as a foreign language. Procedia – Soc. Behav. Sci. **15**(1), 1220–1225 (2011)
23. Ibrahim, K.: Player-game interaction: an ecological analysis of foreign language gameplay activities. Int. J. Game-Based Learn. **8**(1), 1–18 (2018)
24. Rimfeld, K., et al.: Genetic and environmental influences on first and second language achievement at the end of compulsory education in UK. Behav. Genet. 680–680 (2014)
25. Qiang, L.I., Department, F.L.: Paths for ecological classroom construction of college English under the background of education informatization. J. Hubei Correspondence Univ. (2018)
26. Setiawana, B., Iashab, V.: Corona virus disease 2019: the perspective opinion from pre-service elementary education teacher. Educ. Sustain. Soc. **3**(2), 47–50 (2020)
27. Rosa, A.T.R.: Multicultural education system value engineering model in strengthening national identity in the era of the industrial revolution and society 5.0 (R&D study in tebuireng higher education in east java). Educ. Sustain. Soc. **3**(1), 1–4 (2020)
28. Nkemdilim, E.R., Okeke, S.O.C.: Effect of computer-assisted instruction on secondary school students' achievement in ecological concepts. Int. J. Prog. Educ. **10**, 6–13 (2014)
29. Jia, H.: Ecological analysis on college English teaching under the environment of computer network. J. Soc. Sci. Jiamusi Univ. (2014)
30. Zhang, M.L., Ma, .Y., Yang, S.: The exploration of ecological college English teaching mode based on the target of cultivating international talents. J. Jilin Teach. Inst. Eng. Technol. (2012)
31. Simangunsong, E.: Cyberbullying: Identification of factors affecting the quality of higher education in Indonesia. Educ. Sustain. Soc. **3**(1), 15–19 (2020)
32. Haruna, R., Kamin, Y.B.: Application of work-based learning model in technical and vocational education: a systematic review. Educ. Sustain. Soc. **2**(4), 1–4 (2019)
33. Ji, G., Gunasekaran, A.: Evolution of innovation and its strategies: from ecological niche models of supply chain clusters. J. Oper. Res. Soc. **65**(6), 888–903 (2014)
34. Guo, H.: Study on minority bilingual education research achievements based on ecology horizon. J. Qinghai Normal Univ. (2014)

FCST Synergy Education Model Based on Mobile Internet Technology in Chinese Higher Vocational Colleges

Fen Tan[✉]

Institute of Media Art, Hubei Science and Technology College, Wuhan 430074, China

Abstract. FCST collaborative education model is a creative idea for vocational college students cultivating, especially under the background of COVID-19 pandemic. Family, society, school and teacher work together to provide comprehensive education to students. Based on the background of mobile Internet technology, this paper proposes an interactive mode of "FCST" to train higher vocational education students from families, schools, teachers and society. Research methods: literature retrieval method; questionnaire survey; sample interview; logical induction. Research conclusions: Mobile Internet technology is a mature technology and can be widely used in higher vocational education in China. In the process of ideological education for college students, families, schools, teachers and society jointly construct the "1+2+3+N" "FCST" collaborative education model.

Keywords: Mobile Internet Technology · Chinese Higher Vocational Education Institutions · "1+2+3+N" Route · FCST Synergy Education Model

1 Background

Higher vocational education institutions are important positions for cultivating artisans in great powers and are important platforms for carrying forward the spirit of artisans. The Optimization and Innovation of Educational Model in Higher Vocational Education Institutions Is the Necessity of the Intensive Development of Universities and Colleges and an Important Measure to Strengthen the Ideological and Political Work in Colleges and Universities in the New Age. The advent of new media forces ideological and political education workers to change their way of thinking and working methods [1–3]. When the traditional working methods cannot meet the needs of education, they should actively accept the new mode of education. The concrete meaning of "FCST" mode is as follows: F refers to family; C refers to school; S refers to social society; T refers to teacher, that is, society, family and school co-educate people. The way of implementation can be summed up as "1+2+3+N", specifically as follows: For a college student in higher vocational colleges, relying mainly on two main bodies of education relations (schools and families) The school daily management + social life education + family groups and collective sports), through the establishment of N micro-Internet education

E. Bisset Álvarez (Ed.): DIONE 2021, LNICST 378, pp. 447–457, 2021.
https://doi.org/10.1007/978-3-030-77417-2_38

information platform in student ideological education and behavior management to form synergy, co-governance, mutual supervision, and jointly promote higher Adult students become successful, so as to achieve the purpose of Chinese higher vocational education institutions and families to coordinate education [4, 5].

Under the background of the global pandemic of the COVID-19, the safety management of college students has aroused great concern of the government. As the main body of student management, university plays an important role, but students' families and communities should also undertake certain management obligations. In this paper, we establish a college student management model and epidemic prevention and control system with the interaction of families, schools, students and teachers. In order to maintain the normal teaching and scientific research of college students, universities should strengthen the epidemic prevention and control to ensure the health of university teachers and students.

1.1 The Research Methods of This Thesis

Data retrieval method. With the help of internet data, full-text periodical network data and text books, search the relevant information of this paper by data retrieval method to provide research theoretical data support.

Questionnaire method. This thesis draws on the undergraduates of ten vocational schools in central China as the research object, altogether 1000 questionnaires were distributed and 928 were recovered. After invalid questionnaires were removed, 900 valid questionnaires were sent out. The author conducted a statistical analysis of valid questionnaires, and the data obtained will be used as the status quo data in the dissertation.

Interview. In order to understand the use of mobile Internet, such as WeChat, QQ and other parents, the author conducted interviews with 50 parents, mainly about the contact frequency, content and methods of parents and college students.

Logic Induction. According to his own in-depth observation and field visits, this paper summarizes the various materials studied, so that the research results more rigorous, more practical, more application value.

2 The Status Quo of Cooperative Education in Higher Vocational Institutes

Synergy Effects, in short, is the effect of "1 + 1 > 2". In 1971, the German physicist Hermann· Haken proposed the concept of synergy. In 1976, he systematically discussed the theory of synergy and published such works as Introduction to "Symposium Introduction". Synergies hold the view that there is an interaction and mutual cooperation among the various systems in the whole environment. Collaborative effects can be divided into external and internal situations, external collaboration refers to a cluster of enterprises due to mutual cooperation to share business activities and specific resources, and therefore will be more than as a single operation of the business to obtain higher profitability; internal collaboration refers to the overall effect of different parts of the enterprise production, marketing and management, and different phases and different aspects of using the same resource together [6–9].

At present, the factors that affect the construction of institutions of higher vocational education and family co-ordination mainly include the following aspects: First, the responsibility orientation of the co-education between families and colleges and universities in the country is not clearly defined at the national level so that schools and families play synergies Education for a variety of reasons led to "education vacuum." Second, due to economic pressure, work pressure and family burden, parents push the education responsibility to schools and participate in the education of common sense. Thirdly, in the process of giving play to the effectiveness of the leader of collaborative education between families and colleges, the higher vocational education institutes have not taken the initiative and made the coordinated education between families and colleges and universities appear on the surface, making it difficult to make breakthroughs [10]. To sum up, the problems that arise in the co-education between families and colleges are related to the "absence" of the cooperation mechanism of co-education between families and universities. Only when the benign cooperation mechanism is really built can all parties play their initiative, participation and initiative, So as to achieve the optimization of education effect.

The main objective of the interactive model between the family and the college is to make full use of information technology and information equipment so as to solve the problem of lack of effective and smooth communication between parents, students and schools on education issues and to narrow the distance between families and schools. The effect of communication enables parents to recognize the importance of family and school co-education and promote the healthy development and overall growth of students.

2.1 The Main Body Composition of Collaborative Education Model

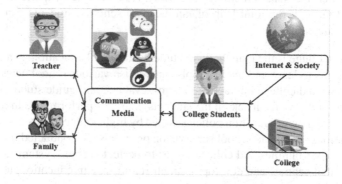

Fig. 1. College, family, society, teacher's collaborative education model.

The 21st century is a century with rapid development of science and technology. The construction of our country needs more talents. These talents not only need professional knowledge but also have sound personality [11–13]. Therefore, the ideological and political education of college students in higher vocational education should be combined with social development, keep pace with the times, follow the pace of the times, take the social road. Higher vocational college students' ideological education, relying on the school's own power is difficult to do, we must from the school, family, society share, to play their respective advantages, to make up their own limitations of perspective, from different angles to promote college student's Moral education to develop and grow, as Fig. 1.

College Students' Parents. Parents play an important role in the interactive system of home and college. The parents' expectations of co-educational model are the main functional requirements of the interactive system of home and college. In China's urban-rural one-child-based society, parents pay more and more attention to their children's education in schools. Parents hope to keep abreast of their children's school conditions, including learning, living and hobbies, through the contact of families and colleges and universities [14, 15]. At the same time, parents also hope that through the mobile communication platform feedback school specific circumstances, including positive and positive, but also negative and negative. Good ideas can be feedback to the school through the mobile Internet technology, hoping the school can further improve and jointly improve the quality of children's education, encourage students to better growth.

Vocational College Students. College students are the objects of common education both at home and at colleges and universities, and also the main body of attention of both parties. Through the interactive platform of home and college, students can see the teachers' evaluation of themselves, including positive praise and negative criticism, and formulate new learning plans based on the assessment so as to find out the gaps and make learning more directional and motivated [16–18]. Students at different stages of growth have different worldviews and values, student assessment is also a way to promote better student growth, mutual evaluation can understand what their classmates in a position, what behavior is more recognized by the students, which it is unwelcome, in order to better understand themselves. Finally, students often experience confusion and loss, and through the joint help of families and schools, students can feel warm during their education.

The College. College is the main body of students in school education, is an important place for students to learn and live. Schools can publish various school news and information on home and college interactive systems, evaluate and guide students 'thoughts and behaviors, and give feedback on the students' learning performance in recent time. They can also make new education policies issued by higher education authorities In the same way, various social and school supervision programs are conducted through interactive platforms of families and colleges so as to perfect various systems and policies in schools so that schools can develop in an all-round way in education, teaching and personnel training.

The Society. At present, China is in a crucial period of reform and opening up. Various kinds of ideology and culture are intertwined and stirred up with each other, leading to the formation of diverse social values. As a special group in society, the main living environment of college students in higher vocational education is school, but their influence by social environment can not be avoided. Poor social environment will lead to the spirit of college students poverty, moral loss, psychological imbalance, therefore, it is necessary to strengthen the social environment. The concept of socialization of ideological and political education among college students in higher vocational education should be analyzed and researched from many angles and many factors.

3   China Mobile Internet Development and College Students' Life

China mobile Internet development status quo. August 4, 2017 China Internet Network Information Center (CNNIC) released the 40th Statistical Report on Internet Development in China (hereinafter referred to as the "Report") in Beijing. The Report shows that as of June 2017, the number of netizen in China reached 751 million, accounting for one fifth of the total number of Internet users in the world. At the same time, China's Internet penetration rate was 54.3%, surpassing the global average of 4.6 percentage points. Mobile Internet users accounted for 96.3% of the dominant mobile Internet, Internet users to further mobile devices centralized.

For social applications, WeChat, QQ space and blog became the top three social application platforms with user usage rates of 84.3%, 65.8% and 38.7% respectively. According to statistics, WeChat and QQ space utilization advantage is very obvious, and the latter has opened the gap.

After 90 college students are accompanied by the development of mobile Internet up a new generation. With the continuous decline in the price of smart phones and mobile traffic tariff reduction, college students have become one of the major mobile Internet. In the classrooms, dormitories, restaurants, bus stations or subway, everywhere with a cell phone access to college students, mobile phone-based mobile Internet has been deeply integrated into the daily life of contemporary college students and profoundly affect and change the students learning and lifestyle.

The advent of new media forces ideological and political education workers to change their way of thinking and working methods [19, 20]. When the traditional working methods can not meet the needs of education, they should actively accept the new mode of education. WeChat relies on these unique features and advantages to attract a large number of users, especially undergraduates, to make college students become one of the major user groups of WeChat and also one of the groups that WeChat needs to pay attention to when it applies and develops software.

College students use WeChat survey. College students use the mobile Internet time. The overall time spent by college students on mobile networks is generally high, with only 33% of students having less than 3 h, 43% of students between 3 and 5 h, and 6% of students even up to 8 h or more. According to the location of the Internet during the day, 84% of college students use their mobile phones for "anytime during the day", followed by "before going to bed" (76%) and "When using the toilet" (56%) and "Wake up in the morning" (51%).

3.1   The Reason and Purpose of Using the Mobile Phone to Surf the Internet

Most of the college students choose to use the mobile phone network because of the "can send a boring time" (75%), "Internet access is free from location restrictions" (74%) and "easy operation" (68%). The main purpose of using the mobile phone network is three. The first is "communication" (89%), the second is "entertainment" (89%) and the third is "access to information" (88%).

3.2 College Students Commonly Used Mobile Internet Applications

The most commonly used mobile phone network applications for higher vocational college students are instant messaging (WeChat, QQ, etc.), personal space like Blog, QQ, search engine, online payment and online music. Among them, every day will use the top five are "instant messaging (WeChat, QQ, etc.)" "Blog, QQ and other personal space" "search engine" "network news" and "online music." From the aspect of each option, instant messaging tools such as WeChat, QQ and so on are used frequently by college students, which is obviously higher than the 91.2% usage rate of instant messaging by nationwide mobile Internet users. Second, the daily use of mobile phones by college students is mainly based on the three types of communication, access to information and entertainment, which is basically consistent with the motivation and purpose of using mobile phone networks for college students.

3.3 College Students Use Mobile Internet to Connect with Family Members

When asked about how often college students and their parents used mobile Internet, 31.03% of respondents chose to get in touch with their parents daily; 26.97% of college students thought they would get in touch with their parents every other day; and 23.40% Students from time to time communicate with their parents using their mobile Internet with their parents. Only 18.60% of respondents choose to use their mobile Internet technology to communicate with their parents rarely. Their contact information is mobile phone.

3.4 College Students Use Mobile Internet Technology and Contact Content

Survey shows that students and teachers to communicate channels, the use of mobile Internet technology accounted for 98.4%. 45.12% of the content is the daily management of the students; 24.21% of the students are learning content; and 19.74% of the exchange is about the students' safety and personal development of the content; 10.93% of the content is the development of party members, social and other content.

3.5 Technical Support and Design Principles

Use the terminal. Android is an operating system designed specifically for mobile devices such as mobile phones, PC and laptop, so the system is designed for simplicity, performance and power savings. Android is largely based on Linux (Linux kernel, currently version 3.0.1) and GNU software. In other words, Android is based on the Linux operating system for mobile terminal development. However, Android is not completely compatible with traditional Linux systems, for example, Android has a proprietary driver, no Glibc support, no native windowing system, and more. Therefore, all Linux / GNU applications can not be ported to Android.

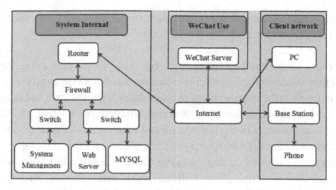

Fig. 2. System physical architecture diagram.

The physical structure of the system. According to the functional requirements of the system, the client can complete the related functions through the WeChat public number and browsing. The server must achieve Internet interconnected, in order to ensure that the server and client network to data transmission and interaction. Users can use system functions from their cell phones, laptops, PCs, etc., and system administrators must use Web applications to manage the system. Therefore, the system physical environment generally includes server devices and network devices. All kinds of client devices access the WEB server through the network. The WEB server interacts with the database server to obtain the data, and then returns the data to the requesting user. Detailed physical architecture (As shown Fig. 2).

Data transmission. In the application architecture, consists of two parts, namely, mobile client and back-end server-side composition. Whether it is a mobile client or a back-end server, the MVC schema is uniformly followed in the technical architecture. MVC (Model View Controller) is a model-view controller model. MVC is a design paradigm. When organizing code, business logic, data, and display are separated. In this way, when improving and optimizing the interface and user interaction data, No need to rewrite business logic code. MVC has evolved independently for importing, exporting and processing traditional input-output (UI) structures into the user interface. MVC ideas and patterns in the background server-side development of the more mature mode, here focuses on the application of MVC in mobile client, as follows:

- Model: The core part of the application to complete the storage and processing of entity data.
- View: An XML layout file is used to describe the interface in Android. The interface is used to generate user interface with the system. Android provides a rich interface layout and flexible interface control technology.
- Controller: The control layer is actually Activity, which is determined by the life cycle of the Activity. Through the Activity to complete the interaction of the entire business logic, where the interaction with both human interaction, but also interact with the system environment through this interaction to control the realization of the entire system business logic. Use the set Content View method in the Activity to display the

specified design interface, the user input data in the displayed interface, the control interface to display and update the model state.

In recent years, China has vigorously developed vocational education, the number of students in higher vocational education institutions has been constantly expanding, and the problems existing in the management of students in higher vocational education institutions have also become increasingly prominent. Under the new situation, strengthening the management of students in higher vocational education institutions and improving the management effect are the needs of promoting the sustainable development of higher vocational education institutions and ensuring the success of adult students. At a 2016 national conference on ideological and political work in universities, Xi Jinping, China's president, emphasized that "ideological and political work should be carried out throughout the entire process of education and teaching so as to achieve full-time education and full-scope education."

4   Construction of FCST Synergy Education System

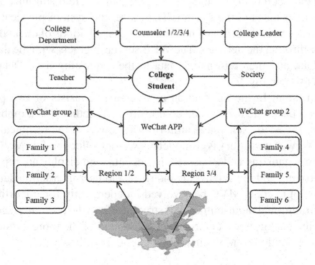

Fig. 3. Construction of "FCST" Synergy Education Model.

In the era of mobile Internet, parents can not only understand the academic status of their children, but also understand the academic status of their classes, colleges and even schools, and fully mobilize the parents' motivation to participate in school education. The mode of collaborative education for families and colleges and universities in higher vocational education institutions is a starting point for strengthening the ideological and political education in colleges and universities in the new era. It establishes a bridge of cooperation between school-based education and family education, and promotes higher vocational education Colleges and universities trained craftsmen. The paper designed the higher vocational education colleges and universities and colleges and universities collaborative education "mobile Internet" model, as Fig. 3.

When applying the FCST system, universities should first classify and code the information of students and their families to form a database, so as to realize the sharing of database resources in the whole university; secondly, classify the students and their families database according to the geographical region, and establish several groups; thirdly, collect the wechat accounts of students and their parents for grouping and coding, filter the error information, and form data The university will input the personal information of teachers into the data matrix, and get the data module of each region. The Internet platform established by colleges and universities is used for constant monitoring and information interaction.

5 Conclusion

With the mobile phone network entering the era of 4G and 5G, higher vocational education institutions can organize teachers and students of the Information Network Institute of our college to develop APP and APP for collaborative education of colleges and universities. Just like Learning Chinese APP software, News, map of our school, student moral education, enrollment and employment, etc. In the column of student moral education, we embed the theoretical content of the ideological and political education and the work experience sharing of the counselors so that the parents can increase the number of students who browse the APP software Understanding of ideological and political education, and enhance their children's moral education. The true effectiveness of information network platform in the development of family and college collaborative education is reflected in the effective interaction between school and family, which can create a one-to-one, many-to-many and many-to-many interaction with each other. Artificial intelligence internet technology should be fully applied in the campus management of college students, especially in the face of public health emergencies, artificial intelligence technology can be used in the behavior management, teaching management and life management of college students. In the process of campus closed management, Internet technology can not only solve the problem of students' contact with their families, but also solve the obstacles of academic discussion between teachers and students. Tencent video conference APP, zoom, DingDing video conference APP and Wechat video conference APP are widely used in Chinese universities.

Family and college co-education need to play the leading role of the school, but also need to mobilize the enthusiasm of parents involved. Compared with the school, parents in the family and college collaborative initiative was significantly less than the initiative. The establishment of a parent committee, hiring high education, coordination ability, with strong ideological and political education theory and practical ability of parents to serve as chairman, responsible for the organization and management of family and college co-education of family-related issues. The result of doing so can effectively reduce the workload of counselors and ensure the smooth connection between schools and families, so as to provide guarantee for higher vocational education institutions to train qualified personnel.

Acknowledgment. Hubei Province Colleges and Universities Practice Education Characteristics Project Fund Project "Study on the Optimization and Innovation of Synergy Education Mode in Hubei Higher Vocational Education Institutions under the Background of "Mobile Internet" (2017SJJPB4007).

References

1. Tang, X.Y.: The ideological and political quality of college students in the perspective of core quality. Theory Pract. Educ. **37**(10), 29–31 (2017)
2. Terry, H., Mark, T.: Moral phenomenology and moral theory. Philos. Issues **1**, 343–351 (2005)
3. Novianto, V., Yogiarni, T., Meidasari, V.E., et al.: Creative economy education-based preservation of gebleg renteng batik culture in Kulon Progo district. Soc. Values Soc. **1**(3), 04–06 (2019)
4. Kurniadi, B., Munggaran, N.R.D.: Political education and political participation of millennial generation in the 2019 Indonesian elections. Soc. Values Soc. **1**(4), 18–20 (2019)
5. Yadav, U.: A comparative study of ancient & present education system. Educ. Sustain. Soc. **1**(1), 160 (2018)
6. Gu, L.M.: Using school websites for home–school communication and parental involvement? Nordic J. Stud. Educ. Policy **3**(2), 133–143 (2017)
7. Mark, M.K.: Community-university engagement: a process for building democratic communities. ASHE High. Educ. Rep. **2**, 45–52 (2014)
8. Kumar, G., Kumar, M., Sandeep Kamboj, S.: Research in higher and technical education by the help of area of applied and innovative ideas for promoting R&D. Educ. Sustain. Soc. **1**(2), 15 (2018)
9. Yin, X.Y.: Research on the application of problem-based teaching in mathematics education. Educ. Sustain. Soc. **2**(1), 8–11 (2019)
10. Shi, T.K., Chengyi Li, C.Y., Zhang, X.H., et al.: A practical study on the construction of high-quality medical humanities gold course under internet background. Soc. Values Soc. **2**(2), 20–22 (2020)
11. Wu, S.L.: Analysis of mental health status and its influential factors of students in vocational colleges. China J. Health Psychol. **01**, 139–142 (2018)
12. Xiao, C., Chen, Z., Meng, W.U.: Investigation on mental health of higher vocational college students. J. Baotou Med. Coll. **78**(5), 382–390 (2016)
13. Wang, L.: Research on the divergence of higher vocational college students' addiction to mobile phones. J. Yangzhou Coll. Educ. **35**(2), 64–67 (2017)
14. Jamin, J., Arifin, N.A.M., Mokhtar, S.A., et al.: Privacy concern of personal information in the ICT usage, internet and social media perspective. Malays. E-Commer. J. **3**(2), 15–17 (2019)
15. Othman, R.B., Rahim, K.F., Kamarulzaman, R.A.B., et al.: Literature review on internet benefits, risks and issues: a case study for cyber parenting in Malaysia. Malays. E-Commer. J. **3**(2), 12–14 (2019)
16. Zeng, X.L.: A study of the innovation of collaborative education in private higher vocational colleges ideological and political theory courses teaching. J. Jingdezhen Univ. **02**, 86–89 (2016)
17. Gretchen, M.W., Michael, W.F.: Historical and contemporary developments in home school education. J. Res. Christian Educ. **18**(3), 303–315 (2019)
18. Paramitha, M., Agustia, D., Noorlailie Soewarno, N.: Conceptual framework of good governance, organizational culture, and performance at higher education in Indonesia. Inf. Manag. Comput. Sci. **1**(1), 18–23 (2018)

19. Huang, W.N., Dai, F.: Research on digital protection of intangible cultural heritage based on blockchain technology. Inf. Manag. Comput. Sci. **2**(2), 14–18 (2019)
20. Unn, D.K.B.: Parental involvement practices in formalized home–school cooperation. Scand. J. Educ. Res. **54**(6), 549–563 (2010)

Reform and Innovation of Financial Teaching in the Environment of Financial Crisis

Pei Sheng(✉)

School of Foreign Languages, Xi'an Shiyou University, Xi'an 710065, China

Abstract. How to apply Internet technology to the classroom in the teaching process of applied undergraduate colleges is a question worth studying. At present, in the context of the domestic financial crisis, it is in this form to create learning videos, students can watch at home or outside the classroom, can study completely at home, or can return to the classroom to study. Face to face communication between teachers and students through the Internet reduces a sense of oppression. And the emergence of the financial crisis has replaced the lack of traditional classroom teaching, highlighting the dominant position and encouraging changes in teaching methods. In the environment of financial crisis, the reform and reform of education is still the same theme of the times, especially the development and innovation of educational methods, combined with the current reform of education and learning forms, in fact, the use of information resources sharing. Based on this, a new model suitable for undergraduate education in China is put forward, which can be used for reference to the innovative research of "Internet+" model of undergraduate education. This paper focuses on the financial teaching reform and innovation path under the financial crisis.

Keywords: Financial Crisis Environment · Financial Teaching Reform · Innovation path

1 Introduction

In the contemporary mode of education, teachers' teaching methods are also constantly improving and improving, in order to make classroom teaching no longer single solidification, the use of modern Internet technology combined with classroom teaching is the most common way so far. Teaching knowledge for many students are more profound difficult to understand, so many students give up learning, to re-engage students to learn confidence, teachers need to combine the current popular Internet technology with classroom content. In today's innovation and reform of education, we should also keep up with the pace of the times and apply convenient and fast Internet technology, which not only makes the teaching method new, but also reduces the time for teachers to give lectures, so that students can learn a lot of knowledge through the Internet [1]. The development of educational innovation is still the most important thing in the educational field, because it can effectively promote the balanced development of compulsory

E. Bisset Álvarez (Ed.): DIONE 2021, LNICST 378, pp. 458–463, 2021.
https://doi.org/10.1007/978-3-030-77417-2_39

education. Prime Minister Li Keqiang has also mentioned strengthening education to speed up the construction of a modern education system. Based on this, we can add the Internet to the existing teaching methods to change the traditional teaching methods of the former single stereotype.

2   Main Problems of Traditional Teaching Mode in the Context of Financial Crisis

2.1   No Comprehensive Training for Students

The traditional mode of teaching in our country can not satisfy the present mode of educational development completely, because the traditional mode of education focuses more on the teaching of students' knowledge, and neglects the cultivation and development of students' various abilities, especially the lack of emphasis on the cultivation of students' innovative ability, as Fig. 1.

Fig. 1. Financial crisis environment.

2.2   Ignoring the Student's Dominant Position

In the traditional teaching mode, teachers have always been the main body, but this kind of education mode is not conducive to students' learning, because students are the "protagonist" in teaching activities, teachers are only "supporting roles", must not confuse the meaning of the two roles. Teachers impart knowledge to students, students learn passively, and subjectivity is greatly inhibited. There is also that most students' learning attitude is relatively negative, not many times will not have a clear plan for their future development, if the attitude of learning, is not conducive to the overall development of students [2]. In particular, the lack of compulsory training and the lag in the construction of the ranks of agricultural teachers have prevented the realization of the current equal education programme. This requires us to think about what we can do and to study the more advanced scientific educational ideas abroad. Strive to improve the current situation of teaching in China, innovate the mode of education and training, and promote the "transparency" and "teaching nature" of teaching resources.

2.3   Significance of Teaching Reform in the Context of Teaching Financial Crisis

2.3.1   Improving Students' Learning Initiative

Fig. 2. Financial crisis environment.

With the continuous progress and development of science and technology, the traditional teaching mode of the past can not meet the requirements of the development society for teaching quality, but also can not train the talents who meet the requirements of today's society. In the process of teaching, put into the use of the Internet, Internet teaching is mainly through the combination of pictures, words, video, sound construction. It can make students feel fresh and urge students to take the initiative to understand learning, so as to improve students' active initiative in learning and improve their learning efficiency, as Fig. 2.

2.3.2   Enabling Teachers to Lighten Their Burden

The Internet is mainly supported by the network, which is convenient and fast in searching data and fast in information transmission. At this time, it can provide good lesson preparation content for teachers, and analyze and summarize students' learning situation through the Internet. Then according to the students' learning situation, make the corresponding teaching plan, so as to reduce the pressure of teachers' lesson preparation, save a lot of time, teachers can spend more time to help students learn.

2.3.3   Innovating and Reforming Modern Teaching Methods to Enable Students to Turn Passive into Active Learning

As a result of the new education model has adopted a large number of innovative Internet network teaching, such as the above micro-class vision, MOOC and flipping classroom and other new learning tools. The greatest advantage of these new media learning tools is that they can break the time limit and space limit, so that students can choose their own online learning tools, which is also the best way to learn freely.

2.3.4   Developing New Educational Innovation Ideas for "Openness" and "Sharing"

Colleges and universities should actively implement the new educational concept of "opening up" and "sharing", and new teaching ideas appear under the background of modern teaching. The main purpose of college education is how to promote and advocate the new concept of transparent sharing of education and how to better use the new educational concept to train students [3] In addition, the current way of education in our country is to find a suitable way for the development of education in our country, because only by finding a suitable way for ourselves can we effectively develop the new concept of "open" and "shared" education and have the opportunity to look forward to the bright future of education [4–6].

3 Concrete Practice of Applied Undergraduate Colleges the Background of Financial Crisis

3.1 Design Rich Teaching Activities to Enhance Students' Interest in Learning

Teachers can download pre-recorded teaching videos to the Internet, students can preview the course through the teaching platform, and then provide feedback to teachers on the problems they face in their own learning process [7, 8]. Teachers allow students to discuss learning problems in groups in groups, and classroom teaching is the driving force to effectively improve students' communicative ability and thinking ability [9–11]. Organize and plan different teaching activities to improve students' interest and enthusiasm for learning, as Fig. 3.

Fig. 3. Financial crisis environment.

3.2 Changing Teachers' Teaching Concepts and Increasing Teaching Videos

Changing teachers' concept is a very important link in educational reform. Teachers' teaching methods should be constantly innovated. In the normal teaching process, we should pay attention to the cultivation of students' subjective initiative, set a good example and model for students, and use people-oriented ideas to cultivate students' learning concepts. Teachers should make it clear that students should mainly improve their learning ability in teaching. In normal teaching to increase the shooting of teaching video, in the grade can also be selected "best learning video" and other activities, so that all teachers and students can participate in the wave of video teaching [11–13].

3.3   Construction of an Information Network Platform and Improvement of the Scientific Evaluation System

Teaching should also build the corresponding network platform teaching, on the basis of using the flipped classroom, first of all, to make use of the characteristics that students like to be in close contact with the Internet. Today's students compare electronic products, through the information network platform can effectively improve students' interest in learning and initiative, effectively enhance the network information teaching learning platform. Can enable students to learn different new knowledge, help them correct their learning thinking and exercise their learning ability, so that students can learn teaching knowledge from various teaching learning software.

4   Conclusion

In the current development of education in China, the traditional mode of high-efficiency education still has a lot of drawbacks, can not meet the current situation of education, but with the rise of the Internet, in the way of education has changed a lot. Therefore, with the development needs of modern education and the background of the times of the Internet, we can develop all kinds of comprehensive qualities and innovative abilities of students in an all-round way, on the one hand, to improve classroom efficiency and students' self-study ability, on the other hand, to carry out the new concept of education and promote the harmonious development of education. In teaching, the success of the application of modern Internet technology in college teaching depends largely on the cooperation between teachers and students. In the context of the financial crisis, the advantages of using this form of teaching to teach students have been shown most vividly in this paper, and the teaching results are certainly self-evident, whether in our country or abroad, the major colleges and universities are already using this method of teaching, in terms of results, it is indeed very effective [14]. Therefore, if our country wants to carry on the education reform innovation, must carry on the teaching method innovation from the actual point of view unifies own development situation, can play the fundamental role, this technology has become the education reform innovation essential tool. In order to do a good job of educational reform and innovation, we also need the joint efforts and progress of educators. In order to provide a strong guarantee for Internet teaching, we should allocate teaching resources scientifically, improve teachers' teaching level and ability, and strengthen investment in human and financial resources [15]. Through the above methods, students can feel the fun of learning, enhance their confidence in learning, so that students have a strong interest in teaching this subject. Internet technology is used to improve teaching methods, mainly to improve the quality of teaching.

References

1. Huang, Y.J.: A study on the education and teaching reform of finance specialty in colleges and universities. Internet Finan. Intell. **4**, 144 (2020)
2. Chen, X.: Behavioral finance course teaching reform and research. Sci. Educ. J. **35**, 94–95 (2019)

3. Wu, W.Q., Zhang, Y.Y.: A Study on the teaching reform of financial mathematics course in mathematics and applied mathematics specialty. Innov. Educ. Res. **6**(3), 181–185 (2018)
4. Zhao, H.Y., et al.: Catalytic reforming of volatiles from co-pyrolysis of lignite blended with corn straw over three different structures of iron ores. J. Anal. Appl. Pyrolysis **144**, 104714 (2019)
5. Guo, J.Y., Xiong, M.J.: The characteristics of inbound tourism in Myanmar. In: Advanced Management Science (AMS), vol. 9, no. 1 (2020)
6. Fadhil, H., Shawi, J.A.: The effect of the four strategic thrusts on improving the performance of educational institutions-An exploratory search in Basra private modern schools. In: Advanced Management Science (AMS), vol. 8, no. 1 (2019)
7. Nketiah, E., Gyamfi, G.A., Adjei, M., et al.: The determinants and the impact of trade openness on foreign direct investment in east Africa. In: Advanced Management Science (AMS), vol. 8, no. 1 (2019)
8. Ding, J., Li, C.B.: Application reliability group evaluation with parameters of big data in energy internet in China. In: Advanced Management Science (AMS), vol. 8, no. 1 (2019)
9. Oluwashakin, A., Oti, A.O.: No poverty vs reduced inequality: partnerships to achieve the goal. In: Advanced Management Science (AMS), vol. 7, no. 1 (2018)
10. Sharma, M.: Women empowerment-an overview of the global context. In: Advanced Management Science (AMS), vol. 7, no. 1 (2018)
11. Ibarra, V.C.: Human resource accounting: its constraints and limitations in the Philippines. In: Advanced Management Science (AMS), vol. 7, no. 1 (2018)
12. Ogbaisi, A.S., Areo, A.B., et al.: Corporate governance and financial performance of firms in Nigeria: Evidence of money deposit banks. In: Advanced Management Science (AMS), vol. 7, no. 2 (2018)
13. Qin, L.: The status quo and summary of China's platform emerging industries based on mobile internet opportunity. In: Advanced Management Science (AMS), vol. 7, no. 2 (2018)
14. Zhu, Z.W.: A preliminary study on the reform of English teaching in higher vocational finance specialty. Sci. Educ. Guide Electron. Ed. **12**, 230 (2019)
15. Liu, X.M.: A study on curriculum reform of higher vocational finance specialty group based on teaching supply-side reform. Shandong Youth **12**, 81–82 (2019)

Research on the Cultural Mission and Path Orientation of School Physical Education Under the Background of Core Values

Daiyong Li[✉]

Xichang University, Xichang 615000, China

Abstract. The humanistic view and socialization of sports are the internal and external factors of the development of school culture, and are the premise and fundamental motive force of the development of school sports . In order to adapt to the development of physical education teaching concept and meet the needs of physical education teaching, school physical education should establish the core values of physical education curriculum suitable for the requirements of contemporary social development. In the process of deepening educational reform, colleges and universities should also recognize the social needs in the new era and consciously undertake their own sports cultural mission. Colleges and universities should further consider the important practical significance of the current sports core values on the basis of re understanding the sports core values. The development of school physical education should be guided by the new concept of health, and the teaching process of college physical education should be people-oriented.

Keywords: Core values · School sports · Cultural mission · Practice path

1 Introduction

School physical education is entering a brand-new period of historical development. Revisiting school physical education's essential function will greatly promote our deeper understanding of the intrinsic value of school physical education. Pushing forward the construction of a healthy China is an important foundation for building a well-off society in an all-round way and realizing modernization. It is also a national strategy to comprehensively improve physical quality and realize the coordinated development of people's health and economic society [1]. Therefore, the college sports culture, which is formed by the integration of college sports and sports culture, leads the fashion of college campus culture with a unique attitude and vigorously promotes the development of college sports. According to the historical development process of social culture, education is an important component [2]. Although various social cultures, including mass culture, exert an irresistible influence on school culture and education, even this influence has many negative elements. Therefore, from the perspective of core values, we should examine the relevant issues concerning the value of school physical education in our country,

E. Bisset Álvarez (Ed.): DIONE 2021, LNICST 378, pp. 464–470, 2021.
https://doi.org/10.1007/978-3-030-77417-2_40

further understand and grasp the status, functions and functions of school physical education, and provide references for promoting the scientific and orderly development of theoretical research and practice of school physical education in our country.

2 Analysis on the Current Situation of Sports Culture Development in Colleges and Universities

In our country's school education system, sports culture has been neglected and its development lags behind. In primary and secondary schools, the first priority of the school is to let students pass the examination and enter key secondary schools and famous universities. In order to carry out scientific research on the core values of Chinese physical education curriculum, we must first sort out the evolution and development process of the core values of Chinese physical education curriculum, so as to excavate the characteristics of the core values of different physical education curriculum in different periods [3]. Influenced by traditional ideas, universities still focus on training students' professional skills and imparting scientific knowledge, and the inheritance of humanistic knowledge and humanistic spirit has not received due attention. School sports culture is a cross-cutting component of social culture and sports culture. It is the product of the interaction and mutual influence between sports culture and campus culture. It is the sum of sports spirit and wealth created jointly by school staff in practice. The current social and cultural value system has seriously affected the development of our school sports, not to mention its leading and positive influence [4]. School physical education does not know itself from inside, but needs to look back at itself from outside. School physical education has been deprived of its right to pursue its own value and meaning, or abandoned voluntarily, while focusing on the practical effect of skill training. School physical education has lost its value of existence due to its neglect of the noumenon value.

3 Cultural Mission of School Physical Education

3.1 Enrich Campus Cultural Life

School sports culture is a combination of sports culture and campus culture, and it is a unique cultural phenomenon with profound cultural connotation and extension. Within the scope of schools, influenced by social culture, many colleges and universities have seen the imbalance of students' sports culture. The level of sports culture and students' cognitive level of sports culture need to be improved. The value of things varies from person to person, and the value of objects varies from person to person, so the inner scale of subject is the fundamental scale of value [5]. The all-round development of human mainly includes the free development of human physical strength and intelligence, the various development of human talents and the high development of personal social relations. The strategic value of school physical education lies in the high consistency between school physical education and social development. Physical education is an important way to realize the complete form of human life. As a practical activity, physical education shapes people from two aspects so as to realize two purposes of education [6].

Of course, while people understand and transform the objective world, these practical activities of human beings, in turn, strengthen people's subjective consciousness, make people have a clearer understanding of their own subjectivity, and further consolidate and strengthen people's subjective position in the objective world. No matter from the position of school physical education in modern social culture or from the perspective of the universality of school physical education for audience groups, school physical education should also serve as the task of leading and reconstructing in the process of social culture.

3.2 To Improve Students' Cultural Accomplishment

Life lies in movement, life lies in balance, and life lies in harmony. These concepts reflect people's cognition of life from different angles and levels. Sports value is the meaning that sports presents to the main body. There are not only static phenomena of sports value, but also dynamic conflicts and changes of sports value. School physical education should strive to create a healthy school atmosphere, actively organize students to carry out sports activities, and comprehensively cultivate students' health literacy. School physical education, as the teaching of physical culture in schools, can be fully displayed only when different levels of culture have a real impact on the educated, and school physical education has the function of cultivating students' external physical quality and improving their internal life value. Campus culture has a subtle influence on students, which can even transform the characteristics of talent cultivation in colleges and universities. The charm of college sports culture is embodied in its own cultural characteristics. Therefore, in addition to improving sports quality, college physical education must also strengthen college students' sports cultural quality [7]. To give full play to the campus cultural characteristics of educating people in sports and become an irreplaceable spiritual strength of campus for other disciplines.

3.3 The Era Mission of School Sports Culture

In the popularization of sports culture, the weak awareness of physical fitness and the lack of educational approaches have greatly affected the development of school sports culture. In physical education teaching, teachers should follow the principle of inspiration and guidance, improve students' initiative and enthusiasm in participating in teaching activities, and enhance students' learning consciousness. The school spirit is a concentrated expression of the value orientation and ideals and beliefs of teachers and students in a school. Therefore, in the practice of the construction of campus sports culture, we should pay attention to the construction of good school spirit, pay attention to the construction of teachers and students' behavior culture, and promote the construction of school spirit with the leading role of example and pioneer. Improving students' own sports cultural accomplishment, deepening their understanding of school sports, and cultivating their sports fitness awareness are important measures to complete the college sports cultural mission and develop lifelong sports, and this is the personal value of school sports. However, according to the laws of education, school physical education is restricted by and serves social politics, economy, science and technology, and culture, which determines that school physical education should have its social value. Through the transmission and learning of sports skills, the effective inheritance of sports skills culture can be realized. At the same time, sports skills exist as a common way of fitness and education. However, the exertion of this educational function should be based on the essential function of physical fitness, i.e. strengthening physical fitness through physical exercise, thus cultivating human spirit, quality and perfect personality.

4 Socialist Core Values Leading the Practical Path of College Sports Culture Construction

4.1 Strengthen the Construction of Sports System Culture in Colleges and Universities

"Democracy" traces back to the basic connotation of "people-oriented". On the one hand, in the modern western educational ideology, it advocates "human nature replaces divinity and human rights replaces divine rights". The core idea is to put the value or status of people in the first place. College sports culture is a specific system in the field of sports culture. It is a comprehensive reflection of college students' cultural quality and sports quality, and reflects college cultural life and spiritual outlook. The conflict in the choice of the value of school physical education is actually the conflict of people's interests and the conflict in the value standard, which is a distortion of the "irrational" function of school physical education. In terms of specific measures, the socialist core value system should be infiltrated into the system construction of campus sports culture, so that the socialist concept of honor and disgrace can become the basic standard for teachers and students' daily learning, life and teaching, and become the basic yardstick for teachers and students' practical learning. As far as sports are concerned, its purpose is to promote and maintain physical health, ensure sufficient physical strength and energy to meet the needs of work, study and life, develop physical strength and skills and balance mental energy, obtain exciting and carefree sports experience, and constantly create and perfect oneself [8]. Students can not only experience the pleasant emotion brought by sports, but also cultivate their life attitude of fairness, justice, consciously abiding by rules and cooperating with each other through the experience of competition and cooperation in sports, so as to promote the formation of their personality and cultivate good sports ethics.

4.2 The Values of School Physical Education Need to Change from Instrumental Rationality to People-Oriented

Value concept is a unique spiritual form of human being, a reflection of people's value life and condensation of practical experience, and a value orientation or value concept in people's heart [9–13]. The function of value concept is that it becomes the evaluation standard system in people's heart. As a part of social culture, the interaction between sports and education has the same reason as social culture [14–18]. The education of sports humanistic spirit to sound personality spirit is embodied in the following aspects: developing people's body and mind, improving people's body and mind adaptability. Cultivate talents with competitive spirit, fighting spirit, team spirit, sense of justice, sense of responsibility and sense of honor. At the same time, they have perseverance, self-confidence, law-abiding and civilized behavior. Therefore, schools should adopt a positive attitude to accept sports culture, combine the situation of physical education in schools, scientifically introduce sports culture, and rationally transform it according to the specific situation of students, so that sports culture can become a powerful tool for shaping students' physical and mental health. In order to achieve the goal of student health, the pursuit of students' psychological, social and moral health of all aspects of the school sports value orientation. In the practice of physical education, scientific physical education curriculum can fully meet the needs of students and instill the correct values of physical education. Gradually establish and improve the school sports management system, so that college sports will become the propaganda base of sports culture and the training base of competitive sports talents.

4.3   The Realization of Teachers' Core Values in Physical Education Teaching

As college physical education teachers, they are faced with high-quality college students. When teaching physical education contents and sports skills, they should be concise and clear, and in line with the theme, so that college students can effectively master and use them. They should also focus on detailed explanation of methodological knowledge and concepts. Teachers and students are encouraged to participate widely, actively discuss and communicate, and effectively guide students' thoughts in sports activities. The lively and novel form has replaced the previous theoretical education, which mainly focused on preaching, and has promoted the enthusiasm and enthusiasm of students to accept education consciously. Under the guidance of socialist values and the ideology of physical education in colleges and universities, we should improve the management system of physical education in colleges and universities and enhance the educational function. Traditional ethnic sports are the crystallization of people's wisdom. With its various characteristics and forms, they carry the cultural accumulation of China for thousands of years. It is not only an important part of China's sports, but also a cultural treasure showing the long history of the Chinese nation. Therefore, the education administrative department should formulate a full-time sports teacher training plan to continuously improve the educational level and professional quality of sports teachers. Increase training efforts, broaden training channels, to create conditions to encourage teachers to go out to study and inspect. As an important position for cultural development, colleges and universities are duty-bound to inherit and develop China's excellent national traditional culture, which is also the basis for highlighting the characteristics of our colleges and universities. The main purpose is to provide direction guidance for PE teachers' self-development, so as to promote PE teachers' professional development effectively and reach the high level of contemporary specialization, thus improving their professional status.

5   Conclusion

At present, building a well-off society in an all-round way and building a socialist harmonious society are the historical missions of our country in the new stage. In the new historical period, college sports should be oriented by the society's demand for talents, based on the macro vision of students' life-long development, carry out sports culture education for students with sports as the carrier, and create a scientific model of school sports. School physical education is geared to the needs of students in school and undertakes the mission of strengthening the nation's constitution, which affects the nation's future direction. Take the students' sports cultural accomplishment as an important task of the school, from enriching the curriculum content to carrying out extracurricular activities, and actively guide students to experience this cultural connotation through various ways. We will truly create a campus sports culture with socialist core values as its guiding feature, so that students will imperceptibly accept the moral education content dominated by socialist core values.

Acknowledgements. The study was supported by "Xichang University "Two High" Talent Research Support Program (Grant No. LGLS201908)".

References

1. Zheng, W.N., Wang, H.: Research on Wushu cultural mission. Sports Culture Guide **000**(003), 72–76 (2017)
2. Zhang, Z.Y.: Research on employment orientation and path of social sports majors in universities under the background of "employment difficulties". J. Fuqing Branch Fujian Norm. Univ. **132**(05), 106–112 (2015)
3. Sun, B.: The value orientation and path selection of China's sports culture development in the new era. J. Sanmenxia Vocat. Techn. Coll. **017**(004), 122–124+138 (2018)
4. Wang, X.D.: Research on the goals and core values of physical education in colleges and universities. Teach. Res. **41**(04), 91–94 (2018)
5. Wang, Y., Yang, Y., Wang, L.: Discussion on the core values and paths of hospital culture construction under the background of new medical reform. Henan Med. Res. **026**(021), 3893–3894 (2017)
6. Zhang, L.: Research on the path and mechanism of improving college students' traditional cultural literacy in the teaching reform of "Humanities Basics". J. Changsha Civ. Adm. Coll. Technol. **3**, 98–101 (2016)
7. Zhang, X.Y.: The cultural mission of the Chinese dream and education. J. Shanghai Jianqiao Univ. **000**(003), 11–18 (2016)
8. Zhai, C.Y., Yang, D.: A probe into the identification and practice cultivation mechanism of socialist core values of contemporary college students——based on the perspective of the construction of the communist youth league in colleges and universities. J. Zhangjiakou Vocat. Coll. Technol. **31**(113), 43–45 (2018)
9. Roopesh, J., Archana, T.: Biological monograph: myristica fragrans. Matrix Sci. Med. **4**(3), 85 (2020)
10. Tanuja, P., Ganesh, D.: Purple urine bag syndrome: an uncommon but noteworthy phenomenon in the ward. Matrix Sci. Med. **4**(3), 86–87 (2020)
11. Roopesh, J., Archana, T.: Monograph: Luteolin. Matrix Sci. Med. **4**(3), 88–89 (2020)
12. Suhail, R., Musaib, D., Mir, U.: Imaging spectrum in patients with nontraumatic ankle pain. Matrix Sci. Med. **4**(2), 35–40 (2020)
13. Liu, Q.: Effect of aerobics combined with strength training intervention on invisible obese college students. Matrix Sci. Med. **4**(2), 32–34 (2020)

14. Anubhav, C., Anchit, W., Deepak, S.: Impacted stone mimicking orbital cellulitis. Matrix Sci. Med. **4**(2), 54–55 (2020)
15. Ferdinand, U., Favour, J.O.: Dangers of organophosphate pesticide exposure to human health. Matrix Sci. Med. **4**(2), 27–31 (2020)
16. Huang, P.: The effect of different sports methods on the body composition of female college students with recessive obesity. Matrix Sci. Med. **4**(2), 51–53 (2020)
17. Dong, K., Xue, X.: A study on the effect of sports intervention based on the energy metabolism on body composition of obese college students. Matrix Sci. Med. **4**(2), 41–43 (2020)
18. Sufiyan, I., Mohammed, K., Bello, I.: Impact of harmattan season on human health in Keffi, Nasarawa State, Nigeria. Matrix Sci. Med. **4**(2), 44–50 (2020)

Author Index

Printed in the United States
by Baker & Taylor Publisher Services